Gender & Technology

PUBLISHING FOR THE WORLD

125 Years

THE JOHNS HOPKINS UNIVERSITY PRESS

Gender & Technology

A READER

EDITED BY

Nina E. Lerman
Ruth Oldenziel
Arwen P. Mohun

The Johns Hopkins University Press
BALTIMORE AND LONDON

The Johns Hopkins University Press
2715 North Charles Street
Baltimore, Maryland 21218-4363
www.press.jhu.edu

Library of Congress Cataloging-in-Publication Data

Gender and technology : a reader / edited by Nina E. Lerman,
Ruth Oldenziel, and Arwen P. Mohun.
    p.   cm.
Includes bibliographical references and index.
ISBN 0-8018-7259-6
1. Sex role. 2. Sexual division of labor. 3. Technology—Social
aspects. I. Lerman, Nina, 1961– II. Oldenziel, Ruth, 1958–
III. Mohun, Arwen, 1961–
HQ1075.G46194 2003
303.48′3—dc21   2003012553

A catalog record for this book is available from the British Library.

To our intellectual foreparents

and to the scholarly community called
Women in Technological History (WITH),
in the Society for the History of Technology

Contents

Acknowledgments

First and foremost, we wish to state plainly that this volume could not have existed without the intellectual examples and the scholarly community provided by members of the interest group WITH (Women in Technological History) of the Society for the History of Technology. Decades of scholarship combining the insights of women's and gender studies with the history of technology are cited throughout these pages; the support and encouragement each of us has received from this group as we developed as scholars ourselves can only be suggested by the dedication of this book to these foreparents, and of any royalties it may accrue to SHOT.

Our early plans for this book included a collection of several pioneering essays from the 1970s and early 1980s. They appear now only as references because the realities of publishing dictated creating a reasonably sized volume. We hope that readers will appreciate as much as we do the continued intellectual presence of this early scholarship in these essays. Work such as Ruth Schwartz Cowan's early essays on domestic technology and consumption, Joan Rothschild's introduction to *Machina ex Dea,* Carroll Pursell's essay on children's toys, Judy McGaw's programmatic review essays, and Autumn Stanley's analysis of patent records can be found in the discussion and citations of the historiographic essay "The Shoulders We Stand On/The View from Here."

Seven of the essays appearing in this volume were first published as a part of a January 1997 special issue of the journal *Technology and Culture.* This project spanned the editorship of both Robert Post and John Staudenmaier, who first encouraged the project and then shepherded it to completion. Joe Schulz at the *T&C* office managed the unenviable task of copy editing in two vocabularies, women's studies and the history of technology. Other essays were published at different times in *T&C*, and three were originally published in *IEEE Technology and Society Magazine, Signs,* and the book *Gender and Archaeology,* edited by Rita Wright (1996). Thanks are due to Raymond Hofman at the University of Amsterdam for administrative support, to the Whitman College History Department for funding, and to Annelise Heinz and Kathryn Krummeck for scanning articles.

The chapter "Gender and Technology: Interrogating Boundaries" and the introductory materials preceding each section and chapter are appearing for the first time in this book. They were improved immeasurably by the careful editing of Andrea Gass. They were further refined thanks to suggestions from Sally Gregory Kohlstedt, Jonathan Coopersmith, and several anonymous reviewers for the Society for the History of Technology and the Johns Hopkins University Press.

Gender & Technology

Introduction:
Interrogating Boundaries

NINA E. LERMAN, RUTH OLDENZIEL,
ARWEN P. MOHUN

I'd like to be just like my Dad,
He's handsome and he's keen;
He knows just how to drive the car,
And buy the gasoline.[1]

—song lyrics, "Mister Rogers' Neighborhood," 1960s

Are cars masculine technologies? What other technologies—rhyming or otherwise—would have made good illustrations of manhood for boys growing up in suburban cold war America? Could Mister Rogers, the helpful and friendly children's TV personality of the years between *The Mickey Mouse Club* and *Sesame Street,* have written his song without reference to technology?

In a different verse, Mister Rogers used knowledge about the material world to encourage female aspirations as well:

I'd like to be just like my Mom,
She's pretty and she's nice;
She knows just how to make the bed,
And cook things out of rice.

Are beds and stoves technologies? Are they feminine technologies? Could Fred Rogers have sung his verse for little girls without reference to material objects and knowledge of how to use them?

In the 1970s, Rogers completely rewrote these lyrics to remove the gender stereotypes. Yet while the material world we inhabit and the gender roles we teach our children may be less blatantly dichotomized now than a generation ago, but Fred Rogers's rhyming stereotypes have modern parallels in childhood activities from baking cookies with mom to the fishing trip with dad. For both grown-ups and children — in the present and in the past — material things (stoves and ovens, cars, fishing rods, bedspreads) can be read as codes for gender. So, too, gender expectations often inform the uses of these objects.

Although we may think of gender as a way of analyzing the social side of human activity and technology as a component of the physical world around us, in fact gender and technology are closely related. Scholars have written of a "mutual shaping" of the social and the technological: each shapes the other.[2] In times of technological change, then, we can expect contests over social categories such as gender; in times of social change, we should look for new kinds of interactions with the material world. As these articles show, such change rarely happens smoothly or automatically. Neither technology nor gender ideology is static or stable.

The articles in this book explore technology through the lens of gender and gender through the lens of technology, and they do so in historical perspective. As a result, the authors shed light on the entwined and reciprocal relationships, in different places and times, between these two important categories of analysis. We hope this volume brings insights and provokes new investigations in gender studies, technology studies, and history.

Technology

In colloquial English, "technology" has come to refer most often to computers and their networks, but the historian interested in discussing telephones or steamboats or blacksmithing uses *technology* in a broader and more anthropological sense. Thinking about technology as people's ways of making and doing things allows the term easily to encompass stone age tools and space age instruments, sewing and cooking and driving cars and programming computers.[3] In cultural studies, meanwhile, technology is often discussed in linguistically grounded terms as a *discourse* or *text*. This approach takes its cue from Michel Foucault, who understood technology to be a whole set of social techniques that become institutionalized.[4] The authors in this book discuss technology historically: they examine new and old material things and people's knowledge about them; the hardening of social facts into clothing, furniture, factories, modes of transportation, energy systems, and so on; the constraints imposed by the physical existence of the built environment once it has been constructed. We argue that the material matters.

All societies have ways of constructing their material worlds, of creating and using artifacts of many kinds. Different cultures—or the same culture in different time periods—accomplish tasks and attach meanings to objects in particular ways. The articles in this book focus on a particular place at a particular time: North America, mostly during the century from 1850–1950, the heyday of industrial capitalism. The authors discuss the technologies and forms of social organization specific to that historical context, including factories and cars, dressmaking and laundry, and bodies and machines. They examine people as well as things, emphasizing the changes and continuities of human technological activities: making, doing, using, designing, producing, consuming, repairing, recycling.

Because these authors understand technology to be situated in time and place, their scholarship explores how the workings, uses, and meanings of technologies are contingent on particular historical contexts. In the past ten years, scholars in gender and cultural studies have marveled at the technological possibilities of biomedical innovations and the Internet to change the sanctity of the body itself. But as some of these essays on the body show, cyborgs have a history, too. Early hair removal technologies, for example, shaped not only gender and racial identities but also modified bodies. The Internet, also, must be situated historically: computers are not the first kind of information technology to allow the masquerades and disembodiment of authors, as the novels of George Sand (b. Amantine Aurore Lucie Dupin, 1804–1876) and George Eliot (b. Mary Ann Evans, 1819–1880) — made possible by nineteenth-century printing, paper, and bookmaking technologies—demonstrate. Throughout the period spanned by U.S. history, technologies have entwined bodies and machines, as studies of early industrial factory mechanization reveal.[5]

Once one considers "technology" in this anthropological and historical way, it becomes clear that its study involves not only material things but also people. Humans' choices, creativity, knowledge, ideologies, assumptions, and values must always be explored along with the objects and machines resulting from their technological activities. After all, technologies do not work without the people who produce, handle, and use them. Furthermore, humans associate their technological activities with the categories by which they divide and define themselves—age, wealth, race, education, work, region, and, of course, gender.

Gender

The word *gender* has come into common usage as one of the many ways by which we sort people into familiar categories. But like many other such schemas, the male/female dichotomy masks complex social and cultural processes. Scholars examining how gender works in society have pointed out that maleness and femaleness do not exist independently but are defined in relation to each other. The boundaries between how people designated male are expected to behave and how people designated female are expected to behave are sometimes redefined, negotiated, or violated. Gender is not only a way to sort people; it is also a way to assign power in particular contexts. The shape of this boundary varies from place to place, from time period to time period, from situation to situation. Elite men once wore wigs and makeup; "painted women," meanwhile, were read as sexually available, not respectable ladies. Flexibility and open communication, once "feminine" characteristics, can now be labeled "good management practice" in the corporate world. Like technology, defining *gender* demands attention to historical processes — gender too is historically contingent.

Keeping track of masculinities and femininities is more easily managed if one thinks of gender as operating at different levels, in layers of function and meaning. At the most personal level, gender is an *identity,* a part of how one sees oneself and presents oneself to the world. People perform their sense of their own gender, not only by words and gestures, but also in material ways: by wearing baseball caps or skirts, ties or jewelry; by tinkering with cars or baking cookies; by shaving with particular colors of razors.

Meanwhile, the larger society makes use of gender in organizational and material ways, so gendered people navigate, create, and modify gender *structures* and *institutions:* Boy Scouts and Girl Scouts, men's departments and women's departments, beauty salons and barbers, men's and women's locker rooms, or, often, corporate typing rooms and board rooms. Finally, gender works in *symbolic and representational* ways, in assumptions about what men and women like, in images of manhood and womanhood, in styles and expectations and ideologies based on portrayals of gender difference. The image of Mr. Rogers's car-driving Daddy is effective not because Mom never chauffeured the kids to piano lessons but because cars were symbolically associated with manhood. Similarly a person toiling over a hot stove represents nurturing womanhood, unless the person wears a tall white hat — in which case this person has been symbolically represented as a professional male of refined taste, a chef.

Interrogating Boundaries: Technology and Gender

Discovering the role of material objects in these layers of gender construction, like considering the role of gender in constructions of technologies, makes clear the importance of treating these two categories together. At every level we find tight connections, technology shaping gender and gender shaping technology. Gender analysis illuminates our understandings of technology, and attention to technology illuminates our understandings of gender. Technology, too, can be analyzed in layers of identity, structures, institutions, and representations.

We can often see this process more clearly by examining the technologies and gender ideologies of the past. For example, early cars were, not surprisingly, machines run by men — a chauffeur, often, who knew the quirks of the machine or an adventurous hobbyist with money for luxury. As cars became more reliable and more popular, their owners constituted a market for a range of gadgets, including now-common items like side-view mirrors and a roof to keep off the rain. One of the most innovative of these was the electric ignition — imagine starting your car from the driver's seat instead of having to crank the engine. But electric "self-starting" ignitions were marketed "for the wife," a specialty item, slow to be adopted as standard equipment: manly men would not mind arm-breaking cranks or trudging around the car on a muddy road to get it started.[6] The long life of the cranked engine makes little sense without considering the expectations of adventurous manhood.

In this case, expected gender identities shaped accessories sold with cars; in turn, the new freedom of automotive transport, now with roof and ignition made safe for the new category of "female drivers," allowed women to extend the reach of appropriate and respectable activity. Similarly the "safety bicycle" was marketed as a machine women could use, not only adding reliable brakes for male cyclists but also redefining cycling as a respectable female pastime. These examples take us from identity to structure, a physical enlargement of where a person could go in a day's travel and also where a woman could go alone. Indeed, physical structures often regulate, reinforce, or impose gender, and gender difference structures the built environment: consider a tree house with a "no girls allowed" sign; a fighter jet off-limits to female pilots; a small electric drill (black case, motor to turn attachment) and an electric mixer (white case, motor to turn attachment); a disposable razor (pastel colors, curved handle, built-in lotion dispenser) and a disposable razor (black and chrome, sharp angles, triple blade). Saloons and bars were traditionally built with a "ladies' entrance" to the dining room, and early hotels were designed with separate parlors for male and female guests. Other dichotomies and categories often intersect and complicate these structures. For example, we expect separate facilities, in the United States called "ladies' rooms" and "men's rooms," to be built in public buildings, while we have no problem with unmarked "bathrooms" in our private spaces at home; for decades in the southern United States, public restrooms were marked "white" and "colored," a racial rather than a gender-based material structure.[7]

These connections become embedded in our vocabulary as well. When typewriters first entered American offices, clerking was a male profession. Soon "typewriters" did not refer to the novel writing machines but the women who controlled them. Similarly, in the early telephone operating systems "switches" referred to the women operating them. Before electronic calculating machines the (female) mathematicians whose work enabled missiles to launch accurately were called "computers"—the new digital technologies were named after the workers.[8]

Finally, representations of gender and technology regularly rely on each other, from the "shop 'til you drop" of female-coded consumerism to the (male) doctor in the white lab coat whose advice in the TV ad is cloaked in the symbols of scientific expertise. Advertising is perhaps the easiest place to find such representations throughout the twentieth century, in images or text. But many other sources—advice manuals, literature, sermons, school reports, Mr. Rogers's song lyrics—use technologies to represent gender categories and gender to categorize technologies.

The connections, indeed, are so ubiquitous they often seem natural and have therefore been hidden from analysis. Treating both categories as constructed, rather than natural or immutable facts, and treating them together, highlights the importance of human agency in understanding historical change and continuity. For example, individual technologies such as the pill, the computer, or the Internet have often been considered causes of women's liberation, but technologies by themselves have no such power. Human beings made these technologies and human beings chose to use them, often based on human desire for (or resistance to) social change.

The stories are often complex, as human stories tend to be. Funding for contraceptive research came out of progressive social ideals including eugenics: "birth control" and "planned parenthood" were labels based on goals of population control and limiting lower-class family size. The research grew out of the assumption that new technological solutions were more promising than teaching men reliable condom use. That the pill became an emblem of changing social mores for unmarried middle-class women was not inherent in the technology but chosen by the women who used it. Conversely, technologies enable social change as well; without reliable contraception, sexual activity is more likely to result in pregnancy. With reliable contraception, sexual activity conducted privately can more successfully be separated, if the participants so choose, from its more public results. But even these "revolutionary" technologies embed continuities: the computer keyboard is based on the typewriter keyboard, which was designed to keep mechanical keys from sticking; packaging for the pill was designed with a week of placebo pills so women could still menstruate "normally" but would not forget to take their pills daily.[9] Social beliefs and practices and technological developments reciprocally shape each other, often with unexpected outcomes, as humans debate and negotiate the alternatives and the constraints.

When we recognize the role of human choice in shaping these categories and their relationships, we must recognize also that both gender and technology are about power: social, cultural, economic, political. Differences are not simply descriptive but shape opportunity and access in industrial capitalism as it has developed in North America. The association of maleness and technological prowess in a society that values technological change camouflages the privileges accorded men; they are labeled privileges of technological knowledge rather than of masculinity. The gendered production/consumption dichotomy, so common in industrial society, labels male-coded activities "production" and camouflages the work and the technological content of activities labeled "consumption" or "reproduction." Consumption and reproductive activities are generally unpaid or underpaid, invisible knowledge and invisible contribution in a capitalist economy.[10]

Gender analysis, scholars have long since pointed out, invites the interrogation not only of the boundaries between maleness and femaleness but also between other categories as well: between public and private, between various racial labels and identities, between animate and inanimate. As you read this book, we invite you to interrogate the boundaries of industrial capitalism: technological/social; production/consumption; skilled/unskilled; expert/user. Outside of this book people usually assume a boundary between the categories "gender" and "technology" just as they assume one between "male" and "female." We seek to examine the gender/technology boundary, to explore its construction, to explain the shapes it has taken in the past and understand its legacy for the present and future.

NOTES

1. This song appeared first on Program #3, produced in 1968 and repeated in several other episodes. Private communications, Elizabeth Mahoney and Laurel Povazan-Scholnick, Mister Rogers' Neighborhood Archives, University of Pittsburgh, January 2002, and Hedda Sharapan, associate director of public relations, Family Communications, Inc., May 2003. The song lyrics were completely revised in 1975; later the song was dropped from the Mister Rogers repertoire. Other revisions in 1975 included attention to whether machines did things on their own: after 1975, machines never initiate interactions with humans. Telephone interview, Hedda Sharapan, November 2002. Lyrics reproduced by permission of Family Communications, Inc.

2. For a full discussion of the literature we rely on in this introduction, see the historiography essay at the end of this volume.

3. Melvin Kranzberg, "At the Start," *Technology & Culture* 1 (1959):1–10; Brooke Hindle, "The Exhilaration of Early American Technology: An Essay" in Hindle, ed., *Technology in Early America: Needs and Opportunities for Study* (Chapel Hill, 1966), reprinted in Judith McGaw, ed., *Early American Technology: Making and Doing Things from the Colonial Era to 1850* (Chapel Hill, 1994). For more recent discussions, see McGaw's introduction to that volume; McGaw, "No Passive Victims, No Separate Spheres: A Feminist Perspective on Technology's History," in Stephen Cutcliffe and Robert Post, eds., *In Context: History and the History of Technology* (Bethlehem, Pa., 1989), pp. 172–191; Merritt Roe Smith and Leo Marx, eds., *Does Technology Drive History* (Cambridge, Mass., 1994); and Donald MacKenzie and Judy Wajcman, eds., *The Social Shaping of Technology* (Birmingham and Philadelphia, 1999).

4. A good example, well known in women's and gender studies circles, is Theresa De Lauretis's book *Technologies of Gender: Essays on Theory, Film and Fiction* (Bloomington, Ind., 1987).

5. A cyborg is a "cybernetic organism," a combination of organic and inorganic parts. On cyborgs in gender studies, see Donna Haraway, *Simians, Cyborgs and Women: The Reinvention of Nature* (London, 1988), which includes her now-classic essay, "A Manifesto for Cyborgs: Science, Technology, and Socialist Feminism in the 1980s," originally published in *Socialist Review* 15 (1985): 65–107. On hair removal, see Rebecca Herzig, "Situated Technology: Meanings," in this reader. Biographical details on Sand and Eliot are from *American Heritage Dictionary of the English Language* (Boston, 1992). For further discussion of the gendered origins of computer technologies, see Paul Edwards, "Industrial Genders: Soft/Hard," and Jennifer Light, "Programming," both in this volume.

6. This discussion is drawn from Virginia Scharff, *Taking the Wheel: Women and the Coming of the Motor Age* (Albuquerque, 1991). In fact, gas-powered combustion engines

were generally associated with masculinity, and electric cars — in their brief heyday — with femininity.

7. For the example of drills and beaters, we thank Deborah Douglas of the MIT Museum. On hotels, see Molly W. Berger, "A House Divided: The Culture of the American Luxury Hotel, 1825–1860," in *His and Hers: Gender, Consumption, and Technology,* eds. Roger Horowitz and Arwen Mohun (Charlottesville, Va., 1998). On bathrooms, see McGaw, "Why Feminine Technologies Matter," in this book and Patricia Cooper and Ruth Oldenziel, "Cherished Classifications: Bathrooms and the Construction of Gender/Race on the Pennsylvania Railroad during World War II," *Feminist Studies* 25, no. 1 (Spring 1999): 7–42. On bicycles, see Wiebe Bijker and Trevor Pinch, "Social Construction of Artifacts," in Wiebe Bijker, Thomas Hughes, and Trevor Pinch, eds., *The Social Construction of Technological Systems: New Directions in the Sociology and History of Technology* (Cambridge, Mass., 1987), pp. 17–50; and Ellen Gruber Garvey, "Reframing the Bicycle," *American Quarterly* 47, no. 1 (March 1995): 66–101.

8. See Margery Davies, *Woman's Place Is at the Typewriter: Office Work and Office Workers, 1870–1930* (Philadelphia, 1982); Sharon Hartman Strom, *Beyond the Typewriter: Gender, Class, and the Origins of Modern American Office Work, 1900–1930* (Urbana, Ill., 1992); Kenneth Lipartito, "When Women Were Switches: Technology, Work, and Gender in the Telephone Industry, 1890–1920," *American Historical Review* 99 (1994): 1074–1111; and Jennifer Light in this book.

9. See notes 29–32 to "The Shoulders We Stand On" in this reader.

10. For further discussion, see "The Shoulders We Stand On."

PART I

Entwined Categories:
Gender Constructs Technology

What is technology?
How has technology been gendered?

What is technology? What is not technology? Is the material world shaped by ideas about gender, and are people's perceptions of the material world shaped by ideas about gender? How? Defining technology broadly—as we have done in the introduction to this volume—as a process of "making and doing things," rather than as a particular set of recent and sophisticated artifacts, allows us to analyze and compare both ordinary and arcane elements of the material world and the ways people interact with it. Because people construct and use technologies, their values shape their choices—but people also deploy the technologies available to them as they construct meaning in their lives.

The articles in Part I allow us to explore the meanings and boundaries of technology in our broad definition and to think about why the word *technology* is often used more narrowly in colloquial English. What parts of the material world have been labeled "technological" and why? In what ways have ideas about gender shaped technological choices? Are technologies sometimes "masculine" and sometimes "feminine"? How do these stories, in turn, shed light on how gender was understood by the actors involved?

ONE

Why Feminine Technologies Matter

JUDITH A. McGAW

Bras, closets, collars, bathrooms: Judith McGaw begins with a series of artifacts not often included in common definitions of technology. She calls these "feminine" technologies because they are either used by or predominantly associated with female people. This piece was originally written for archaeologists, who are often left to draw conclusions about entire cultures from surviving objects and house plans. McGaw argues that the issues raised by studying these sometimes invisible modern technologies will shed light on technologies we may think of more readily— cars, electricity, and so forth—as well as on our understandings of technologies of earlier times. How does gender analysis influence McGaw's perspective on technology? In what ways have gender ideologies shaped the technologies she discusses here? McGaw is deliberately provocative in this piece—what do you think about her conclusion that "feminine technologies matter because they disclose the system's fatal flaws"?

From Judith A. McGaw "Reconceiving Technology: Why Feminine Technologies Matter," pp. 52–75 in *Gender and Archaeology* edited by Rita P. Wright. Copyright © 1996 University of Pennsylvania Press. Reprinted with permission.

What can we learn about gender and technology by subjecting contemporary artifacts to the close material and cultural scrutiny we generally reserve for prehistoric and precapitalist objects? Can the gender conscious study of modern, Western technology offer archaeologists analyzing gender and technology in earlier eras a salutary comparative perspective? This essay demonstrates the utility of such an approach. It carefully examines several objects—the brassiere, the closet, the white collar, and the bathroom—selected to represent what I call "feminine technologies." I argue that unearthing the cultural context and artifactual precursors of such feminine technologies exposes pervasive aspects of our society's relationship to technology—aspects that as scholars and as citizens we ignore at our peril.

Although informed by perspectives drawn from historic and industrial archaeology and physical and cultural anthropology,[1] what follows is the work of a historian of technology rather than an archaeologist. . . . Archaeologists face the perennial challenge of being sufficiently self-critical to avoid reading personal cultural conditioning into the artifacts of other times and places. Turning to a new area of concern—the archaeological study of gender and technology—creates a new responsibility to become aware of the intertwined assumptions about gender and technology that inform our modern, Western perspective. Ideally, the historical study of recent technology promotes such cultural awareness. And research such as mine—shaped by the desire to address general public concerns—is especially suited to provoke personal reflection among Americans who are also archaeologists.

My concern to reach an audience that includes the literate lay public also raises issues germane to this volume's emphasis on teaching about technology and gender. Like most Americans, our students often feel disenfranchised in the realm of technological decision-making because they lack the requisite expertise. In fact, my commitment to research and write so as to communicate with a broader audience stems from my growing conviction that, in asserting their claims to expertise as befits a rapidly professionalizing group of scholars, historians of technology merely deepen the central dilemma of modern technology—the pervasive conviction that most of us do not know enough to participate in technological decision-making; that we have to trust technology to the experts.

My guess is that, like most of us, when it comes to understanding technology and gender, archaeologists and their students know too much rather than too little. For example, one of my principal arguments is that when we consider modern American technology, we are most constrained by our narrow preconceptions of technology. We start off on the wrong foot because we hear the word "technology" and immediately envision complex mechanisms, sophisticated electronics, or mysterious molecular combinations. In any event, our initial image is of technology as hardware. We may, as historians of technology generally do, notice the "-ology" in technology and concede that it includes knowledge as well as tools, but, again, we tend to imagine most of that knowledge as the purview of engineers and their corporate and government managers.[2] Hence my title: "Reconceiving Technology." My largest contention is that we will continue in the squirrel-wheel of current technological thinking as long as we construe technology so narrowly.

My subtitle — "Why Feminine Technologies Matter" — is meant to point the way toward a broader outlook. There are, of course, lots of reasons why the technologies associated with women should matter to scholars and citizens. Feminist scholars have already articulated many of them.[3] At the very least, a full picture of technology has to include the tools, skills, and knowledge associated with the female majority. Moreover, in societies such as ours in which women do most child rearing and tending, our subconscious, unarticulated technological convictions must derive principally from our early preverbal experience around technologies selected and manipulated by women. It is also true that until we began to study women and technological change, we were able to remain unaware and ignorant of technology's masculine dimensions — we studied inventors, engineers, and entrepreneurs as though they were simply "people," oblivious to the ramifications of the overwhelming masculine predominance, both numerically and politically, in the so-called technological professions.

Thus, there is an abundance of good reasons to study women and technology. In what follows I emphasize yet another reason: looking at feminine technologies makes visible precisely those aspects of technology that we need to examine if we seek alternatives to a modern, Western technology that appears to be self-destructive, self justifying, and self-perpetuating. By feminine technologies, I mean those technologies associated with women by virtue of their biology: tampons, brassieres, and IUDs, for example. And I mean those technologies that almost all American women use by virtue of their social roles: kitchen utensils, household cleaning products, and sewing needles, for example. Beginning with a look at such items, I hope to suggest a broader view of feminine technologies as well.

Archaeologists, anthropologists, and historians of non-Western technology might readily argue that studies of prehistoric and non-Western technologies serve equally well to make visible aspects of technology that we fail to see when examining more familiar examples. Although this is certainly the case, archaeologists also recognize how inextricably intertwined are technologies and their cultures. Indeed, many have learned from their own experiences how formidable is the challenge of disentangling their personal cultural sense of technology from the assumptions they deploy when associating prehistoric or non-Western technology with its culture. For Westerners generally and for archaeologists who are also Westerners, then, the more idiosyncratic features of our technological beliefs and practices may best be observed by working with Western examples. Feminine technologies have the advantage of already being domesticated, in both senses of the word: they are Western, but outside the masculine mainstream.[4]

Moreover, an accessible study of feminine technologies promises an immediate practical payoff. It can help to persuade the people in our society most convinced that they lack technological expertise — namely women — that, to the contrary, they know more than enough to contribute intelligently to any discussion of technology policy. Simultaneously, it can challenge archaeologists and other scholars with an interest in gender to recognize unexamined and unsupported assumptions about the sexual division of technological expertise in the past.

In what follows, I discuss briefly several of the case studies that form the backbone of my current work. In so doing I will delineate some of those neglected aspects of technology that we suddenly see clearly when we look at feminine technology. In general, by calling our attention to the technological knowledge associated with the selection and use of products, the study of women's technologies reveals technological choice and technological knowledge to be pervasive, not confined to a corps of experts. It shows that neither tools nor professional technologists determine the ultimate form of a technological activity: purchasers and users do.

Looking at feminine technologies also means looking especially at the relations of technology and biology. In the process, it suggests how extensively modern technology has been driven by the impulse to obscure the reality that we are animals as well as intellects. And, because feminine technologies are often the technologies of the private sphere, bringing them to light exposes how interwoven with modern technologies are novel conceptions of privacy, conceptions very much at issue in current political debates. At the same time, examining women's technological activities helps expose the deep and obscure ties between public and private, linkages that ultimately make the distinction problematic, linkages encapsulated in such an oxymoron as the "right to privacy."

Finally, and perhaps most importantly for archaeologists and other students of the past, studies of feminine technology give us a novel perspective on that most troublesome of notions in technological history: progress. Although historians of technology have made a major commitment to eradicating this value-laden concept, what they have mostly done is to reveal technology's social construction.[5] They have demonstrated, that is, that technology has no inherent logic of its own, but embodies the perspective of its creators: that it gets "better" only in the sense that it better serves the interests of those empowered to make technological decisions. This is a long and laudable step away from the "march of progress" story that historians generally made of technology several decades ago. Nonetheless, it leaves untouched the notion that there is such a thing as a "better" technology: one that would materialize if only a better social system governed technology's development.

Mechanizing the Biological: The Brassiere as High-Tech

Close scrutiny of feminine technologies reveals that our whole notion of "better" may be specious. Consider, for example, the brassiere. The brassiere is one of the many feminine technologies about which technological history tells us virtually nothing. Yet in other respects—other than being used by women, that is—the brassiere falls into precisely those categories that scholars have deemed most worthy of study. Judging from the word *brassiere*'s initial appearance in our language—1911 according to the OED supplement—this undergarment is a thoroughly modern invention.[6] Like many of the technologies historians of technology have deemed especially worth studying—automobiles, industrial research labs, and electric light and power systems, for example—it evidently originated during the late nineteenth and early twentieth centuries, the era often designated the Second Industrial Revolution. As clothing goes, the brassiere qualifies as "high tech," incorporating relatively early on such sophisticated products of the chemical industries as rubber and synthetic fiber. And its inclusion of such components as wires to stabilize it and adjustable straps and fasteners to modify it makes it visibly more complex than we expect of mere clothing.[7] Indeed, early brassieres frequently list patent numbers on attached labels, underscoring their technological character.

Those of us who lived through the early years of the contemporary women's movement, when feminists were often dismissed as "bra burners,"[8] will immediately recognize one way in which the history of the brassiere renders progress a profoundly dubious notion. The avowed purpose of the brassiere is to support the breast tissue, yet there is no convincing evidence that breasts need support. Even breasts so large that they quickly droop in the absence of support cannot be said to "need" support; at best we are discussing a preference. Clearly, the brassiere serves essentially cosmetic purposes; changes in it are closely associated with changes in fashion, changing definitions of feminine beauty, and changing prescriptions for feminine behavior.

Our first impulse, then, might be to dismiss brassieres as not "real" technology, like such functional technologies as automobiles, industrial research labs, and electric light and power systems, for example. Yet if we are willing to take this feminine technology seriously, we may be pushed to ask important neglected questions about "real" technologies. How much, for example, do automobiles serve a cosmetic purpose, especially a purpose associated historically with enhanced masculinity? Do they in fact provide the freedom and speed of travel often considered as their function, or merely the illusion of freedom and speed? Likewise, how much of the role of industrial research labs is "real" and how much part of projecting the image that DuPont wants us to experience "better living through chemistry" and that GE devotes itself to bringing "good things to light"? For that matter, although anyone who knows the history of industrial fires or of domestic cleaning can recognize important safety and labor-reducing functions of electric light and power systems, how much of their early and continuing output has gone to uses best described as decorative, cosmetic? How bright do our homes and offices "need" to be, for example?

Looking at the brassiere more closely makes the notion of technological progress problematic at yet another level. The heart of my study of this technology has involved talking with women, focusing on a single question: How do you know that your bra fits? When women's answers are combined with the accounts of saleswomen and gynecologists, and with prescriptive literature on the subject, articles in girls' and women's magazines, for example, one large conclusion emerges: it doesn't. No one's bra fits. When asked, women talk in terms of making the best of a limited array of choices or of finding something less unsatisfactory than their previous choice. They may even wax enthusiastic about the latest innovation: a more comfortable fabric, less irritating strap width, more accommodating cup size, less inconvenient fastener. But although no one says it directly, it is implicit in everyone's responses: brassieres don't fit. At best, the real experts in this field — individual consumers and saleswomen — learn to know what brands come closest in particular cases.

The central problem with this feminine technology goes beyond issues of capitalist exploitation of the consumer or patriarchal disregard for women's concerns, arguments to which analysis through social construction readily leads. The central problem is that you cannot make a bra that fits. You cannot because breasts are living things, not standardized commodities. Any given woman's two breasts are never exactly the same size and shape. And the size and shape of any given woman's breasts change continuously — as she ages, as she gains or loses weight, as she goes through pregnancies, as she experiences her monthly hormonal cycles. Added to the underlying problem with all ready-to-wear clothing — that people do not come in standard sizes — these considerations make the brassiere an especially stark example of modern technology's inherent inadequacy in the area to which it has increasingly turned its attention: standardizing the biological.[9]

Myths of technological expertise to the contrary notwithstanding, ordinary women are the great experts in this area. They make the compromises and create the knowledge that permits a deeply flawed system to work. In the case of brassieres, they begin in adolescence to build and maintain a highly personalized body of information (technology as knowledge) that serves to adjust a standard system to individual difference. Thus far, I have found little evidence that women have ever gained much of this expertise from their mothers, probably because adolescence — the time when a woman would be most likely to need it — is, in this culture, a time when communication between the generations is especially difficult — and nowhere more so than in decisions about personal appearance. Peer exchanges and individual experimentation serve as the principal sources of technological development, with professional expertise — either from saleswomen or from written materials — playing a distinctly minor role.[10] The resulting body of knowledge about the best brands and styles for the individual; the necessary adjustments, including structural modifications and additions; and the variations that suit particular overgarments and activities needs continuous modification because the options available for purchase change frequently. And, as women who want or need to avoid synthetic fabrics make especially clear, there is little to suggest that, even within the inherent limitations of the technology, brassieres exhibit technological progress.

Brassieres are only one of many technologies intended to suit the mechanical to the biological. And women's knowledge plays a crucial role by compensating for the inadequacies of many of these technologies. Women do a disproportionate share of the society's clothes shopping and clothing modification, an area in which their labor is economically and socially invisible and is further obscured by jokes about their propensity to shop.[11] Likewise, women work hard to offset the deficiencies of the various foods suited to mechanical harvesting and long-term storage, devising ways to create flavorful meals while relying on products such as canned vegetables, iceberg lettuce, and Delicious apples, to name but the most obvious examples, and acquiring the continuously changing knowledge of how to assess food's probable freshness, potential longevity, and possible safety.[12] And they have developed the skill that has modified several generations worth of new and standardized diapering products to variously shaped and rapidly changing infants and toddlers.

Thus, beginning with a simple technology like a bra and asking a simple question like, "How do you know it fits?" leads us rather quickly to some important reconceptualizations of technology. First, we find that, although we usually associate technology with utility, its actual role is often decorative or cosmetic. Its kinship to the visual arts, heretofore recognized mostly in studies of inventors,[13] may turn out to be its most important relationship. Second, we recognize that, where living things and people are concerned, apparently functional technology is inherently flawed; that the services which the producers of goods apparently render actually serve us only because goods are made serviceable through women's invisible bodies of knowledge—the technology of consumption and modification. In sum, these observations reveal technology to be far less solid, substantial, and straightforward than we commonly assume in our discussions of technological progress or technology policy. Assessing whether technology is "better" seems a pretty dubious enterprise once we see how much of it is smoke and mirrors, how illusive are its boundaries and intentions.

Inside the White Box: Closets, Cupboards, and the Technology of Filing

Feminine technology proves equally provocative when we turn to an older, simpler, and more clearly utilitarian example: the closet. For those acquainted with early modern Anglo-American material culture, the story of closets begins as a familiar one. In fact, its history offers an abbreviated account of technological development from the late medieval and early modern era through the Industrial Revolution. Judging from surviving structures, house plans, inventories, and the linguistic history of terms such as "closet," "cupboard," "dresser," "bureau," "pantry," "larder," and "cabinet," closets and other storage spaces were comparatively rare until the century or so just before the onset of the British and American Industrial Revolutions.[14] Earlier usage of terms denoting the technologies of storage expresses that closets and other built-in or wooden containers were limited to the dwellings of the extremely affluent or privileged: royalty, the church, and those with plate, jewelry, or similar goods requiring specialized storage.

Beginning in the seventeenth century, this situation began to change. Words such as *closet* and *cupboard* came increasingly to be used in their modern sense, as storage spaces for utensils, provisions, and other, more common goods. Inventories confirm that storage furniture such as cabinets and dressers grew more pervasive. They also reveal that storage technology had developed at least in part because more and more people had more and more goods to store. Thus, the growing commonness of closets and cabinets highlights something histories of industrial technology too often ignore: the Industrial Revolution was, in the first instance, a response to rising consumption—to a growing market among people of the middling sort for goods such as ceramics, woodenware, textiles, and iron implements. Consumer initiative directed the activity of those who came to invent, develop, and invest in the new manufacturing technology.[15]

By the early nineteenth century the Industrial Revolution was in full swing, lowering substantially the real cost of most of the items people had come to store in closets. Not surprisingly, closets proliferated. Indeed, the abundance of closets —built-in storage spaces—is one key technological change evident in Victorian house plans.[16] Building technology abetted this innovation. By the nineteenth century, Americans generally employed balloon-frame construction, a system using light two-by-fours and nails in lieu of heavier timbers of irregular dimensions that had been specially notched and pegged together. The new system dramatically reduced building costs, permitting house purchasers not only to have more rooms, but also to have more ells, bay windows, closets, and similar small nooks set apart within larger rooms.

Linguistic clues show that, within the new context of abundance, storage technology was also transformed—we might even say reinvented. Historically, the various sorts of storage spaces had been highly specialized. The "-ology" part of the technology—the rules governing use of closets, larders, and pantries—had circumscribed their contents. Thus, larders, as the word echoes, originally housed pork products in particular and came to serve as meat storehouses. Pantries, as the word's French derivative might signal, were bread rooms, a use broadened to include other foodstuffs by the seventeenth century. Cupboards originally held cups and other vessels, as one might guess from looking at the word. Cabinets, historically meaning "little cabins," were repositories with locks for storing valuables such as jewels or documents.

By the eighteenth century these distinctions were clearly breaking down. Cupboards had come to hold meat, money, bread, and books; larders stored food of all sorts; and pantries might contain plate and linen, as well as provisions generally. Nineteenth-century usage suggests even more overlap in function. Nor was the reinvention of storage technology confined to its software: the rules governing what might be stored where. At the close of the eighteenth century, that mistress of precise word use, Jane Austen, wrote of "A closet full of shelves . . . it should therefore be called a cupboard rather than a closet."[17] Her attempt to reassert traditional usage signals its decline. Clearly, closets—separate tiny rooms—now sometimes contained shelves, whereas formerly shelf storage had characterized cupboards, pieces of furniture often placed in a room's corner or recess.

The most pervasive change in the hardware of storage was the shift from open display to total enclosure. Originally the cupboard was a board or table on which plate and ceramics were displayed. Likewise, "dresser" designated the kitchen table on which food was dressed and the hall or dining room table from which dishes were served and on which empty dishes were arrayed. Both word usage and surviving artifacts show dressers and cupboards becoming more cabinet-like, more enclosed. Simultaneously, the terms "closet" and "cabinet," originally signifying separate rooms, increasingly referred to mere storage enclosures.

Although the axiom "necessity is the mother of invention" might explain the proliferation of storage spaces, the abundance of new things to be stored did not necessitate their enclosure, much less their increasingly promiscuous arrangement. Indeed, one might more easily imagine that having more things to keep track of would encourage people to place them where they could be seen and to array them in specialized containers.

Making sense of the new storage technology begins by remembering who performed the work of storage. In late medieval and early modern England, where larders and cupboards and closets were confined to palaces and monasteries, specialized servants supervised and manipulated the technology of storage. Terms such as "butler's pantry" or "housemaid's pantry" indicate the linkage of specialized technology with specialized laborer. By contrast, as household goods grew more numerous in common people's homes on the eve of industrialization and, later, as they gradually filled the residences of the new industrial middle class, a single, unspecialized worker — the housewife — supervised disposition of the entire array and performed much of the manual labor of storage and retrieval as well.[18]

Housewives, because of their multiple concerns, brought new considerations to decisions about how to array goods. For example, eighteenth- and nineteenth-century housewives could expect to spend a substantial part of their careers supervising small children while performing their work. Storing goods so as to keep them away from infants and toddlers must have been important to such women, whereas the butlers of the noble or affluent could expect nursemaids to keep children away from ceramics, foodstuffs, and linen.

Similarly, as unspecialized domestic workers, housewives performed cleaning as well as storage tasks. Storing goods in enclosed spaces greatly reduced the work of cleaning by separating linen and utensils from the soot and ash pervasive in houses containing the innumerable fires necessitated by eighteenth- and nineteenth-century cooking, heating, and lighting technologies. Indeed, as Ruth Schwartz Cowan has observed,[19] modern notions of household cleanliness are only conceivable where most items are no longer left on the table but have separate places where they can be stored. Only where goods and implements can be moved to cupboards or closets can table and counter surfaces be kept free from food debris and dust.

So much for the hardware of the closet, what of the software? Why wouldn't it still make sense to keep the contents of closets homogeneous? Judging from household inventories, the answer, once again, lies in the fact that closets had become women's technologies. And what housewives needed most from their storage technologies was to find the implements of their work proximate to their workplaces. Food, cooking implements, and seining dishes belonged together in the kitchen pantry or dresser. Dining linen was handiest when placed with china and glassware, whereas bed linen was most accessible if stored in bed chamber closets alongside the clothing and personal goods of the chamber's occupant. Whatever their original uses, broken goods, obsolete tools, and worn-out items could be consigned to attics or spare rooms.

This new technology of filing did not emerge overnight. Household inventories from the eighteenth century suggest a diversity of arrangements although, as Robert St. George has noted, rules of appropriate order become increasingly evident.[20] By the mid-nineteenth century, the process was sufficiently complete, at least in the mid-Atlantic states, that inventory takers could merely list "Sundries in the Kitchen," "Sundries in the Pantry," and "Sundries in the Chamber" in the certainty that people knew roughly what items belonged where.

But only the housewife knew precisely what belonged where. As with all filing systems, the knowledge part of the technology had to be idiosyncratic to function successfully. No two households had the same array of goods, users, or storage spaces, or the same chronology of acquisition. So only the housewife knew where to find and where to replace all of the family's diverse and growing inventory of possessions. By the late nineteenth century virtually all middle-class girls must have received domestic training in this sort of filing. Thus, it is hardly surprising that, once government and corporate bureaucracies came to require extensive filing systems, they could find an abundant supply of female employees to perform the labor skillfully, yet at low pay,[21] just as manufacturers had experienced little trouble finding cheap female labor to perform supposedly unskilled sewing tasks.

Ironically, then, like the complex household appliances of the twentieth century, the eighteenth- and nineteenth-century closet began by promising to be labor saving—to reduce cleaning, that is—but ended up creating more work.[22] Anyone who has kept house knows how much time is spent in picking things up and putting them away. And, like most twentieth-century household appliances, the work of nineteenth-century storage technology came to be used not to reduce cleaning, but to permit higher standards of cleanliness.

Anyone who has performed housework also knows how hard it is to delegate tasks such as cooking to someone who does not know where to find the tools and ingredients, or tasks such as laundry to someone who does not know where to put things away. In other words, comic strips and sitcoms featuring the inept husband dependent on his wife to find necessary items of dress or grooming point to real skills housewives need to have. The extensiveness and social invisibility of these household filing skills help explain why housework has remained so resistant to the division of labor, even where husbands and wives have attempted more equitable arrangements.[23]

White Collar: On Cleanliness and Class

Since women's work often employs apparently simple technology, such as closets, that actually entails extremely complex bodies of knowledge, such as filing, study of feminine technology makes us especially aware that much resistance to technological change derives from our investment in the invisible software that is generally the largest aspect of technology and the aspect we most frequently miss when attempting to comprehend technology. Not only does the knowledge and skill part of technology help account for the persistent gendered division of labor that remains inexplicable in terms of economic rationality, it also helps explain many other apparently irrational aspects of technological choice. Why, for example, has the desire for greater cleanliness inspired extensive innovation in household technology and household labor for more than two centuries? The answer to that question is multifaceted and largely beyond the scope of this essay. A full answer will certainly require that we examine the profound influence of evangelical Protestantism in our avowedly secular culture. It is no coincidence, for example, that the belief that "cleanliness is next to godliness" was first articulated on the eve of industrialization. And it is no surprise that the originator of the axiom was John Wesley, writing in a tract entitled "On Dress" that was especially directed to women.

Understanding the driving force of cleanliness also involves tracing the association of cleanliness and class, especially in America where geographic mobility and ethnic diversity made external markers more essential than in most other Western societies.[24] Ruth Cowan's work conveys with particular clarity the dynamic linkages between class, cleanliness, and twentieth-century household technology.[25] My window on the process is a single artifact — the white collar — whose history encapsulates central themes in the history of women's domestic work.

Although specific features of the white collar's history differ from anything archaeological evidence is likely to reveal, at least two aspects of its history are germane to interpreting archaeological evidence. First, the case of the white collar directs our attention to a prerequisite for technology's adoption and diffusion that scholars rarely consider: maintenance. The diffusion of the white collar—a clothing technology employed to signify class—cannot be understood without comprehending the nature of laundry technology—the maintenance tools and procedures that assured the white collar's social utility. Second, study of the white collar indicates that we cannot simply write off changes in a garment's form as superficial matters of fashion. As in this instance, fluctuations in form may be firmly linked to a garment's shifting social function. Thus, where archaeologists have reason to suspect the presence of social hierarchy, engendering archaeology may call for close attention to technology's possible symbolic functions. Or it may require imaginative reconstruction of maintenance procedures that were the work of very different people from those who deployed the technology.[26]

The history of the white collar may be recounted briefly. Among early modern English people and colonial Americans, almost no one wore collars, or anything we would consider white. The starched white ruff or collar band had long symbolized gentility, because, unlike the rest of the shirt, its quality and condition were visible.[27] Keeping a collar starched and white required considerable labor, despite the general practice of making ruffs and collars detachable, a practice that at least limited the laundress's most heroic efforts to collars alone.[28] A white collar signified that someone, usually a female servant supervised by the mistress of the house, had performed the following tasks: making soap, hauling water, filling and emptying washtubs a number of times, building a fire and heating the water, soaking the clothes, rubbing the clothes against a washboard, wringing them out several times, soaping dirty spots several times, boiling the clothes, rinsing them, making starch, dipping collars and other parts to be stiffened in the starch, hanging the clothes to dry, dampening them, heating irons on the hearth, testing the irons, and ironing the garments.

During the late eighteenth and early nineteenth century when industrial and political revolutions were transforming Anglo-American class structures, the white collar reflected that social instability in its exceedingly diverse forms. Among the elite, the collar was now generally attached to the shirt, underscoring the owner's ability to pay someone to render the whole garment white. Its size varied considerably. In the early nineteenth century, fashionable men's collars might be so large that before folding they hid the face entirely. Contemporaries clearly recognized what large white collars represented in women's work. One early nineteenth-century periodical commented that when Beau Brummell first sported the new fashion, "dandies were struck dumb with envy and washerwomen miscarried."[29] At the same time, many men eschewed the display of the collar as undemocratic, wearing neckcloths that hid it completely. This was also the era when the "dickey" or false shirtfront with attached collar appeared to accommodate those who couldn't afford to launder entire shirts. Although out of favor among the genteel, detachable collars also persisted. In sum, during an era of enormous social change and unrest, collars reveal the guardians of gentility struggling to maintain "the thin white line which cut the community in two, separating the gentleman of leisure from the manual worker."[30]

Part of the difficulty, of course, was the proliferation of ambiguous jobs whose occupants were neither leisured nor manual laborers: professionals, factory owners, and, especially, clerks and salesmen of various sorts—people who came to be designated white-collar workers. By mid-century the vast majority of those sporting white collars fell into the latter category. What distinguished white-collar workers was that, despite their modest incomes, their work entailed close proximity to employers or customers who required assurance of their social respectability. White collars continued to testify that the wearer could afford their maintenance, which remained considerable. The technical aspects of keeping white collars white had changed little by the mid-nineteenth century save that women now often purchased ready-made soap and they increasingly used cast iron stoves to heat their water and irons. They also generally subjected white garments to an extra rinse in water containing bluing, requiring an extra wringing as well. In general, then, wearing a white collar showed that a man paid a laundress, but also supported a full-time housewife who supervised the labor. In other words, white collars signaled adherence to the basic tenets of Victorian middle-class life.

As the nineteenth century wore on, the tenuous hold many white-collar workers had on middle-class status was recognized by the increased acceptability of detachable collars, although the existence of disposable paper collars reveals that maintaining white-collar respectability remained a struggle for many. It also remained a struggle for their wives, although by the late nineteenth and early twentieth centuries increasing numbers of women had indoor running water, eliminating the heaviest washday chore. At the same time, however, burgeoning white-collar demand made good servants harder to find. One response was to send at least some laundry, especially the breadwinner's valuable white collars and cuffs, to one of a growing number of commercial establishments.[31]

Only in the post–World War I era did detachable shirt collars disappear, meaning that someone had to keep the whole garment white. Initially that someone was likely to be a woman tending a machine in a commercial laundry; men's shirts were the one item even hard-pressed white-collar households managed to send out. After commercial establishments had underwritten the development of a washing machine technology that could be modified and marketed to middle-class households, however, the someone in question was increasingly the housewife who washed, starched, and ironed the family's shirts, using new electrical appliances to compensate for the dearth of domestic servants, appliances whose cost discouraged white-collar families from sending any laundry outside the home. We all know the end of the story. By the 1960s and 1970s, household technology and clothing technology made ironing and starching relatively unnecessary. Instead, housewives laundered many more shirts and blouses in an increasingly diverse array of fabrics and hues, requiring more extensive software in the form of washing procedures and chemical agent selection. The progressive extension of standards of cleanliness is nicely summarized in our ability to concern ourselves with the hidden rings inside our collars.

For me, the persistent extension of the white collar and its associated technological and labor system nicely captures the powerful, largely unexamined role of women's maintenance labor in defining who qualifies for white-collar work. Perhaps part of our current preoccupation with "the decline of the family" stems from our inability to rely on such external evidence to assess and avoid those whose family structures we deplore. After all, we cannot really see inside people's collars. Like the mass-produced calicoes that made it possible for early nineteenth-century working girls to dress like ladies — a phenomenon their contemporaries found simultaneously attractive and unnerving — modern household technology had rendered traditional markers of class increasingly obsolete. Although our culture generally applauds cleanliness, we are made uneasy by its pervasiveness. How can we tell whose homes our children can safely visit? Who can we trust in front of our classrooms? Is there a way to use databases and electronic technology to screen potential life partners?[32]

From Closet to Water Closet: The Technology of Privacy

To comprehend more fully this aspect of technology, let us return to the closet. Or, to be more precise, let us consider what returning to the closet has come to signify. Clearly, in modern parlance, the notion of people being in closets conveys the sense of something amiss. People are being forced to hide something, usually their sexuality, that should not need to be hidden and, thus, in a real sense, to hide themselves. The very choice of the phrase announces that among the many acceptable uses of closets, the housing of people is not one.

Originally, by contrast, that is precisely what closets were meant to do—assure people privacy. All of the early uses of the word refer to small enclosed spaces where people performed acts that early modern English people deemed best carried out in private: sleeping, dressing, praying and meditating, studying and speculating. Indeed, so strongly were closets associated with privacy that in the seventeenth century "closet" served as an adjectival synonym for "private," in the full and positive sense of that word. "Cabinet," the other general term for an enclosed storage space, has a similar history. Only in the seventeenth century, on the eve of industrialization, that is, did closets and cabinets become places for things rather than for people. And part of the software of enclosed spaces that was transposed from people to things was the emphasis on privacy, a linkage succinctly conveyed in one of the industrial era's favorite phrases: "private property."

Neither the limits of the essay form nor the state of my researches permits an attempt to explain how the shifts in who and what deserved privacy played out. One thing is clear: the increased relegation of women to the private sphere was only one of several redefinitions of privacy's role that coincided with the great technological revolutions of the eighteenth and nineteenth centuries. Again, paying attention to feminine technologies matters because the association of women and privacy has been so visible and because women have often been especially associated, rhetorically at least, with other increasingly private aspects of life.

One especially arresting embodiment of the technology of privacy, and one of more than passing interest to archaeologists, is the bathroom. The bathroom ranks as a truly remarkable technological development because of the rapidity with which this utterly new room assumed its standard modern form—a process completed within a few decades in the late nineteenth and early twentieth centuries.[33] The invention of the bathroom meant more than just privatizing excretion by moving it inside the house. It also meant use of a new piece of hardware, generally denominated a "water closet," a term that also applied to the room in which the toilet, to use more modern parlance, was housed. After a few decades in which closets had been deemed appropriate receptacles mostly for things, new standards of privacy made it appropriate to return at least some human activities to the closet.

Moreover, the original hardware of the bathroom makes clear that all, excretion—sweating and menstruation as well as urination and defecation—grew increasingly private. The growing social value assigned to hot water baths to remove sweat and sebaceous excretion was the other principal motive for creating the new room, a motive embodied in the sink and bathtub installed together with the toilet in the standard American bathroom. Simultaneously, the early development of disposable menstrual products permitted the gradual confinement of menstrual blood to the water closet rather than permitting it in the laundry.

The other significant cultural function of the water closet was to "eliminate waste," a phrase worth paying attention to. Environmentally, of course, there was nothing particularly novel about Americans' using the continent's abundant pure water to carry away waste; early American factories were situated on rivers as much for their utility in removing industrial byproducts as for their water power. What was remarkable about the bathroom was the growing conviction that humans could be rendered immaculate, divorced from the aromas and physical evidence of their status as animals. It is both ironic and unsurprising that a people intent on denying their biology created a system of waste removal that helps to threaten our survival as animals, both through its excessive use of and its extensive pollution of water.

Nor is it coincidence, I think, that the era in which Americans committed themselves to eliminating evidence that humans excrete ushered in the era in which engineers became obsessed with eliminating industrial waste. What they meant, unfortunately, was not the elimination of industrial pollution, but the reduction of production workers to automatons through time-motion studies, and the elimination of white-collar and managerial discretion through the wholesale introduction of new record-keeping forms and quantitative assessment techniques. As in the bathroom, so also in scientific management, eliminating waste meant denying the full range of human behavior.

Here, as elsewhere, feminine technology matters, because the technology of women's work made both odor-free bathrooms and paper-filled offices possible. In the process, of course, women, at least, remained fully cognizant that family members excreted and that proliferating forms did not make bosses scientific. Of course, women were creatures of the private sphere—schooled through the decades to launder, rather than air, the dirty linen. It seemed safe to trust them with the secrets.

No wonder women's growing unwillingness to stay in the private sphere or keep the customary secrets has created such a backlash. Far more is at stake here than who changes dirty diapers, although this act, too, has wider technological, social, and environmental implications. What is at stake is an entire economic, techno-logical, cultural system in essence, the society erected through the Industrial Revo-lution. That revolution has been widely touted as a great miracle: creating abun-dance, leisure, longevity, and a host of other good things. Certainly, critics have noted the inequity of the system, but by confining their focus to masculine, public technology they have tended to limit their critique as well. They have mostly been concerned with who runs the machines versus who manages, and with how the spoils are divided. They have generally missed such important and central issues as the qualitative inadequacy of the system's products—such as brassieres or ice-berg lettuce; the inherent tendency of the system to create more work—such as putting everything away in closets or rendering whole garments spotlessly clean; or the frightening prospect that a system so divorced from biological reality in-evitably threatens the biological—as do water closets and scientifically managed factories. Feminine technologies matter because they disclose the system's fatal flaws: it threatens and reduces natural abundance, necessitates increased labor, and promises longevity at best only to the current occupants of the planet.

For archaeologists, I expect that the lessons of these cases will be as diverse as their researches. Certainly many will find my concern with distinctive feminine attire, storage technology, class-specific household labor, and technological activi-ties shrouded in privacy directly applicable to the era and culture they study; they are certainly technological features common to many times and places. Certain themes that surface in my research also seem broadly relevant.

First, because software, the knowledge component of technology, predomi-nates in our society's feminine technologies, it requires a special effort to make the nature and implications of that technology visible. It is especially easy to dis-miss women's technological knowledge—filing items in closets and cupboards or selecting and modifying garments, for example. Although it is certainly possible to recognize in other cultures what we fail to see in our own, our tendency will be to miss precisely those aspects of the past that we miss in the present. Given the nature of artifactual evidence, seeing the software will always require greater com-mitment than seeing the hardware. Gender bias can only exacerbate our blindness.

Second, because the public/private dichotomy in our society is pervasive and emotionally charged, and because technology is culturally associated with public, masculine endeavor, it takes an extra effort to identify as technological those artifacts associated with the private, feminine aspects of culture. Neither the brassiere nor the closet comes to anyone's lips when asked to list some technologies. Rather, we put up a bit of a struggle before conceding their relevance to the topic. Given the nature of the archaeological enterprise, this challenges us to avoid reading such associations into the remote past. Even where we recognize the irrelevance of our culture's public/private dichotomy, we may still tend to see activities currently denoted private as inherently less technological.

Third, the persistent association of Western, feminine technology with human biological functions also tends to render these technologies less visible—to naturalize them in a sense. Whereas we know that we don't need automobiles, industrial research labs, and electric light and power systems, for example, we assume some level of need for food, clothing, shelter, and hygiene and we inadvertently extend the aura of necessity to the technologies that serve those functions. We are less likely, that is, to view human creativity and social choice as just as free ranging when it comes to producing technologies tailored to biological needs. We reason as though the need dictated the shape of the technology. Coming from a Western perspective this may make us especially blind when a culture gives low priority to technologies closely linked to the body, especially the female body.

Finally, if, as I think they do, these case studies of feminine technologies raise fundamental and troubling questions about the entire modern, Western technological enterprise—questions very different from those raised by the technologies we usually study—the lesson for archaeologists and other students of technology seems especially clear. Paying attention to technologies associated with women can neither be dismissed as a luxury nor written off as a concession to political correctness. Some might argue that neglecting feminine technology means telling only half of the technological story. I submit that it means missing the most important parts.

NOTES

I wish to thank Ruth Schwartz Cowan, Mary Kelley, Patrick Malone, Jean Silver-Isenstadt, Fredrika Teute, and Rita Wright for reading and commenting on earlier versions of this essay. I am also grateful to Krishna Haugland, assistant curator of costume and textiles at the Philadelphia Museum of Art, for sharing her expertise in the history of underwear and for showing me a collection of historic brassieres.

1. Like many social historians of women, in framing my early research I depended on scholarship in physical and cultural anthropology, guided by my then colleague Jane Lancaster. See Judith K. Brown, "A Note on the Division of Labor by Sex," *American Anthropologist* 72 (1970): 1073–78, and "An Anthropological Perspective on Sex Roles and Subsistence," in *Sex Differences: Social and Biological Perspectives*, ed. Michael S. Teitelbaum (New York, 1976); Jane Beckman Lancaster, "Sex Roles in Primate Societies," in Teitelbaum, *Sex Differences*; Michelle Zimbalist Rosaldo and Louise Lamphere, eds., *Women, Culture, and Society* (Stanford, Calif., 1974); Nancy Tanner and Adrienne Zihlman, "Women in Evolution, Part I: Innovation and Selection in Human Origins," *Signs* 1 (1976): 585–608; Adrienne Zihlman, "Women in Evolution, Part II: Subsistence and Social Organization among Early Hominids," *Signs* 4 (1978): 4–20; and Rose E. Frisch, "Population, Food Intake, and Fertility," *Science* 199 (1978): 22–30. Many of these are the same works Gero and Conkey cite as current resources for the construction of an engendered archaeology: Joan M. Gero and Margaret W. Conkey, "Tensions, Pluralities, and Engendering Archaeology: An Introduction to Women and Prehistory," in *Engendering Archaeology: Women and Prehistory*, eds. Joan M. Gero and Margaret W. Conkey (Oxford, 1991). My training and practice of the history of technology has also consistently emphasized the importance of the artifact, relying on scholarship in historic and industrial archaeology, and material culture. See, for example, James Deetz, *In Small Things Forgotten: The Archaeology of Early American Life* (New York, 1977); Ian M. G. Quimby, ed., *Material Culture and the Study of American Life* (New York, 1978); Brooke Hindle, ed., *Material Culture of the Wooden Age* (Tarrytown, N.Y., 1981); Robert B. Gordon and Patrick M. Malone, *The Texture of Industry: An Archaeological View of the Industrialization of North America* (New York, 1994); and Judith A. McGaw, "'Say It!' No Ideas but in Things: Touching the Past through Early American Technological History," in *The World Turned Upside: The Reorientation of the Study of British North America in the Eighteenth Century*, ed. William G. Shade (Bethlehem, Penn., 1996).

2. This "high-tech" bias is evident in the history of technology as well as in popular thought. See Judith A, McGaw, "Introduction: The Experience of Early American Technology," in *Early American Technology: Making and Doing Things from the Colonial Era to 1850*, ed. Judith A. McGaw (Chapel Hill, 1994), 2–7; see also John M. Staudenmaier, *Technology's Storytellers: Reweaving the Human Fabric* (Cambridge, Mass., 1985). Needless to say, an emphasis on technology as the purview of engineers and managers means telling a story with an overwhelmingly male cast of characters.

3. Ruth Schwartz Cowan, "From Virginia Dare to Virginia Slims: Women and Technology in American Life," *Technology and Culture* 20 (1979): 51–63; Judith A. McGaw, "Women and the History of American Technology: Review Essay," *Signs* 7 (1982): 798–828; McGaw, "No Passive Victims, No Separate Spheres: A Feminist Perspective on Technology's History," in *In Context: History and the History of Technology*, eds. Stephen H. Cutcliffe and Robert C. Post (Bethlehem, Penn., 1989), 172–91.

4. Here and hereinafter, in the interest of brevity, I use the term "feminine technology" to mean "modern, Western, predominantly American, feminine technology."

5. Staudenmaier, *Technology's Storytellers* (n. 2 above). The popularity of the rhetoric of social construction becomes apparent from even a cursory reading of recent volumes

of *Technology and Culture,* the journal of the Society for the History of Technology. For an outstanding example of such scholarship, see Donald MacKenzie, *Inventing Accuracy: A Historical Sociology of Nuclear Missile Guidance* (Cambridge, Mass., 1990).

6. R. W. Burchfield, ed., *The Compact Edition of the Oxford English Dictionary: Volume III, A Supplement to the Oxford English Dictionary, Volumes I–IV* (Oxford, 1987). Feminine technology throws into high relief the gradual, incremental nature of most technologies' development and, hence, the relative meaninglessness of precise dates of origin. Here and in what follows I will often rely on word origins to supply a good rough chronology. The virtue of this approach is that word use signals the time by which a technology not only was in existence but also had sufficient cultural visibility to warrant linguistic innovation to permit discussion of it. By contrast, other frequently used measures such as patenting or initial production do not guarantee that anyone other than the patentee or the manufacturer had any awareness of the technology.

7. Rosemary Hawthorne, *Bras: A Private View* (London, 1992); C. Willett Cunnington and Phillis Cunnington, *The History of Underclothes* (New York, 1992); Jennifer Craik, *The Face of Fashion: Cultural Studies in Fashion* (London, 1994), 129.

8. Although it is beyond the scope of this discussion, one of the more fascinating aspects of this technology is the mythology that has grown up surrounding it. That feminists burned bras is one such myth. Virtually everyone I discuss my research with knows that feminists burned bras in protest. Indeed, I also knew it. Nonetheless, there is no evidence that anyone in the women's movement ever burned a bra. The power of this image, probably "coined by some feature writer searching for a clever phrase," says volumes about gender and technology in America. Susan Brownmiller, *Femininity* (New York, 1992), 45–46. See also Susan Faludi, *Backlash: The Undeclared War against American Women* (New York, 1991), 75. Of course, brassieres are not unique among technologies in the elaborate mythology surrounding and partially obscuring them, an aspect of technological history that has attracted little serious scholarly attention. Here and in what follows I draw on the rich and growing body of literature derived from probate inventories, described more fully in my own study of such materials. See Judith A. McGaw, "'So Much Depends upon a Red Wheelbarrow': Agricultural Tool Ownership in the Eighteenth-Century Mid-Atlantic" in McGaw, *Early American Technology,* 328–57. I also draw on several decades worth of visits to historic structures, principally in New England, the mid-Atlantic states, and Virginia. Sources cited here form the basis for many of the observations that follow.

9. Brownmiller, *Femininity,* 40; Daphna Ayalah and Isaac J. Weinstock, *Breasts: Women Speak about Their Breasts and Their Lives* (New York, 1979). In the history of technology, most scholarship on this topic treats agricultural technology. See Deborah Fitzgerald, "Beyond Tractors: The History of Technology in American Agriculture," *Technology and Culture* 32 (1991): 114–26; Jim Hightower, *Hard Tomatoes, Hard Times* (Cambridge, Mass., 1972).

10. As with mother-daughter exchanges of information, young women's distrust of older women as a source of information about dress and appearance makes them unlikely to seek advice from saleswomen or to accept advice when it is proffered. It is easy to see how American culture helps young women develop their suspicion. To take but one example, "wardrobe engineer" John T. Malloy, in his first *Dress for Success* book, offered only a few

comments on women's dress, under the heading "How to Set Dress Codes for Women Employees." After noting that men's and women's perceptions of what is good and beautiful in clothing are "diametrically opposed," Malloy adds, "By the way, fifty-five-year old female executives are no better or only slightly better qualified to choose the clothing for young female employees than are their male counterparts." John T. Malloy, *Dress for Success* (New York, 1975).

11. Susan Porter Benson, *Counter Cultures: Saleswomen, Managers, and Customers in American Department Stores, 1890–1940* (Urbana, Ill., 1986).

12. Cowan, *More Work for Mother.*

13. Brooke Hindle, *Emulation and Invention* (New York, 1981).

14. McGaw, "Red Wheelbarrow"; Furnivall et al., eds. *The Compact Edition of the Oxford English Dictionary* (Oxford, 1933); Dell Upton and John Michael Vlach, eds., *Common Places: Readings in American Vernacular Architecture* (Athens, Ga., 1986). As archaeologists certainly know, storage technology has a long history and prehistory. Thus, it is important to note that my phrasing here is not meant to assert the novelty and/or rarity of storage technology generally. Certainly, many early modern houses had attics or cellars where an array of items was stored, and they held barrels, ceramic containers, and other items used to store food in particular. My focus here is on a particular type of storage technology that consisted either of a separate small room built into the living areas of the house or of substantial pieces of furniture that fulfilled similar functions. In the interest of succinct exposition, I will sometimes refer to this particular class of items by the more general terms "storage technology" or "storage spaces."

15. Carole Shammas, "The Domestic Environment in Early Modern England and America," *Journal of Social History* 14 (1980): 3–24; Neil McKendrick, John Brewer, and J. H. Plumb, *The Birth of a Consumer Society: The Commercialization of Eighteenth Century England* (Bloomington, Ind., 1982); Colin Campbell, *The Romantic Ethic and the Spirit of Modern Consumerism* (Oxford, 1987). Here, let me note in passing that most new consumer goods raised people's living standards only after domestic processing. Yard goods, for example, had to be sewn into clothing and linen. Many of the newly emphasized processing tasks—laundering and baking, for example—combined to comprise a new form of labor—an activity for which a new word, "housework," was coined. See Cowan, *More Work for Mother*, esp. pp. 16–18.

16. Clifford Edward Clark Jr., *The American Family Home* (Chapel Hill, 1986); Sally McMurry, *Families and Farmhouses in Nineteenth Century America* (New York, 1988). Despite its increasing evidence in the floor plans illustrating most histories of American family houses, the closet has evoked virtually no comment from historians of American building. The word is, for example, absent from all of the indexes I consulted, offering additional testimony to the relative invisibility of technologies associated with women's work.

17. Furnivall, *Compact Edition of the OED*, 440.

18. Cowan, *More Work for Mother.*

19. Ibid., 162–63.

20. Robert Blair St. George, "'Set Thine House in Order': The Domestication of the Yeomanry in Seventeenth-Century New England," in Upton and Vlach, *Common Places.*

21. It is still true, as I noted more than a decade ago, that we lack a history of filing as a technology. The overwhelming predominance of software over hardware, combined with its feminization, readily explains its neglect. See McGaw, "Women and the History of American Technology," 811.

22. Cowan, *More Work for Mother.*

23. Joann Vanek, "Time Spent on Housework," *Scientific American* 231 (1974): 116–20.

24. This is an especially challenging question to answer for *American* domestic technology. Whenever I talk about the history of domestic technology for an audience that includes Europeans, someone invariably asks about the apparent American obsession with cleanliness, submitting their personal experience as evidence.

25. Cowan, *More Work for Mother,* 152–219.

26. On the importance of imaginative reenvisioning to an engendered archaeology, see Gero and Conkey, "Tensions, Pluralities, and Engendering Archaeology," 20–21.

27. Historically, the shirt was a male undergarment, so displaying the whole garment was not customary. Attitudes informed by the shirt's history still clearly undergird rules of male business and formal attire. See Cunnington and Cunnington, *Underclothes.*

28. In general, my account here and below of changes in shirt form relies on Cunnington and Cunnington, *Underclothes,* modified to take account of American variation as necessary. Given the considerable gulf between early modern behavior and our own, it is probably worth stating that there was no need to keep the shirt body particularly clean. It was not visible, and any odors that permeated it would hardly be perceptible since no one bathed very often. Indeed, the practical function of the shirt was to protect the outer garments from bodily filth.

29. Quoted in ibid., 100.

30. Ibid., 99.

31. Susan Strasser, *Never Done: A History of American Housework* (New York, 1982), 104–24; Cowan, *More Work for Mother.*

32. Here and elsewhere, my research reflects my conviction that the historian's task is to look to the past for insight on contemporary social, political, or cultural concerns; that history is never merely about the past; antiquarianism is. Although not peculiar to feminists, this is an approach especially suited to feminist scholarship because of our commitment to the relevance of personal experience. At least some archaeologists committed to the study of gender share my perspective. See Gero and Conkey, "Tensions, Pluralities, and Engendering Archaeology," 22–23.

33. Siegfried Gideon, *Mechanization Takes Command: A Contribution to Anonymous History* (New York, 1948), 682–711; Strasser, *Never Done,* 96–103.

Why Masculine Technologies Matter

RUTH OLDENZIEL

In contrast to McGaw's non-obvious technologies and female perspective, making the invisible visible, Ruth Oldenziel begins with a
very visible kind of technology: the automobile. She argues, however, that the fondness of boys for cars and the nature of male
technophilia in the twentieth century are anything but obvious, that
boys learn to love their toys with the help of auto manufacturers
and others who have mobilized extensive economic and cultural
resources in the interests of shaping what is partly a consumer
relationship. Like McGaw, Oldenziel insists that we do not assume
boys should like machines any more than girls should like putting
things away in cupboards, cabinets, and closets. In what ways
has technological knowledge been transmitted and nurtured?
How does Oldenziel treat the gendered associations of production and consumption categories in an age when consumers were
increasingly being coded female?

"Boys and Their Toys: The Fisher Body Craftsman's Guild, 1930–1968, and the
Making of a Male Technical Domain," *Technology and Culture* 38 (1997): 60–96.
Reprinted by permission of the Society for the History of Technology.

In 1931, an advertisement for the Fisher Body Craftsman's Guild in *National Geographic* invited teenaged boys to participate in a model-making contest. It showed a boy offering a girl a miniature version of a "Napoleonic Coach" — an image that had been chosen as the emblem of the Fisher Body Company in 1922 to convey luxury, comfort, and style. The emblem had been modeled on the coaches Napoleon I of France used for his wedding and for his coronation as Emperor. Fisher Body, the organizer of the Guild, was the world's largest manufacturer of automobile bodies, which it supplied principally to General Motors. The Fisher Body Craftsman's Guild aimed to train "the coming generation" and to secure "fine craftsmanship" (Fig. 2.1). Intended to appeal to boys of high school and college ages between 12 and 20, the ad portrays "the Fisher boy" as fatherly: mature and responsible, ready to take a bride — a far cry from the boisterous bachelor or daredevil hot rodder. Opposite the Fisher boy stands a girl, positioned as the passive and grateful but critical recipient of his Napoleonic coach and suggesting the kind of future that such a gift seems to promise. The illustration implies that the Fisher boy is not only a builder of coaches, but also a builder of families and security as a future husband and breadwinner.

The Fisher Body Craftsman's Guild (1930–1968), the organization that sponsored the ad, marks one of the most playful by-products of the very successful partnership between Fisher and cosponsor General Motors (GM). At first glance, the Guild invites us to view the world of boys' toys hidden in attics, basements, barns, and backyards as whimsical, playful, and innocent, but a second reading reveals an intricate web of institutions that defined and maintained a male technical domain. The fascinating but now-forgotten history of the Guild suggests that the definition and production of male technical knowledge involved an extraordinary mobilization of organizational, economic, and cultural resources.[1] The Guild, "an educational foundation devoted to the development of handiwork and craftsmanship among boys of the North American continent," directly appealed to boys and relied for recruiting on the Boy Scouts, the YMCA, and the public school system.[2] Girls found themselves excluded as a matter of course.

This explicitly male technical domain came into existence at precisely the same time that "the consumer" became more and more explicitly gendered female, as

THOUSANDS of boys all over America are completing miniature model Napoleonic coaches in the first year's activity of the Fisher Body Craftsman's Guild. These models they will shortly submit in a nationwide competition for four university scholarships of four years each, 98 trips to Detroit, and 882 other valuable awards.

The Fisher Body Corporation sponsored this inspiring movement, believing that this exercise of creative talent, this quickening of the hand of youth, are essential steps toward the development of high ideals—that only by training the coming generation can fine craftsmanship be perpetuated and superior coachcraft be assured.

CADILLAC · LA SALLE · BUICK · OAKLAND · OLDSMOBILE · PONTIAC · CHEVROLET

Figure 2.1. "Fisher Boy Offers Girl His Napoleonic Coach,"
National Geographic, June 1931. This advertisement for the
Fisher Body Craftsman's Guild portrays a boy between
the ages of 12 and 20, of eligible age for the contest.
(From the Collections of Henry Ford Museum &
Greenfield Village, neg. 91.303.2027.)

scholars of consumer culture have argued.[3] Through various means such as the "Body by Fisher" ad campaign, GM and the Fisher Body Company aligned their companies with women as their potential consumers.[4] To consider a single example among many, the same Fisher Body Company that created the Craftsman's Guild ran an advertisement in *Life* magazine in 1927 in which we find a different

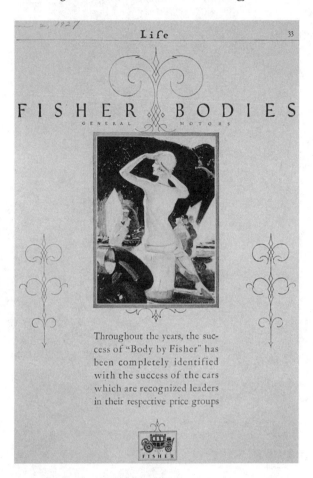

Figure 2.2. "Fisher Bodies," *Life,* June 1927. Advertisement
for "Body by Fisher" campaign, one example among many
created by illustrator McClelland Barclay for the Fisher
Body Company. The ad draws semiotic and graphic
parallels between the automotive and female bodies.
(From the Collections of Henry Ford Museum &
Greenfield Village, neg. 64.167.657.531.)

Fisher girl, a flapper whose body sensuously replicates the curves of an automobile (Fig. 2.2).[5] Seen side by side, these two Fisher promotional campaigns exem-
plify the complementary ways in which we have come to portray men and women
in their stereotypical relationships with the technological world—a world where
men design systems and women use them; men engineer bridges and women cross
them; men build cars and women ride in them; in short, a world in which men are

considered the active producers and women the passive consumers of technology. Both ads point to a specific historical moment in which these roles were being articulated and shaped by GM and the Fisher Body Company. Considered in this light, the exclusion of girls from the Craftsman's Guild was not so much a culturally determined oversight as it was an expression of the need to shore up male identity boundaries in the new world of expanding consumerism precariously coded as female.

The case study of the Fisher Body Craftsman's Guild also suggests that an exclusive focus on women's supposed failure to enter the field of engineering is insufficient for understanding how our stereotypical notions have come into being; it tends to put the burden of proof entirely on women and to blame them for their supposedly inadequate socialization, their lack of aspiration, and their want of masculine values. It also runs the risk of limiting gender, as an analytical tool for historical research, as merely an issue affecting women.[6] An equally challenging question is why and how boys have come to love things technical, how boys have historically been socialized into technophiles, and how we have come to understand technical things as exclusively belonging to the field of engineering. The focus on the formation of boy culture is not to deny that women often face formidable barriers in entering the male domain of science and engineering; they do. The story of the Fisher Body Craftsman's Guild introduces one episode into the institutionalized ways in which boys, male teenagers, and adult men have been channeled into the domain designated as technical.[7]

This article considers one side of the gendering processes. The most substantial part focuses in detail on the male gendered codes in the Fisher Body Craftsman's Guild and its miniature world of model cars to show how from the 1930s to the 1960s the Guild helped socialize Fisher boys as technophiles and sought to groom them as technical men ready to take their places as managers or engineers in GM's corporate world. If the first Guild's advertisement points to the making of a corporate male identity, the second ad suggests, as the Fisher Body Company explained, that the making of the "technical," "hard," and "male" coded world of production has also been produced by and produced its opposite: a world of consumption coded as nontechnical, soft, and female.

Building Model Cars and Male Character

Between the 1920s and 1940s boys' toys developed into a booming consumer market.[8] Wagons, sleds, scooters, bicycles, aeroplanes started to clutter boys'

Figure 2.3. Example of Napoleonic Coach model built for the Fisher Body Craftsman's Guild contest during the period 1930–47. Built from scratch, each coach demanded on average 960 hours to complete. (GM Media Archives, neg. 20134-L-1. Copyright General Motors Corporation, used with permission.)

rooms, while chemistry and Erector sets were sold because "every boy should be trained for leadership."[9] Girls also acquired toys from their parents, of course, but theirs were less varied and not aimed to help smooth a career path. Toys were intended not only to amuse and entertain but also "as socializing mechanisms, as educational devices, and as scaled-down versions of the realities of the larger adult-dominated social world."[10] Many toy companies such as the Gilbert Company, the Wolverine Company, or Toy Tinkers, Inc., exploited the new passion, but none of these companies turned play with toys into the totalizing experience that the Fisher Body Craftsman's Guild managed to create. Under the auspices of GM, the Guild combined the appeal of toys and the modelmaking tradition with corporate needs for training new personnel while crafting consumers' tastes.

The annual Fisher Body Craftsman's Guild contest awarded a $5,000 scholarship at an engineering school to the American or Canadian teenaged boy who managed to build the best miniature Napoleonic coach (1931–1947) or car (1937–1968) (Figs. 2.3 and 2.4). One recruiting sign in 1930 read "Boys!! Enroll here in the FISHER BODY CRAFTSMAN'S GUILD. No dues . . . no fees. An opportunity to earn your college education or one of the 980 other wonderful awards" (Fig. 2.5). When the Guild was founded in 1930, $5,000 was an average worker's income for three years and would buy eight Chevrolets or Fords; in 1940 Americans could buy a house for that price.[11] With a college education perceived as an avenue for upward mobility, young men and their families could gain a great deal from participating in the Guild. GM's investment in the organization was not trivial either: beyond the $20,000 to $100,000 spent on actual awards, the company budgeted at least twenty times more for organizational expenses and publicity each year.[12] Promotional literature boasted that the Guild had the largest membership of any young men's organization in the United States except for the Boy Scouts of America (established in 1910) and claimed that by 1960 over eight million male teenagers between the ages 12 and 20 had participated in the Guild through national, state, and local contests and clubs. Whether these figures are trustworthy or not, it is clear that through its recruitment efforts alone the Guild influenced num-

Figure 2.4. A 1962 example of Free Design Model for the Fisher Body Craftsman's Guild contest. It took on average 275 hours to complete a model after one's own design. (GM Media Archives, neg. X42321-28-A10. Copyright General Motors Corporation, used with permission.)

bers of male adolescents much larger than the high school students who actually managed to finish and submit the complicated models each year.[13]

If the stakes were high, so were the requirements. The teenaged boy who built a miniature coach or car had to be willing to invest an extraordinary amount of time, possess a large measure of patience, and acquire a high level of skill. The Guild's officials apparently realized that a completion of a coach would be extremely challenging without substantial corporate encouragement. Hence, they ensured that replicas would be prominently displayed in department store windows and that color prints and scale drawings were printed in local newspapers and in the Guild's newsletters. To be sure, displays of the Fisher Coach served pro-

Figure 2.5. Applicant to the Fisher Body Craftsman's Guild in 1930 in the Fisher Building in Detroit. The sign behind the woman official reads: "Boys!!! Enroll here in the Fisher Body Craftsman's Guild; No Dues . . . No Fees; An opportunity to earn your college education or one of the 980 other wonderful awards. Join the Detroit Times Chapter." Smaller sign on the desk reads: "For First Instructions in Coach Building see the Sunday Times." (GM Media Archives, neg. 19206-2. Copyright General Motors Corporation, used with permission.)

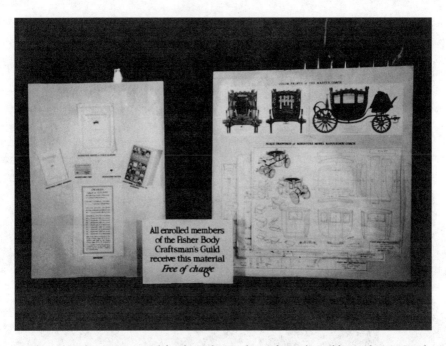

Figure 2.6. Application material for the Fisher Body Craftsman's Guild Napoleonic Coach contest: "Scale drawings, contest rules, guide book to parents, membership button and card" (1930). (GM Media Archives, neg. 19206-1. Copyright General Motors Corporation, used with permission.)

motional purposes as well. Contest rules demanded that all parts be handmade, which necessitated the ability to build a miniature Napoleonic Coach (measuring 11 × 6 × 8 inches) from scratch, to read complicated patterns, to draft accurately, carve wood painstakingly, work metal, paint, and make upholstery with utmost care (Fig. 2.6). Boys of high school and college ages had to construct functioning mechanical parts: windows that could slide, steps that could be folded away, spoked wheels and cambered axles that could turn, and a working leaf-spring suspension. The interior also needed painstaking attention to evoke the proper royal texture of lush upholstery, silk covers, rabbit fur carpets, and brocade curtains. Harking back to the time-consuming labor of craft traditions, the completion of a miniature Napoleonic coach to specification demanded an extraordinary amount of dedication and time—about 3 hours a day for over 10 months—not to mention the investment in materials.

The craft theme presented the organization with a full range of medieval sym-

Figure 2.7. Fisher Body Craftsman's Guild contestants who had won in their home states were treated to a four-day "Fisher Body Convention" where they competed in sports and games. Photograph from 1937. (GM Media Archives, neg. 37633-2. Copyright General Motors Corporation, used with permission.)

bols tailored to contemporary corporate needs. These were smoothly mixed with the most up-to-date technologies of the time: during the 1930s live radio broadcasts announced the winners to parents, family, friends, and neighbors; after World War II, airplanes carried the boys to GM's headquarters in Detroit for the festive four-day Fisher Body Convention. Here, GM officials staged events ranging from essay-writing contests to matches in swimming, golfing, and other athletic events in order to foster the boys' competitive spirit (Fig. 2.7). Finally, the teenaged boys toured carefully selected industrial sites and GM laboratories that served as windows through which they could view their possible future in the corporations (Fig. 2.8).

The evocation of the medieval theme found its culminating moment during the last day of the convention when the organization offered the contestants a banquet and an initiation rite at a candlelit table against a Gothic backdrop. Clad in medieval costumes, the state finalists entered into the corporate world as apprentices under the blare of trumpets. In 1939 Embury A. Hitchcock, a Guild judge and engineering educator, vividly described the spirit of the ritual and showed how the ceremonies marked the transition from apprentice to master craftsman and from boyhood to manhood. He fondly recalled how "the light of flickering candles shows the ornate walls, the heavy-beamed ceilings, and shields and draperies much as they were in the guild halls of Brussels . . . The trumpeteers [*sic*], dressed in doublets, breeches, and buckled shoes . . . lead the procession of contestants, each man carrying his own coach. After the seating, a casement window on the second floor swings open and a representative of the master workmen of the guild days addresses the group on what is required in the way of long years of service to qualify as a craftsman." Most of all, the Guild succeeded in updating the old "cor-

Figure 2.8. Fisher Body Craftsman's Guild contestants, wearing their Guild berets, with corporate officials inside a Fisher Body Plant. Four contestants lift a body top. One worker looks on. (GM Media Archives, neg. 37633-4. Copyright General Motors Corporation, used with permission.)

porate" world of medieval guilds to modern times. Hitchcock described how—after the evocation of European Guild traditions—the medieval ornaments served as a backdrop for GM's American corporate modernity: "a picture of the modern boy, using power-driven tools in building his coach, shows the contrast between work in the Middle Ages and today."[14] By deftly wedding medieval motifs to symbols of the modern age, then, the ritual trumpeted the past and broadcast the future, reaching millions through radio shows, news bulletins, department store displays, photographs, short films, and advertisements.

The Fisher Guild did more, however, than just update the medieval values of apprenticeship for the modern corporate world. As the Guild's ads suggested, the company sought to create a future generation of corporate workers while also expanding consumer markets. During the guild festivities organizers allotted time for shopping trips in downtown Detroit, suggesting that in the expanding consumer society men were no longer just breadwinners and producers but were also expected to take on new roles as consumers. At the same time, the Guild's advocates and GM officials explicitly encouraged Guild winners to seek GM jobs after graduation. During a 1931 radio broadcast announcing that year's winners, GM President Alfred Sloan Jr. extended "to all you boys the opportunity to become employees of the corporation as soon as your schooling is completed."[15] In the depths of the Depression, this was a powerful message indeed.

Sloan's 1931 invitation turned out to be more than a mere public relations ploy, for it was sustained by the corporation's active recruitment policy. As many participants later testified, the sumptuous banquet offered the teenagers easy access to key GM officials, and indeed the event was designed to encourage the boys to converse with men held up as successful role models and potential mentors. Local business leaders, GM chief designers and upper management, and the presidents and deans of major engineering colleges all fraternized with the contestants. GM's attention to the male teenagers went beyond fleeting moments of attention at banquets. By sponsoring a special Alumni Organization, the Guild held winners of past contests up as examples to others. Each year all members of this exclusive club were invited back to the banquet as guests of honor, giving GM ample opportunity to monitor their advances as they grew up. The Guild newsletter, *The Guildsman,* printed biographical narratives next to instructions on how to design the miniature coaches and cars. Working together, these narratives and technical instructions advised simultaneously on building perfect models and proper male character.

The corporation's recruiting efforts paid off handsomely: many of the winners later became chief designers and high-level managers at General Motors and elsewhere in the corporate world. In 1968, for example, 55 percent of the creative design staff at GM had been involved in Fisher Body's Guild, while many other former contestants occupied key positions in other large corporations.[16] "These cars," a design director for Walter Dorwin Teague Associates, Ken Dowd, recalled in 1985, "were truly the beginning of my design career."[17] The Guild, another alumnus remarked, "was very much part of my teenage years. I was a scholarship winner in 1963 . . . The Guild and the Soap Box Derby got me started on the road to being an industrial designer."[18] Over the years, the winners constituted a true fraternity of designers, still voicing deep emotions when recalling the pleasures of joining the Guild.[19] Although the story of these winners fails to account for the many who never managed to complete a model or lost the contest, the Guild's carefully planned efforts reached many more teenaged boys than those who actually submitted models for the contest.[20] Even those who merely learned about the Guild's existence through friends or at school praised its impact. Carroll Gantz, for example, recently recalled its lasting influence when he wrote, "I was not a participant, but I certainly recollect the program's introduction in 1948 as inspirational to my career choice."[21]

The public narratives in the media stressed individual merit and preached "rugged competition," faithful as they were to the middle-class American ideal of the self-made man, but personal recollections suggest that the efforts were often collaborative. Many of the entrants came from the lower middle classes and small towns, and building the coaches and models fitted into the family economy and ambitions for upward mobility. After all, college education was the prize, a potential reward difficult to ignore for a teenaged boy and his family. In 1930, Raymond Doerr's father, for instance, allowed his son to postpone entry into the job market after high school graduation. Young Doerr lived off the family's income for about a year to devote all of his time to the competition. This family decision indeed paid off, because Doerr won the 1931 competition. Other fathers assisted their sons with advice, tools, capital, or skills. Mothers helped with the complicated and elaborate work on the majestic upholstery that adorned the Napoleonic Coach. Myron Webb recalled that his mother "had at one time worked in a millinery shop designing and making hats and did beautiful handiwork. She did the sewing on the inside trim [of the Coach]." Brothers assisted by exchanging skills and sharing earlier experiences in the competition. The Pietruska brothers, Richard, Ronald, and Michael, were all national winners, "needless to say we were very proud of our accomplishments, individually and as a family."[22] Thus, while the contest pushed a masculine identity of autonomy, individuality, and honor in building the cars, actual practices suggest that modelmaking was embedded in the family economy, in which family members shared their talent, capital, and time. Such pooling of family resources is perhaps not surprising given the promise of a scholarship, but it contrasted sharply with the Guild's representations of building proper male character as a lone, individual effort.

These family strategies developed in tandem with GM's search for personnel. The company sought to socialize male teenagers not only as future corporate employees but also as breadwinners, and consumers. As one contemporary observer close to the automobile industry remarked, the goal of General Motors's sponsorship of the Guild "was to build good will, rather than to sell automobiles," but also considered "the boy's influence in automobile selling . . . a very powerful factor."[23] In another appreciative assessment an advertising trade journal stated that the Guild served to whet the boys' appetite as prospective consumers.[24] If this trade-literature assessment is correct, it is particularly significant that the Guild presented the boys in their new consumer roles as knowledgeable producers and builders—a portrayal that stood in marked contrast to the passive roles mapped out for girls in GM's advertisement campaign "Body by Fisher," initiated a few years earlier, and the craft's recruiting literature during this period.

Building Male Institutional Networks

The Guild owed its remarkable success to more than the luster of banquets and the promise of substantial scholarships, however important they might have been in motivating the Fisher hopefuls. In an age of increasing marketing sophistication, the Guild's promoters succeeded particularly well because General Motors's organizational apparatus enabled the company to reach and recruit young men from across the United States and Canada, in a manner so convincing that the contest appeared to be an integral part of the life of the teenaged boy and his family. This was due both to GM's deft mobilization of leading economic, social, and cultural institutions to support the competition and to the intimate organizational parallels between the Guild and its corporate parent.

The organizational shape of the Fisher Body Guild's contest closely resembled General Motors's business organization and followed the company's strategy of multidivisional management structure in various ways. In Sloan's formulation of the GM corporate strategy, the company sought "decentralized operations through coordinated control." While centralized control played an important role, the Guild, like its parent company, invested in local economies and communities around the United States and Canada. By 1933, the Guild's organization had covered over 600 major cities and many more other communities.[25] As is well known, GM's management approach contrasted with Ford's hierarchical and centrally organized structure, which sought to integrate production vertically.[26] Ford and GM differed not only in their internal management structure but also in their views on the world outside the confines of their companies. For one thing, Sloan sought to manage the reproduction of skills and the succession of people through calculated and predictable bureaucratic means. Promoting the virtues of the "Organization Man" as the model of the new corporate worker, Sloan detested idiosyncratic personalities such as Ford's and Durant's.[27] If the Ford company emphasized vertical and backward integration of production, Sloan's strategy stood out because it also crafted a consumer framework for GM's products by seeking to integrate both personnel and consumers forwardly into the organization in a more planned and organic fashion. GM's sophisticated advertising campaigns, such as the "Body by Fisher," represented one means of accomplishing that integration. The Guild represented another.

The Guild also marked an important alliance between the corporation and educational institutions. Its judging system, for example, cemented GM's collaboration with educators by integrating the school system into its ranks. Teams of judges enlisted from local and national educational elites evaluated the models for faithfulness to the original and level of craftsmanship (Fig. 2.9). On the national level, General Motors recruited a group of judges that reads like a roll call of engineering's educational elite. In 1937, for example, five presidents of engineering schools and seven deans of engineering colleges participated. An advisory committee included heads of secondary public school systems and leaders in manual arts teaching. These leading educators had ample opportunity to fraternize with GM's high-level managers and to exchange views with Harley H. Earl, head of the GM Art and Color Section, or Daniel C. Beard, president of the Boy Scouts, while they were in Detroit to judge the many models and participate in the festivities.[28]

How did a boy get involved in such an institutional mobilization in 1930? How did the Guild succeed in becoming such an integral part of the life of the teenager and his family that he would be willing to spend at least three hours a day

Figure 2.9. Fisher Body Craftsman's Guild judge examining 1960 entrants in the free model design contest at the GMTech Center. (GM Media Archives, neg. X36699-11. Copyright General Motors Corporation, used with permission.)

after school on the Guild?[29] Based on information supplied by nearly 200 contestants, a composite biography emerges of how a boy got drafted smoothly into the Guild and learned to nurture his passion for cars as if it were his second nature.[30] There were at least three all-male institutional settings where the teenage boy might be introduced to the Guild: the YMCA, which organized local Guild chapters, provided the first avenue; the Boy Scouts, which also participated in recruitment and integrated the contest into their merit badge program, was the second; and finally, the high school, where the vocational counselor's advice to participate often received further endorsement from the high-school principal's active support, offered a third entryway.[31] GM secured the sponsorship of high school principals, rewarding that collaboration by presenting a trophy not only to the boy who had won the contest but also to the school he attended.[32] To further wed these networks with the appropriate educational message, GM arranged for some thirty-two renowned athletes to narrate stories about enduring difficulties and overcoming initial failures on the path to ultimate success. Finally, GM organized promotional teams that visited 1,200 high schools each year. Some even visited the contestants at home.[33]

Once introduced to the Guild and encouraged to participate in it, the teenager enrolled by submitting his name to the local Chevrolet, Buick, Oakland, Cadillac, or Oldsmobile dealer in his area well in advance of the deadline; officials calculated that on average seven boys enrolled at each GM dealership throughout the country. After enrolling, they received a membership card, a bronze Guild button, a guide for their parents, a detailed manual with plans and instructions, and a quarterly newsletter called *The Guildsman,* which counseled them on how to proceed and paraded previous winners who showed off their successful careers as additional trophies.[34] To cap it off with a show of personal attention, members of the Guild received greeting cards wishing them a Merry Christmas with the compliments of GM.[35] In some cases the support was much more substantial than that: GM divisions such as Delco Remy in Anderson, Indiana, the Packard Electric Company, Mansfield Tool and Die Company, and the Fisher Body Plant in Hamilton, Ohio, organized local guild clubs in their own communities under management supervision to help guide and encourage the boys making their models.[36] James Barnett recalled that community and industrial support for the Guild program in his home town of Anderson, Indiana, was substantial: "it is only in retrospect that I can fully appreciate that guidance we guildsmen received."[37] The judging system, as has been mentioned already, brought GM's corporate management together with the Boy Scouts, high school teachers, and engineering educators. All these institutional networks and rituals helped reproduce distinctly male patterns of paternal mentoring.

The emerging social and economic network extended beyond the coalition between the corporation and the engineering education elite to include the active support of the media. More than twenty national and local newspapers participated in weaving these intricate social and economic networks together into a seamless web. Among several newspapers, the *Detroit News* directly sponsored the Guild by providing weekly instructions on how to plan, design, and build a model; other newspapers faithfully helped to build suspense by carrying accounts of deadlines, events, displays, or announcements of winners throughout the year.[38] The Guild's advocates built the annual cycle of each contest in such a way that reports on the Fisher Body Craftsman's Guild appeared in the press monthly and sometimes even weekly.

Figure 2.10. Above: Winners of the state level contest with their Napoleonic Coach awaiting the announcement of the four national winners at the banquet against a backdrop of the Brussels Guild. *Below:* Two senior and two junior winners for the year 1932. (Smithsonian Institution, National Museum of American History, Division of Transportation, neg. 9988.)

In other words, the contest was as much about building media events and suspense as about building models and male character to which girls had no access. A photograph taken at the annual banquet in 1931 just moments before the winners were announced symbolizes these close parallels most graphically. The photograph shows rows of straight-backed boys identically clad in Guild attire: jacket, beret, tie, and pin. Facing the camera with similar expressions of suspense on their faces, each boy clings to his exact miniature replica of the Napoleonic Coach (Fig. 2.10). We can read this 1931 photograph as a perfect rendition of the emerging male corporate ideal. The contest's demand for exact imitation of the original Coach model is neatly replicated in the demand for identical male character, something that would come to symbolize the ideal of the "Organization Man." As propagated so eloquently by Sloan, GM's corporate male ideal demanded patience, hard work, and a willingness to conform to the rules and regulations of a large organization, the very antithesis of the behavior associated with unpredictable and colorful personalities.[39]

Although many of the contest's features remained the same throughout the years, over time organizers gradually introduced one important change. Until the outbreak of World War II, the contest required that entrants build a miniature replica of the coach featured in the Fisher Body logo, but after the war the Guild's organizers decided to change this requirement, and began to ask for an original design instead of the faithful imitation (Fig. 2.4). This change occurred neither suddenly nor in straightforward fashion, but reflected the contradictions and challenges General Motors faced. If the "free model" design seemed a radical departure from the straightjacket of careful imitation of the craft as represented by the 1931 photograph, a closer look at the change also shows continuities between the values at work in the coach and in the free model contests and between the idealized nineteenth-century culture of production and the twentieth-century culture of consumption.[40]

The Napoleonic Coach versus the Free Model: (Dis)continuities and Contradictions

Why would an automotive giant such as GM and a body company such as Fisher sponsor an organization that harked back to the European Middle Ages and their craft traditions? Why would Fisher Body go to such extraordinary lengths to instill "craftsmanship" in a younger generation when auto manufacturers changed their production methods so thoroughly? How did the Guild fit into the company's overall strategy? The premium put on skilled craftsmanship and endurance entailed a historic irony. At first glance, the Guild's emphasis on craft seemed at odds with the growing economic trend toward a Fordist mode of mass production that sought to eliminate workers and replace them with machinery.

The Fisher Body Guild celebrated the craft ideal and demanded undivided labor (from the purchase of raw materials through tool-making, design, execution, and finishing) at the same moment that production in the Fisher Body plants moved toward the assembly of parts by semiskilled workers. Because of the extraordinary degree of difficulty, many teenagers who started the process never finished; others negotiated the craft challenge by competing year after year; some competed for as long as seven consecutive years—from age 12 when they were first allowed to join the contest until the maximum age of 20 when they lost eligibility. The Guild's initial emphasis on craft as a path of male socialization disguised the emasculating nature of corporate America that produced it. The borrowing of Guild past attempted to recapture and remake a masculine culture in the context of a twentieth-century society looking for new resources. In time, the Guild evolved to fulfill the dual purpose—one concerned with the crafty imitation of existing models, the other based on the inspiration of new designs.[41] This tension of near opposites reflected an often uncomfortable transition within the company and the automotive industry as a whole. The Guild expressed the contradictions, tensions, and solutions of GM's conflictual world of corporate culture that the Fisher family confronted as it moved into the corporation. As historian Roland Marchand has shown, many of GM's corporate strategies during the 1920s not only reflected an "outward quest for prestigious familiarity" but also sought to promote internal loyalty and corporate centralization.[42] The Guild and Fisher's promotional campaigns were no exception.

From a thriving German-American family firm in Detroit during the second half of the nineteenth century, Fred Fisher and his six brothers swiftly built their company into the world's largest manufacturer of automobile bodies when they began to mass-produce closed bodies for various automobile companies during the first two decades of the twentieth century. Before the first World War, combustion-engine cars had been mainly associated with utilitarian farmers or upper-class male adventure and racing.[43] Soon thereafter, automotive design changed dramatically as manufacturers sought to broaden its appeal and market to include women.[44] Closing the automobile's body on all sides did just that. The automobile's closed body moved motoring away from an exclusively sporting, summer, and leisure-time activity to a practical mode of transportation all year round, in all weather conditions. The Fisher brothers simultaneously stepped into and created this new market.[45] "The Fishers kept their eyes on closed car possibilities from the start," one chronicler of the firm explained the Fishers' particular need for women as the company's market niche. "They saw that motoring would remain a summer sport until drivers and owners could be comfortable in the winter months. Women would never be really pleased with the automobile so long as their gowns and hats were at the mercy of wind and weather. After pressing these points on car manufacturers they were at last rewarded . . . for the first 'big order' for closed car bodies."[46] Thanks to the closing of the car's body, the mass of middle-class and urban women could venture out on the road in all weathers. More importantly perhaps, the car's body became the selling point of the automobile as a whole, over its technical specifications. The body of the car "is emphasized by thousands of successful automobile salesmen as an introduction to their selling effort and as an easy and sure way of having the buyer accept the entire car."[47] In this technical and marketing transformation of the automobile, the Fisher Body Company played a critically important role as the world's largest producer of closed bodies and became the key company in GM's marketing strategy to beat Ford and other competitors.

Not only was the Fisher firm phenomenally successful in carving out a powerful new niche in the market, it also succeeded in making a smooth transition from a traditional, craft-oriented nineteenth-century family firm to a twentieth-century division of GM, despite rapid changes in product and modes of production this new market strategy entailed. The Fisher brothers, all seven of them, were brought into GM's managerial structure and without exception became leading corporate managers. At first the brothers successfully negotiated for their continuing control within General Motors. Holding onto their craft-inspired past and playing a crucial role in helping to bring about General Motors's success at styling, the Fisher clan moved into GM's managerial command during the 1930s. Fisher Body remained a family firm tightly embedded into the corporate structure; eventually, however, the brothers became the victims of their own success precisely because their very effective incorporation into the corporate structure rendered them obsolete.[48]

The Fisher Body company's choice of the handmade Napoleonic Coach as its logo illustrates the emblematic ways in which Fisher reworked the discontinuities and contradictions with the corporate world. Fisher Body's Napoleonic Coach did not draw on an old family trademark but represented an invented tradition. "This symbol" read an announcement in 1922, "will appear, from this time forward, on all finished products of the Fisher Body Corporation [and] records the care which the motor car manufacturer has exercised in providing your car with a body of the very best quality obtainable."[49] The imperial coach harked back to an old craft tradition and symbolized comfort and luxury—values believed to appeal to women in particular. Registered in 1922 and officially introduced as a trademark in 1923, the Napoleonic Coach logo began to circulate in the commercial and visual domain in 1926. That year also marked Fisher's incorporation into GM and the surrender of its autonomy as a coach-making firm.[50] Ironically, the Napoleonic Coach—symbol and model for the Guild—represented the craft tradition that the Fisher family was about to lose to GM. The Napoleonic Coach perhaps breathed nostalgia for a Fisher that was long gone, but in the hands of GM it became less a symbol of the past than a malleable and invented tradition suitable for present and future use.[51] The trademark proved so successful that the coach became a stand-in for GM's own logo well into the 1980s. Inside GM, Fisher and its Napoleonic logo mitigated the contradictions between the worlds of craft and of mass production; outside, they carved out a new market. The Fisher slogan "Body by Fisher" in advertisements, featuring the suggestive curves of the female body, sought to convey an image of beauty, elegance, luxury, and craftsmanship associated with European royalty, and held it as a promise to the newly emerging middle classes. Strictly speaking, of course, the European handcrafted Napoleonic Coach was out of reach for American middle classes, but the emperor's mass-produced coach by Fisher beckoned consumers to enter its fantasy world through the illusion of a custom-made body—something a Ford could and would not provide. Most importantly, the emphasis on comfort, luxury, and safety aimed to appeal to women.

However successfully wedded with GM's market strategy, the contradiction of the Fisher craftsman's world within a modern world came to a head in the Guild's contest itself. The image of a Napoleonic coach might have served marketing, public relations, and corporate organization strategies very well indeed, but it failed to fulfill the needs of the participants in the Guild, who labored relentlessly at a task far beyond a world available to them. For a few years between 1930 and 1937, the world of the old crafts reigned supreme in the contest. Then, for another ten years, the old world of craft and the modern world of design were brought together in a delicate balance that worked as a compromise for a short time, but after World War II the modernist design ideal triumphed completely.[52]

The changes in the contest punctuated the shifts in the company and in the fate of the Fisher family firm. The announcement of the free model design in 1937 symbolized the imminent erasure of the Fisher family's past as coach builders and its direct claim to European lineage. While the Fisher Napoleonic Coach trademark continued to figure prominently in successive decades, the decision of the organizers to permit the design of free models announced both the integration of the Fisher family firm into GM's management structure and the firm's growing loss of autonomy. For years, the Fisher brothers clung to their belief in the superiority of wood frames that were sheathed in steel and tried to hold onto their tradition of craft in the production of automotive bodies, but they lost out to GM's increased overall control in 1937.[53] In the same year that the Guild adopted the free design competition based on plaster and synthetics, GM eliminated all wooden parts from their cars. That change in policy also marked the growing confidence of General Motors in a new marketing strategy, where women and style occupied center stage, pushing craft knowledge and technical innovations to the background and marketing strategies to the fore.[54]

Many other analogies existed between the requirements in the contest and the manufacturing of automotive bodies. When Harley Earl joined the Guild as an official in 1937, it was also the first year the Guild allowed the boys to enter a car model after their own design. As the head of GM's Art and Color Section, Earl stood at the center of the new marketing strategy emphasizing style. Funded by the Fisher Body Division, managers expected Earl "to direct general production body design and to conduct research and development programs in special car designs," in his newly established department.[55] He pioneered many new techniques in automobile design that made styling an institutionalized and closely coordinated activity and crucial strategy to GM.

The contest's post–World War II requirements squarely reflected the policy shift from the craft tradition of coach building to streamlined design, personified by Earl. While in both versions of the contest (the Napoleonic coach and the free model), the emphasis fell on building from scratch and using raw materials with no prefabricated parts, the free model mirrored GM's focus on "style," one that came to dominate the automotive industry as a whole.[56] The decision to eliminate the Fisher's Napoleonic Coach from the contest altogether in 1947 effectively consolidated and completed the erasure of the Fisher family's coach-building past. It emphasized style rather than structure. If the lush and majestic interiors and the proper mechanical functioning of parts rendered the points essential for a boy to win the contest in the Napoleonic coach competition, in the free-model contest, mechanically accurate movement had no bearing whatsoever on the outcome. After World War II, smooth exterior finishing and an eye-pleasing style formed the sole criteria for winning a scholarship or money award. "I was impressed" Tristan Walker Metcalfe recalls of a conversation he had with one of the top designers as a young boy at the Guild's banquet in the 1950s, "[at] how unimportant the efficiency and performance were relative to appearance and style, in their attitude then."[57] Typical of the ideology of streamlined design, the Guild' s instruction books and newsletters reflected the change and defined bad design as those models that appeared "slow, boxy, heavy, or square," and good design as "graceful, light, fast, flowing," cultural values that projected the new design curve of femininity.[58]

The parallels between the boy's world of modelmaking and the internal workings of the corporation did not end here. During the 1950s and 1960s, the contest instruction manuals closely followed the first twelve months of the planning stages of GM's celebrated annual model change: sketching, drawing to scale, making a clay model, casting plaster model, and finishing. The manuals recommended that contest entrants use clay modeling as an essential part of the design process just as GM designers employed the clay modeling technique pioneered by Earl to create fluid car lines.[59]

It is hard to say how far GM planned the nexus between the Guild and the corporation, but the linkages forged were powerful indeed. Providing an easy source of design ideas might not have been the original intent of the Guild's advocates when the contest first was introduced in the 1930s, if only because the Napoleonic Coach did not lend itself easily to that purpose, but sometime in the 1950s the Guild's promotional literature cast it as explicit contest policy in no uncertain terms: "it is possible that a submitted model may include a design or idea which General Motors Corporation may use at some time, and it is understood that the Corporation and its licensees are entitled freely to use any such design or ideas." Furthermore, officials warned those who participated in the competition that GM retained the right to "freely use, for advertising or publicity, reproductions of likenesses, statements, names and addresses of Guild members and the models or reproductions of models submitted by them."[60] The most successful model cars generated enough interest for GM's Design Department to go to the trouble of buying them, and in some cases displaying them at GM's headquarters.[61] James Garner recalls, for example, that "my 1955 model sports car entry was purchased by G.M., design rights included. It was on display at their headquarters . . . It was in a traveling show throughout the U.S."[62] And James Sampson remembers seeing, "in the Fisher Body Office Building in Warren, Mich., storerooms on the lower level [containing] a number of models that they had purchased from Guildsmen," which were then loaned to Sampson when he went on a promotional tour for the Guild in 1956.[63] Whatever the Guild's initiators had in mind at first, these statements suggest at the very least that the nexus between the contest organization and GM's design had become unambiguous by the 1950s.

If the Guild's instructions and practices copied or extended the adult world of automotive design, they also revealed major differences with GM and Earl's Art and Color Section. Styling at GM had become a coordinated, controlled, and institutionalized strategy. Fifty designers were employed in this department alone and worked at the introduction of a new model that took two years of planning. While the Guild's manuals instructed teenagers to make models by themselves — a process including tool making, designing, modeling and execution — in Earl's hands this was no longer left to the "haphazard activity of engineers or salesmen as the need for a new model arose," but involved teams of specialists working for over a year.[64] In contrast to the adult world — where the actual shaping and manufacturing of automotive bodies shifted from the hands of Fisher Body's engineers, foremen, and skilled workers to the creativity appropriated by GM designers in the styling department — the boys were instructed to accomplish the same results in seclusion in attics, cellars, and bedrooms at home.[65] No wonder many teenagers who entered the contest never completed it.

Other organizations tried to emulate the Fisher Body Guild but never attained quite the same success. Ford's Industrial Arts Awards Program, for instance, sought to establish a similar program through high school industrial arts and vocational education classes in the 1960s, but organizers soon discontinued it due to lack of enthusiasm.[66] In 1962, the Bank of Dearborn and the Ford Motor Company Design Center established The Greater Dearborn Automotive Design Competition, known as the Thunderbird Design Contest in the period between 1962 and 1971. The distribution of literature, dependent on school cooperation, proved to be a major obstacle because attempts to lure teachers into the program failed. Ford decided to abandon the competition because the "number of participants, public relations and advertising value [it] did not warrant the necessary investment in time, manpower and money."[67] These failures show by contrast just how successful the Fisher Body Craftsman's Guild had been in aligning the school system to the corporation and how complicated the organization, maintenance, and reproduction of male technical knowledge and skills — all neatly lined up with a particular corporate ideal — actually was.

The success of the partnership between educators and GM during the thirties, forties, and fifties is perhaps best illustrated by the dramatic way in which the Guild unceremoniously unraveled in 1968 when the company was forced to terminate the project. Standard promotional procedure had GM officials visit the high schools of the winners, but after John Jacobus won the contest at the state level during the 1960s, he recalled that "when GM came to my high school principal and requested permission to make a presentation to an assembly of 2,000 in my honor, the corporation was turned down."[68] Going against well-established expectations, such official neglect would be startling enough for a teenaged boy who had spent many spare hours on his model. For another, it announced that the coalition between GM and the educational institutions could no longer be taken for granted. More dramatic signs flagged the ending of the once successful alliance between corporations, educational institutions, the teenager, and his family. By the sixties, male teenagers no longer projected their future careers into the corporations, as canvassing corporate representatives were shocked to find out. Someone close to the organization remembered that "in the late sixties, [GM's] presentations at inner-city high schools were not that well-received." He thought that "often the disillusioned, turned-off young of that era felt little motivation to exercise the kind of self-discipline required for the creativity and craftsmanship it took to win even a college scholarship" and concluded, "I hate to say it, but I think a few of our Field Representatives felt fortunate to escape from some of those school assemblies in one piece—it got that bad."[69]

By the stark contrast they provide, these examples illustrate the sheer organizational resources, capital, and good will that had sustained the Guild for over three decades. Even though it had grown to appear quite natural, the values of this corporate, male coalition could no longer be taken for granted by the late 1960s. Other and earlier signs signaled that all was not well in the corporations. Since World War II, for example, the immense popularity of a marginalized subculture of hot rods and drag racing, where young men with ties to the automobile and aircraft industries souped up and stripped down mass-produced cars and raced them illegally, announced the emergence of an exuberant, youthful male rebellion at a grassroots level. As historian Robert Post characterizes these early racing aficionados, they were "unmarried males, many of them ex-GIs, with plenty of spare dollars, enhanced mechanical skills, an assertive bent, and a love of speed." One could read their tinkering as a rebellion of sorts against gender roles mapped out by the corporations. Their illegal activities implied a rebellion against the modern corporate male identity of the "Organization Man" promoted by Sloan and by GM and against what they considered the frivolous effeminate designs coming from Detroit.[70]

In the post–World War II era, the Guild's introduction of a free design model sought to simultaneously convey and plan consumers' freedom of expression. In contrast with the young men, who cherished the buoyant grass roots and autodidactic culture of hot rods and drag racing, the Guild promoted a sense of individual expression that was essentially corporate and adult-sponsored; it was carefully managed, supervised, and circumscribed from above. Moreover, if hot rods and drag racing emphasized high performance, mechanical ingenuity, exposed interiors, and adventure, GM and the Guild merchandised smooth surfaces, lush interiors, convenience, and comfort associated with women and family values. Hot rodders relished stripped-down bodies; GM's automotive style relied on enhanced car bodies as a marketing strategy. As the advocates of the Fisher free design contest instructed, the car's body, not its interior, constituted the pièce de résistance in this new design configuration. The miniaturized world of the Guild's free design contest replicated GM's famed annual model changeover that cosmetically altered the car's exterior irrespective of interior technical specifications. GM's emphasis on enhanced bodies, smooth surfaces, and female models had not been a matter of casual choice but was part and parcel of an elaborate effort to beat Ford's successful formula of Spartan utilitarianism associated with male virtues of thrift.[71] The Fisher firm and its parent company General Motors confronted and produced the female-gendered consumer through several institutional means such as their establishment of the Art and Color Department, their engagement of the advertising agency Batten, Barton, Durstine, and Osborn, and their recruitment of famed illustrator McClelland Barclay for the Fisher Girl campaign, as we have seen. The strategy was so successful that when McClelland's Fisher Girls appeared for the Fisher Body Company in *Life* magazine, *Motor* magazine editor Ray W. Sherman described the change from Fordism to Sloanism as follows: "the automotive business has almost overnight become a feminine business with a feminine market."[72] The strategy of smooth surfaces, convenience, and comfort consciously projected female values, however. As a Fisher advertisement told retailers and other interested readers, "for years Fisher Bodies have been built with feminine tastes in mind."[73] In the new design curve of the automotive bodies that Fisher produced, women and the female body played a prominent if not essential role, as the hot rodders sensed. The 1927 *Life* magazine "Body by Fisher" advertisement (Fig. 2.2) encapsulated GM's and Fisher's deliberate marketing creation of women as consumers.

However successful, the automotive style as a consumable female-coded product proved to be a precarious enterprise, indeed.[74] If women were all surface and cosmetics, the automotive industry's content and operations, the company insisted, were to be left to men.[75] Establishing all-male organizations such as the

Fisher Body Craftsman's Guild was one of many ways to reestablish firm, strict, exaggerated safeguards against possible female incursions. They helped reestablish clear boundaries between designers and users, men and women, and between producers and consumers. While the corporation sought to tease out women as consumers, it recruited boys through the conduit of the YMCA, the Boys Scouts, and other all-male organizations.

The 1931 *National Geographic* advertisement (Fig. 2.1) suggests, as has this article, that the playful world of model cars was not merely light-hearted, diverting, and amusing, nor was it inconsequential: it was a very serious business indeed. While Fisher boys found themselves making model cars with a view to an engineering scholarship, the future proposed to girls cast them as receivers — consumers — of what the boys produced. The world of Fisher Body provides us insights into the institutionalized ways in which boys, male teenagers, and adult men were channeled into the domain designated as technical and, conversely, the ways in which girls, female teenagers, and adult women were positioned as consumers, as the seemingly natural antitheses of the productive, masculine domain. If the gentle path to the showroom was open to all, the hard road to the design department demanded manly virtues acquired through boys' rites of passage, carefully constructed by just such coalitions as the Fisher Body Craftsman's Guild.

The smooth change from the Napoleonic Coach to the free model showed how the craft ideal could be playfully adapted to the new requirements of mass production in a new phase of the automotive industry and consumerism associated with women's new role: style was supposed to recapture notions of freedom and individuality in mass-produced and standardized consumer goods. But attempts to remake the past into the future by employing medieval rituals obscured pertinent facts about the present — including the changing representations of masculinity and femininity. The contest recruited boys to make models, casting them as knowledgeable producers, while Fisher girls were groomed to be models for the new consumer goods. These seemingly clear-cut roles nevertheless obscured a new truth about gender in the new phase of the consumer society. Boys, and for that matter men, were potential consumers as well, as even the supporters of the Guild seemed to have acknowledged. The Guild recruited not only at schools but also through automobile showrooms and department stores such as Macy's and Hudson's, while the organizers made sure to reserve ample time for contestants to shop in downtown Detroit. As we have seen, an industry trade journal considered the influence of boys in automobile purchasing significant enough to argue that, by way of the Guild, General Motors meant to extend goodwill through the recruitment of boys for the modelmaking contest. By the same token, women and girls entered the buyer's market not merely as passive actors but as the essential builders of an expanding consumers' society. Yet in the tales of modern gender mythology told by organizations like the Guild, boys' technophilia, whether as consumer or producer, was born of this role as potential designer.

NOTES

The author is indebted to John Jacobus, Roger W. White (Curator, Smithsonian Institution, National Museum of American History), and Amy James (Archivist, GM Media Archives) for making various primary documents available to her, and to Eileen Boris, John M. Staudenmaier, and Robert Post for careful and insightful readings of earlier drafts of this article. She thanks Nina Lerman and Arwen Mohun for ongoing conversations on the subject.

1. So far, the Fisher Body Craftsman's Guild has not been the subject of any scholarly treatment. John Jacobus generously shared his sustained childhood passion for the Guild with the author and has generated most of the primary source material. Unless otherwise noted, all citations of primary source materials pertaining to the Guild are to Fisher Body Craftsman's Guild Papers (FBCGP), Smithsonian Institution, National Museum of American History, Division of Engineering and Industry, Division of Transportation, donated by Jacobus to the Smithsonian. Jacobus also gave access to his personal collection on the Guild.

2. Letter W. A. Fisher to Walter S. Carpenter, W. S. Carpenter Papers, Series II Part 2, Box 821, Hagley Museum and Library, Manuscripts.

3. On the creation of the consumer as female by corporations and professional groups, see Roland Marchand, *Advertising the American Dream: Making Way for Modernity, 1920–1940* (Berkeley and Los Angeles, 1985); Virginia Scharff, *Taking the Wheel: Women and the Coming of the Motor Age* (Albuquerque, N.M., 1991); by women professionals, see Carolyn Goldstein, "Mediating Consumption: Home Economics and American Consumers, 1900–1940," (Ph.D. diss., University of Delaware, 1994); Jackie Dirks, "Righteous Goods: Women's Production, Reform Publicity, and the National Consumers' League, 1891–1919," (Ph.D. diss., Yale University, 1996).

4. On GM's advertising and marketing strategy, see Roland Marchand, "The Corporation That Nobody Knew: Bruce Barton, Alfred Sloan and GM," *Business History Review* 65 (Winter 1991): 825–75.

5. For examples of the "Body by Fisher" campaign, see the many advertisements from 1926 through the 1960s in such magazines as *Vogue, Life, Saturday Evening Post, National Geographic,* and *Woman's Home Companion.* Created by McClelland Barclay, Fisher Girl ran for nine years and established the genre for Fisher Body company and GM. On Bar-

clay, see *National Cyclopedia of American Biography,* vol. 34, p. 351; W. Blackman, *Facts and Faces by and about 26 Contemporary Artists* (n.p., 1937); *Art Digest,* May 1939, 28; and *Current Biography* (1940): 50–51. GM aligned itself also with the quintessential flapper of the movies, Colleen Moore, who appeared in "Perfect Flapper" and "Flaming Youth"; see *International Dictionary of Films and Filmmakers III—Actors and Actresses* (London, 1992), p. 698, and Colleen Moore, *Silent Star* (New York, 1968), pp. 231–45. For a general history, see Marchand, *Advertising the American Dream.*

6. The classic statements are Joan W. Scott, "Gender: A Useful Category of Historical Analysis," *Journal of American History* 91 (1986): 1053–75, and Sandra Harding, *The Science Question in Feminism* (Ithaca, N.Y., 1986). On gender and technical knowledge, see Ruth Schwartz Cowan, foreword to Cynthia Cockburn, *Machinery of Dominance: Women, Men, and Technical Know-how* (Boston, 1985).

7. Margaret W. Rossiter's *Women Scientists in America: Struggles and Strategies to 1940* (Baltimore, 1982) and *Women Scientists in America: Before Affirmative Actions, 1940–1972* (Baltimore, 1995) are the best analyses of barriers to women's entering the fields of science and engineering. On engineering, see Martha Trescott Moore, "Lillian Moller Gilbreth and the Founding of Modern Industrial Engineering," in *Machina Ex Dea: Feminist Perspectives on Technology,* ed. Joan Rothschild (New York, 1983), pp. 23–37. See also "Women Engineers in History: Profiles in Holism and Persistence," in *Women in Scientific and Engineering Professions,* ed. Violet B. Haas and Carolyn C. Perrucci (Ann Arbor, 1984), pp. 181–205, and Ruth Oldenziel, "Gender and the Meanings of Technology: Engineering in the U.S., 1880–1945," (Ph.D. diss., Yale University, 1992), chap. 6.

8. On the history of toys and consumer culture, see Lawrence Frederic Greenfield, "Toys, Children, and the Toy Industry in a Culture of Consumption, 1890–1991" (Ph.D. diss., Ohio State University, 1991). More specialized information is provided by Erin Cho, "Lincoln Logs: Toying with the Frontier Myth," *History Today* 43 (April 1993): 31–34; Richard Saunders, "Pedal Power: The Kiddie Car," *Timeline* 6, no. 6 (1989): 16–25; "A Glimpse into the Magical World of Old-Time Toys," *American History Illustrated* 19, no. 8 (1984): 22–29; Robert K. Weis, "To Please and Instruct the Children," *Essex Institute Historical Collections* 123, no. 2 (1987): 117–49; Janet Holmes, "Economic Choices and Popular Toys in the Nineteenth Century," *Material History Bulletin* 21 (1985): 51–56; and Mark Irwin, "Nineteenth-Century Toys and Their Role in the Socialization of Imagination," *Journal of Popular Culture* 17, no. 4 (1984): 107–15.

9. Carroll W. Pursell, "Toys, Technology and Sex Roles in America, 1920–1940," in *Dynamos and Virgins Revisited: Women and Technological Change,* ed. Mary Moore Trescott, (Methuen, N.J., 1979), pp. 252–67.

10. Donald W. Ball, "Toward a Sociology of Toys: Inanimate Objects, Socialization, and the Demography of the Doll World," *Sociological Quarterly* 8 (1957): 447, quoted in Pursell, p. 254.

11. Wick Humble, "The Fisher Body Craftsman's Guild: GM's 34-Year Talent Search," *Special Interest Autos,* February 1981, 28; John L. Jacobus, "Once & Future Craftsmen: A Fisher Guild Scrapbook, 1930 to 1968," *Automobile Quarterly* 15, no. 2 (1987): 206.

12. These figures do not reflect the costs incurred by other organizations, such as the Boy

Scouts' recruitment efforts on behalf of General Motors. Jacobus, p. 206; Arthur Pound, *The Turning Wheel: The Story of General Motors through Twenty-Five Years, 1908–1933* (Garden City, N.Y., 1934), p. 300.

13. "The Story of Fisher Body," (Detroit, 1952), General Motors Research Laboratories Library, Warren, Mich., p. 10. All figures come from GM. Pound, p. 300; "Fisher Guild Will Submit Car Designs," *Automobile Topics,* March 1, 1937, p. 157; Embury A. Hitchcock, *My Fifty Years in Engineering* (Caldwell, Idaho, 1939), p. 262; *Fisher in the News* (1966 pamphlet) and clippings, FBCGP; Jacobus, p. 206; Skip Geear, "The Fisher Body Napoleonic Coach," parts 1–3, *Generator and Distributor,* June 1988, 15–19; July 1988, 24–30; August 1988, 15–19.

14. Hitchcock, p. 262.

15. Raymond Doerr Scrapbook (1931) Collection, Archives Center, National Museum of American History, Smithsonian Institution (hereafter, Doerr Scrapbook); Humble, p. 33.

16. Jacobus (n. 11 above), p. 207.

17. Ken J. Dowd to Jacobus, February 21, 1985.

18. Ray Peeler to Jacobus, May 6, 1985. For similar reactions and assessments of the Guild's influence on their career paths, see letters to Jacobus from Raymond Doerr, April 4, 1985; Leo C. Peiffer, April 9, 1985; Bert E. Ray, July 17, 1985; L.W. Jacobs, April 2, 1985; James Garner, February 23, 1985; David Rom, March 6, 1985; Gilbert McArdle, March 20, 1985; Art Russell, undated; Randall Wrington, March 8, 1985; M. B. Antonick, March 4, 1985; James Barnett, undated; David P. Onopa, undated; Albert W. Brown Jr., February 1, 1985; Anthony Simone, undated; Anthony Simone, undated; Lane Prom, February 8, 1985; Dale Gnage, January 22, 1985. The Soap Box Derby was established in 1934 by GM.

19. The sense of male fellowship can be gleaned from the extensive correspondence between Jacobus and other ex-Guildsmen. See Leo C. Peiffer to John Jacobus, April 9, 1985; Jacobus, telephone interview by author, September 28, 1991; Jacobus, conversation with author, August 6, 1996.

20. The records primarily concern winners. Future research will concentrate on the differences and similarities between the winners and the rank-and-file members of the Guild. The Guild's organizers were self-conscious about the difference. "Guildsmen are cautioned not to compare their own model cars with the cars in the exhibit. They must keep in mind that the models in the exhibit are some of the best from among the top winners in past Guild competitions and are the products of four, five, even six different attempts at the project." *The Guildsman* 4, no. 4 (1951): 3.

21. Carroll M. Gantz to Jacobus, November 13, 1984.

22. Throughout Doerr Scrapbook we find evidence of family cooperation in the families of other participants. Michael Pietruska to Jacobus, November 26, 1984. See also "Craftsman Remembered," in *The Arkansas City Traveller,* August 16, 1983; Myron Webb to Jacobus, June 18, 1985; David Rom to Jacobus, March 6, 1985; Jacobus interview. In a still broader context, sisters also contributed to this household economy as they went to work to contribute to the family income, allowing their brothers to go to college. Upholstery was also women's job at Fisher; see "General Motors II: Chevrolet," *Fortune,* January 1939, pp. 36–46, 103–4, 107–10; and Sidney Fine, *Sit-Down: The General Motors Strike of 1936–1937* (Ann Arbor, Mich., 1969), p. 156.

23. Humble (n. 11 above), p. 29.

24. "Building Tomorrow's Customers: How Fisher Body is Securing the Good-Will of Boys Through Its Craftsman's Guild," *Printers' Ink,* November 20, 1930, 11–12; "Guild Wins Goodwill," *System and Business Management,* March 1934, 135–36.

25. Pound, (n. 12 above), p. 300.

26. Alfred P. Sloan, *My Years with General Motors* (1963; reprint, New York, 1986); Arthur J. Kuhn, *GM Passes Ford, 1918–1938: Designing the General Motors Performance-Control System* (University Park, Pa., 1986).

27. James J. Flink, *The Automobile Age* (Cambridge, Mass., 1988), p. 232.

28. Presidents of engineering schools: Thomas Baker, Carnegie-Mellon; M. L. Brittain, Georgia Tech; S. W. Stratton, MIT; P. R. Kolbe, Brooklyn Polytech; R. A. Millikan, Cal Tech. Deans of engineering colleges: M. E. Cooley, University of Michigan; George J. Davis Jr., University of Georgia; W. F. Durand, Stanford University; E. A. Hitchcock, Ohio State University; D. S. Kimball, Cornell University; R. L. Sackett, Pennsylvania State College; T. A. Steiner, University of Notre Dame. Hitchcock (n. 13 above), p. 264; See also Mortimer E. Cooley with Vivien B. Keatley, *Scientific Blacksmith* (Ann Arbor, 1947), p. 137; Dexter S. Kimball, *I Remember* (New York, 1953); Pound (n. 12 above), p. 300; and various issues of the *Guildsman.*

29. In the 1960s the teenagers needed 275 hours to finish a model of their own design.

30. This composite is based on biographical information supplied by 200 participants collated from a variety of sources, including Jacobus's correspondence with a great number of former participants, several issues of *The Guildsman,* various Guild pamphlets describing winners throughout the period between 1930 and 1968, and the rich descriptions contained in local newspaper clippings. In some cases a nearly complete biography emerges; in others only the mere outlines are generated in this manner.

31. Doerr Scrapbook; Humble (n. 11 above), p. 29. The exact class background of the Guild participants and the role the Guild played in their class aspirations remain in question, but some inferences may be drawn on General Motors 1962 statistics on occupations of the fathers of Guild entrants, second, the preponderance of immigrant names and the place of residence in the sample of 200 members collated, and personal conversations with former participants Nil Disco, John Jacobus, Art Mollela, and Rudi Volti. The involvement of the Boys Scouts included more than active recruitment efforts since the organization also helped to formulate the Guild's dress code and chaperoned winners of the state competitions from their home states to Detroit. Pound claimed that the Guild was the only boys' organization sponsored by the Boys Scouts; see Pound, p. 300.

32. Robert D. Smith to Jacobus, November 8, 1984; Humble, p. 29; Jacobus, "Once & Future Craftsmen," (n. 12 above), p. 206.

33. Doerr Scrapbook.

34. Geear (n. 13 above), p. 15; Humble, pp. 27, 29.

35. John Rempel Jr. to Jacobus, August 20, 1986.

36. David O. Burnett to Jacobus, March 10, 1988; *Fisher in the News* and clippings (n. 13 above).

37. James Barnett to Jacobus, undated.

38. Doerr Scrapbook; Humble (n. 11 above); Skip Geear to Jacobus, April 7, 1992; Herbert Lozier to Jacobus, August 10, 1984.

39. Despite his aversion to colorful personalities, Sloan himself did not fit the character of the organization man, as Peter Drucker emphasizes in his introduction to Sloan's autobiography, *My Years with General Motors* (n. 26 above); see also Sloan and Boyden Sparkes, *Adventures of a White-Collar Man* (New York, 1941), and Flink (n. 27 above), p. 232. For contrast, see Warren Susman, "Culture Heroes: Ford, Barton, Ruth," in *Culture as History: The Transformation of American Society in the Twentieth Century* (New York, 1984), pp. 122–49.

40. Susman, *Culture as History;* Marchand, *Advertising the American Dream,* (n. 3 above).

41. Fisher Body Craftsman's Guild Pamphlet, n.d., Henry Ford Museum & Greenfield Village Library and Archives, Dearborn, Michigan; see also *The Guildsman* for the year 1951 and thereafter.

42. Roland Marchand, "The Inward Thrust of Institutional Advertising: General Electric and General Motors in the 1920s," *Business and Economic History* (1989): 188–96, and "The Corporation That Nobody Knew" (n. 4 above).

43. Reynold M. Wik, "The Early Automobile and the American Farmer," pp. 37–47, and Michael L. Berger, "The Great White Hope on Wheels," pp. 59–70, both in *The Automobile and American Culture,* ed. David L. Lewis and Laurence Goldstein (Ann Arbor, 1983).

44. Scharff (n. 3 above), chaps. 3–4. For a more general account, see David A. Hounshell, *From the American System to Mass Production, 1800–1932: The Development of Manufacturing Technology in the United States* (Baltimore, 1984), chap. 5; Flink, (n. 27 above); and Kenneth T. Jackson, *Crabgrass Frontier: The Suburbanization of the United States* (Oxford, 1985), chap. 6. Electric cars had been catering to upper-class women at an earlier date, but failed. Rudi Volti, "Why Internal Combustion?" *American Heritage of Invention and Technology,* Fall 1990, 42–47; Mark Schiffer, *Charging the Wheel: Women and the Electric Car* (Washington, D.C., 1995).

45. Roger B. White, "Body by Fisher: The Closed Car Revolution," *Automobile Quarterly* 29 (August 1991): 46–63, and "Fisher Body Corporation," in *The Automobile Industry, 1896–1920,* ed. George S. May (New York, 1990), pp. 187–92. Surprisingly, White's articles are the only scholarly treatment of the subject. I am grateful to him for sharing his research on the subject. See also Michale Lamm, "Body by Fisher," *Special-Interest Autos,* May–June 1978, 18–25, and Donald Finlay Davis, *Conspicuous Consumption: Automobiles and Elites in Detroit, 1899–1933* (Philadelphia, 1988).

46. Pound (n. 12 above), p. 289.

47. "The Origin of the Emblem 'Body by Fisher'" (July–August 1928), Fisher Body Papers, Division of Transportation, Division of Engineering and Industry, National Museum of American History, Smithsonian Institution.

48. White, "Body by Fisher" and "Fisher Body Corporation."

49. "Fisher Bodies," *Saturday Evening Post,* June 24, 1922, 37.

50. From 1926 until 1984, Fisher was a division of the GM company.

51. On the uses of history by major companies, see Susman, *Culture as History* (n. 39 above).

52. The annual competition ceased for several years because of the war.

53. White, "Body by Fisher," p. 63.

54. Richard Tedlow, *New and Improved: The Story of Mass Marketing in America* (New York, 1990), chap. 3; Marchand, "The Corporation That Nobody Knew" (n. 4 above) and "The Inward Thrust of Institutional Advertising" (n. 42 above).

55. Flink (n. 27 above), p. 235.

56. *The Guildsman* explicitly stated that interiors were no longer important but that style was; *The Guildsman* 5, 1 (1957). In the 1960s, controversy arose over GM's emphasis on design and styling and would soon become the focus of consumer advocate Ralph Nader's indictment, *Unsafe at Any Speed: The Designed-In Dangers of the American Automobile* (New York, 1965). In it, Nader accused GM of callous negligence in the design and manufacture of the rear-engine Corvair. For further criticism, see Jeffrey O' Connell's *Safety Last* (New York, 1966).

57. Tristan Walker Metcalfe to Jacobus, November 25, 1984.

58. For an example, see Fisher Body Craftsman's Guild, "How to Build a Model Car" (1957), General Motors Research Laboratories, Library Collection, pp. 2–3.

59. Flink, p. 201.

60. Fisher Body Craftsman's Guild, "Designing and Building a Model Car" (1958), General Motors Research Laboratories, Library Collection, p. 26.

61. Harold C. Krysak to Jacobus, n.d.; James Garner to Jacobus, February 23, 1985; Art Russell to Jacobus, n.d. Norman Law sold his model to GM; see Fred C. Schollmeyer to Jacobus, February 21, 1985.

62. James Garner to Jacobus, February 23, 1985.

63. James T. Sampson to Jacobus, January 1, 1985.

64. Flink (n. 27 above), p. 236; Jeffrey L. Meikle, *Twentieth Century Limited: Industrial Design in America, 1925–1939* (Philadelphia, 1979), p. 12.

65. For some insights into the role of foremen and skilled workers in styling of cars on the shop floor at Fisher, see White, "Body by Fisher" (n. 45 above); on the rise of designers over engineers in the Earl's styling department, see "General Motors II: Chevrolet," *Fortune,* January 1939.

66. David L. Lewis to Jacobus, January 10, 1986.

67. H. Benson to Jacobus, March 13, 1974.

68. Jacobus interview (n. 19 above).

69. Humble (n. 11 above), p. 34.

70. For a rich description of this subculture, see Robert C. Post, *High Performance: The Culture and Technology of Drag Racing, 1950–1990* (Baltimore, 1994). See also Gerald Silk et al., *Automobile and Culture* (New York, 1984), pp. 177–250. For an unsurpassed journalistic account, see Tom Wolfe, *The Kandy-Kolored Tangerine-Flake Streamline Baby* (New York, 1972).

71. Meikle (n. 64 above), p. 12.

72. Scharff (n. 3 above), p. 115. On Batten, Barton, Durstine and Osborn, see Sloan, *My Years with General Motors* (n. 26 above), and Marchand, "The Corporation Nobody Knew" (n. 4 above) and "The Inward Thrust of Institutional Advertising" (n. 42 above). On the

Art and Color Section, see Meikle. On the subject of Gibson Girls, flappers, and fashion, see Jennifer Craik, *The Face of Fashion: Cultural Studies of Fashion* (London and New York, 1994).

73. "A Coach for Cinderella," *Literary Digest,* February 27, 1932, 25.

74. For a general discussion on the precarious nature of the female market for the advertising profession, see Marchand, *Advertising the American Dream* (n. 3 above), chaps. 1 and 2, and Scharff, *Taking the Wheel.*

75. "A Coach for Cinderella."

THREE

Situated Technology: Meanings

REBECCA HERZIG

McGaw and Oldenziel both demonstrate the range of objects falling under the broad rubric of technology and highlight the extent of technological knowledge necessary to make such objects useful, desirable, or profitable. Their discussions remind us of the importance of human agency—users as well as designers—in understanding technological change, and thus of the need to examine various human ideas about what they (and others) want to accomplish in the material world.

Rebecca Herzig asks us to consider not only what technology is but also what things people choose to make and do, in which contexts, and why. Her article explores a now-familiar medical technology—the x-ray machine—sought out by women for what is now an unexpected purpose: facial hair removal. Radiation effectively causes hair loss, albeit with some risk. Hair loss— or hair removal, depending on one's perspective (and generally on the location of the hair in question)—carries highly gendered meanings, and highly racialized ones as well. For the women she studies, these meanings were potent enough to overcome the risks involved in x-ray hair removal. Why, according to Herzig, was hair removal so important to these women at this time? How, in Herzig's portrayal, does gender shape technology? How does technology in turn shape and maintain gender ideologies and gendered identities?

"Removing Roots: 'North American Hiroshima Maidens' and the X-Ray," *Technology and Culture* 40, no. 4 (1999): 723–45. Reprinted by permission of the Society for the History of Technology.

On 25 February 1940, an officer with the San Francisco police department's homicide detail reported a "rather suspicious business" operating in the city. At 126 Jackson Street sat an old, three-story rooming house, recently leased by Dr. Henri F. St. Pierre of the Dermic Laboratories. As Assistant Special Agent J. W. Williams later described the scene, "women had been seen entering the place from the Jackson Street side at various times of the day, subsequently leaving by . . . an alley at the rear of the building. Following the arrival of the women, cars would arrive with a man carrying a case resembling . . . a doctor's kit. They would also enter the building for a short time, come out, and drive away. . . ."[1] At first sight, the medical kit, the furtive departures, and the seedy locale all signaled to Williams that St. Pierre was running a "new abortion parlor."[2] As it turned out, however, "the so-called 'Dr.'" was offering a somewhat different service to these women: the removal of their unwanted body hair through prolonged exposure to X rays.

At the time of Williams's writing the practice of removing hair with x-radiation was thoroughly disreputable — if not illegal — in most of North America. By 1940, health officials and X-ray clients had long since realized that intense doses of ionizing radiation not only removed hair, but also led to other, dangerous physiological changes. Articles lambasting the potentially lethal practice had been appearing regularly in medical journals and popular magazines since the early 1920s, and these articles only became more graphic with the passage of time. In 1947, an article in the *Journal of the American Medical Association* described in gruesome detail dozens of cases of cancer resulting from depilatory applications of x-radiation.[3] In 1970, a team of researchers found that more than 35 percent of all radiation-induced cancer in women could be traced to X-ray hair removal.[4] In 1989, two Canadian physicians suggested a new name for the widespread pattern of scarring, ulceration, cancer, and death that affected former epilation clients: "North American Hiroshima maiden syndrome."[5]

Although physicians might now liken X-ray hair removal to international atomic attack, the analogy obscures several of the most crucial aspects of the history of this technology. To begin, unlike the famous "Hiroshima Maidens" (twenty-five young bomb survivors brought to the U.S. for plastic surgery in 1955), epilation clients were anything but the unsuspecting targets of foreign military action.[6] As Williams's 1940 description of back-alley hair removal makes clear, these individuals were not "passive victims" but willing participants in the diffusion and persistence of a controversial technology.[7] Moreover, unlike the atomic explosions that punctuated the summer of 1945, X-ray hair removal was not a dramatic aberration in the history of technology. Rather, from its first use among professional physicians in 1898 through its slow demise in the late 1940s, X-ray epilation enjoyed nearly fifty years of continuous practice. One 1947 investigation concluded that thousands of Americans visited a single X-ray hair removal company, the Tricho Sales Corporation.[8] Since Tricho was just one of dozens of similar X-ray epilation companies in operation in the 1920s and 1930s, one can conclude that tens of thousands—if not hundreds of thousands—of other American women also irradiated themselves in order to remove unwanted body hair.[9]

Rather than focusing on the eventual physiological impacts of X-ray epilation, a tragic story already told in meticulous detail by medical reports, this article explores the circumstances in which prolonged, repeated self-irradiation seemed appealing to its myriad users and promoters. Understanding the practice's allure requires consideration of two questions: first, why did so many early-twentieth-century American women wish to remove their body hair? Second, why would some women choose the X ray over other available hair removal technologies? The answers to these questions, we will see, quickly lead from X rays and hair to larger problems of race, sex, and science in the interwar period.

The Emergence of "Superfluous" Hair

The manipulation of human body hair has a long and rich history. Like teeth and foreskin, hair has been removed according to prevailing social custom for millennia.[10] Yet if unwanted hair and techniques for its removal have been around—in the words of one commentator—since "hoary antiquity," the years after 1870 saw an increasing fascination with "superfluous" hair in the United States.[11] In the wake of Darwin, body hair became newly invested with evolutionary significance and attendant questions of racial and sexual difference.[12] Changing patterns of immigration stimulated further attention to comparative physiognomy, while shifting economic and political roles for white, middle- and upper-class women provoked particular interest in "woman's" proper physical appearance.[13] The establishment of the American Dermatological Association in 1877 created a professional medical niche for the study and treatment of *hypertrichosis,* the condition of excessive body hair thought to be most troubling among young white women.[14] As excessive hair became increasingly identified with individual pathology, the threat of transgressed sexual roles, and racial atavism, new medical and popular energy was devoted to "remedying the evil" of superfluous hair.[15]

Scientific and popular fascination with excessive hair expanded still further between the two world wars. In the years between 1914 and 1945, as historian Christine Hope has demonstrated, popular U.S. women's magazines increasingly promoted models of hairless, white feminine beauty.[16] As ideals of smooth, white skin spread through the nation's periodicals, some American women developed new anxiety about hair growth. A number of these conveyed their mounting apprehension about body hair to popular magazine editors, beauty experts, and health advisors.[17] Medical practitioners also noted the intensity of women's concern, describing the severe depression, self-imposed seclusion, and nausea common to those "afflicted" with superfluous hair.[18] Physician Paul Bechet recalled one typical patient who "gave up a very lucrative position, shunned all her acquaintances, refused to go out unless heavily veiled, and slowly drifted into true melancholia" due to her hairy condition.[19] Others described patients who considered (or accomplished) escaping their misery through suicide.[20] Struggling to accommodate intensified ideals of hairlessness, many American women began to strip their body hair, removing the troubling growth from arms, legs, faces, armpits, breasts, buttocks, and pubic regions. By 1938, one expert could declare without sarcasm that any hair not on a woman's scalp was rightly considered "excessive."[21]

Such newly stringent norms of feminine hairlessness expressed changing cultural understandings of race. In the United States, where questions of racial identity have long focused on perceived morphological characteristics, discussions of physical appearance carry particular histories of race and racism.[22] Like skin color, body hair has long been central to perceptions of human variation, and concern with hair, hair growth, and hair manipulation often mirrors more general attitudes toward race and racial difference.[23] In the nineteenth century, for instance, most published discussions of human hair reflected a larger interest in the recognition and classification of distinct races, as evident in Peter A. Browne's 1853 *Trichologia Mammalium.* In this elaborate treatise, Browne delineates "three distinct species of human beings" based on differences in the quality of hair types.[24] Browne describes new instruments — the trichometer, the discotome, and the hair revolver — designed to divulge and quantify these fixed somatic differences. Other

North American and European naturalists similarly measured human hair growth in order to elucidate racial taxonomies. The measurement, classification, and manipulation of human body hair, in short, enabled and upheld nineteenth-century taxonomic understandings of race.[25]

By the 1920s and 1930s, nineteenth-century efforts to define stable, distinct racial categories had shifted to apprehension about racial mutability and blending. "White" was no distinct and fixed category of human being, but an identity that required maintenance and display. Concerns about hair shifted accordingly. Amid new discussions of immigration, racial degeneration, and "passing," women's body hair came to signify both the promise and threat of slippery racial boundaries. Some beauty advisors began to underscore the racial significance of body hair, emphasizing the importance of a pale, hairless complexion.[26] Physicians treating hypertrichosis similarly began to focus their attention on the racial characteristics of the disease. While debating environmental, gonadal, and psychological influences on "excessive" hair growth, most concluded that the ailment pointed to subtle, underlying racial differences. As Ernest L. McEwen concluded, "heredity is the most frequent etiological factor. The condition is seen very often in several members of the same family, and may be traced in the lateral branches of the same stock. A racial tendency is observable, notably in those of Jewish and Celtic extraction."[27] Emerging in a period of widespread anxiety about racial mixing and racial discrimination, commercial hair removal salons echoed popular and medical opinion on the value of "smooth, white, velvety skin," linking the eradication of hair to the eradication of "troubling" racial markers.[28]

While the emergence of new norms of hairless, white femininity helps account for some women's increasing interest in hair removal, these norms alone do not explain the prevalence of the specific technique of X-ray hair removal. Although dozens of other feasible hair removal technologies were readily available during the interwar period, thousands of American women sought this particular method of epilation. Indeed, clients continued to pursue X-ray hair removal even as the practice was denounced in popular women's magazines and assailed by medical and legal authorities.[29] While not all women remained committed to the technique once learning of its dangers, X-ray epilation persisted even after the practice's formal prohibition, as Assistant Special Agent Williams's report on the surreptitious activities at 126 Jackson Street demonstrates.[30] In the context of a widespread cultural demand for feminine hairlessness, how might we account for the peculiar popularity of hair removal with the X ray?

Selling the Imperceptible

To understand the enduring and widespread appeal of X-ray epilation, we must begin where consumers of the time would have: with a comparison between the X ray and other available hair removal technologies. During the 1920s and 1930s, a woman with unwanted body hair had a variety of temporary solutions at her disposal. Abrasives such as fine-grained pumice stone, sugar-paste solutions, or so-called "Velvet Mittens" made from sandpaper all depilated hair at the surface of the skin, but were expensive and likely to lead to skin irritation and scabbing.[31] Razors of various types could also be used to remove hair at the skin's surface, yet several commentators noted the distaste women had for shaving themselves, particularly for shaving their facial hair.[32] Since tweezing hairs individually frequently proved too time-consuming and too painful for densely covered areas, modified shoemaker's waxes were used to rip off large patches of enmeshed hair in a single motion. Waxes, too, remained expensive and painful to use.[33]

Chemical depilatories were also available, ranging from ineffective peroxides through irritating, foul-smelling sulfides to the outright lethal thallium acetate solution known as Koremlu.[34] These hair removers were not only hazardous but also caused significant discomfort to the user. One physician, H. L. Baer, speculated that the pain associated with chemical depilatories was not limited to the location of hair removal, but might also be transmitted "along the branches of the facial nerve to the teeth and the tongue," where metallic fillings and bridgework may help "to conduct the nerve current of the pain." Baer advocated the placement of a wooden barrier between the teeth of the patient during chemical epilation, thus "interrupting the transmission of this nerve current."[35] As a final method for the temporary removal of hair from the root, at least one specialist recommended a procedure known as "punching." With this technique, a slender cylindrical knife was inserted through the cutis around the hair shaft and immediately withdrawn, leaving a severed column of skin containing the hair-root. Punching was never one of the more popular methods of epilation.[36]

Each of these methods of epilation, although generally effective, provided only temporary relief from superfluous hair, since the follicle itself often remained intact and capable of still further hair growth. For a more lasting effect, women of means might try diathermy or electrolysis—complicated, expensive procedures typically practiced outside the home by experienced specialists.[37] Technically, diathermy and electrolysis were quite similar. First, a slender needle charged by a galvanic battery was inserted directly into the hair shaft. The electrical circuit was then closed and a slight bubbling at the skin's surface was produced as both the hair root and surrounding tissues were blanched. The roasted hair was then easily removed with "epilation forceps" (more commonly known as tweezers). Unlike razors, waxes, or other hand tools, diathermy and electrolysis both involved contact between the woman, a separate, skilled operator, and sophisticated, spark-snapping machinery (fig. 3.1). The techniques required not only meticulous attention and skill on the part of the operator but also extreme patience and tolerance for pain on the part of the client. Indeed, since the client was often responsible for completing the electrical circuit by grasping an electrode of the machine, her fortitude enabled the very functioning of the apparatus.[38] As one physician described the Kafkaesque procedure: "When the needle is inserted the patient is directed to touch the electrode. She naturally touches it first with the end of one finger, adding gradually another one, till all five fingers are in contact with the handle, ready to grasp the handle with the full palm, if necessary, thus controlling by her own action the degree of intensity of the current."[39] In order to ensure the full intensity of the current and thus the full efficacy of the treatment, occasionally anesthetics such as cocaine were necessary to help the client become "hardened" to her task.[40]

Compared to these methods of epilation, the new rays offered several distinct advantages. To begin with, they were undeniably effective at removing hair.[41] Even the American Medical Association's Bureau of Investigation—one of the staunchest opponents of X-ray hair removal—grudgingly admitted the amazing results achieved with this technique.[42] More importantly, the rays bypassed the inescapable physicality of all other hair removal technologies. Appropriately named "X" by Roentgen in recognition of their enigmatic nature, the rays were alluringly imperceptible. Gone were the noxious smells of depilatories, the root-ripping pain of hot waxes, and the frightful appearance of multiple electrolysis needles. In the most popular commercial hair removal chain, Albert C. Geyser's Tricho salons, the client was seated before a large mahogany box containing the X-ray equipment (fig. 3.2).[43] A metal applicator the shape and size of the area to be treated was adjusted to the box's small front window, and the turning of a switch started the operation. The X-ray equipment itself was visible to the user only through a small window in the front of the machine, and the machine shut off automatically after the appropriate period of exposure, usually three to four minutes.[44] The

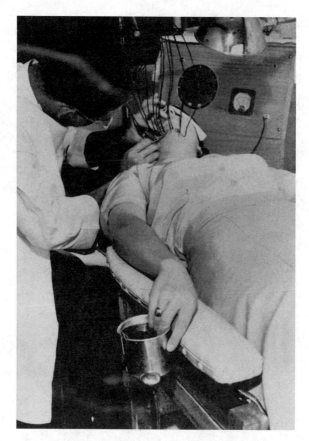

Figure 3.1. Electrologist and client, demonstrating
the multiple needle technique. (Delmar E. Bordeaux,
Superfluous Hair, Its Causes and Removal [Rockford, Ill.,
1942], 48. Reproduced with permission of the American
Medical Association.)

client might hear the sounds of electrical generation or detect an odd ozone smell,
but the epilating rays themselves were clean, quiet, invisible, and mysterious.

Commercial salons seized on the imperceptible nature of the rays as their primary selling point, and eagerly perpetuated the promise of effortless self-fashioning available with the X ray. As one epilation pamphlet assured, "The woman being
treated absolutely does not feel, hear, or see the action of the ray."[45] Or, as a number
of advertisements summarized the experience, "Nothing but a ray of light touches
you."[46] Not only would the gentle new light banish hair, the promotions further

Figure 3.2. Albert C. Geyser's Tricho Machine, with automatic timer. (Reproduced with permission of the American Medical Association.)

emphasized, but also the messy, smelly, time-consuming labor once required for its removal. Hand tools such as tweezers and sandpaper were described as "Antiquated Methods," rudimentary vestiges of a painful, toilsome past happily superseded.[47] One Boston newspaper advertisement urged prospective clients to think of the "joy of freedom from depilatories or razors" possible with the X ray.[48] The theme of women's emancipation from routine work was reiterated in a pamphlet from a Tricho salon in Detroit, which concluded simply: "It is no longer necessary for any woman to resort to the old makeshifts, since science has shown the perfect way."[49]

The "Science" of X-Ray Epilation

The ray's imperceptibility augmented an equally crucial aspect of its popularity: its association with "science." Like countless other early Progressive-era businesses, hair removal salons learned that references to "science" both piqued customers' interest and provided the enterprise with an aura of legitimacy.[50] Rhetorically appropriating this aura, salon operators made frequent references to the "scientific" methods and equipment used in their establishments, and to the "scientifically-sound principles" on which their epilation process was based. Yet unlike many other early-twentieth-century businesses, the salons' references to science were not simply spurious. To consider the relations between science and X-ray epilation, we must return to its origins in nineteenth-century physical research.

The hair-removing properties of X rays were discovered accidentally by two Vanderbilt University researchers in 1896, within weeks of Roentgen's first public announcement of the "new kind of light."[51] In March of that year, researchers John Daniel and William L. Dudley were asked to locate a bullet in the head of a wounded child. As Daniel later recalled the chain of events, Dudley, "with his characteristic devotion to the cause of science," agreed to lend himself to an experiment with skull X rays. Twenty-one days after Dudley's head had been exposed for an hour with the tube placed half an inch from his scalp, Daniel reported that all hair had fallen out from the area held closest to the tube.[52] Word of the researcher's new bald spot spurred substantial "editorial merriment. . . . [T]here were even suggestions in the newspapers and technical journals that the X rays might render daily shaving obsolete."[53]

Soon after these merry speculations, professional physicians began to experiment with the X ray in the treatment of hypertrichosis. In 1898, two Viennese dermatologists, Eduard Schiff and Leopold Freund, published the first good results of this medical therapy.[54] On the heels of their success, scores of dermatologists, roentgenologists, and other physicians in both Europe and North America adopted the effective new treatment.[55] By 1910, one specialist declared that the "electric needle, formerly so prevalent, but tedious and painful in operation, has largely given way to the X-rays."[56]

In fact, medical use of the electric needle never actually gave way to X-ray epilation. Some European and North American physicians continued to use X rays well into the 1920s to remove hair from skin intended for grafts or on skin affected by ringworm, but most U.S. physicians had grown reluctant to use the X ray for other hair removal even before the First World War.[57] Why would physicians decide to abandon the highly effective (and lucrative) practice of X-ray epilation?

To be sure, American physicians' abandonment of X-ray epilation stemmed partly from their increasing recognition of radiation risk. As time provided further evidence of the latent effects of X-ray exposure, physicians grew loathe to expose their patients to the hazards of the ray.[58] Concern for radiation risk alone, however, does not account for physicians' eventual disavowal of X-ray epilation. Scientists, technicians, and physicians had grown wary of radiation "burns" prior to widespread experimentation with X-ray hair removal, and their support for the technique continued even as they witnessed the destruction wrought on the bodies of X-ray "pioneers."[59] It would be difficult, therefore, to attribute physicians' increasing reluctance to treat superfluous hair with X rays solely to their sudden recognition of the effects of radiation on human tissue.

The decline of medical X-ray epilation in the 1910s reflected not only increasing awareness of radiation risk but also physicians' growing unwillingness to apply the prestigious new ray to "minor" concerns such as hypertrichosis. Definitions of "excessive" hairiness were maddeningly fluid, and the intractable ambiguity of the disease's diagnosis moved hypertrichosis to the contested border between "cosmetic" and "medical" concerns. Although most practitioners agreed that excessive hair was a matter of widespread public interest, hypertrichosis remained on the margins of professional therapeutics, the country cousin of more stately concerns such as cancer and tuberculosis.

The X ray, on the other hand, enjoyed unquestionable sovereignty as one of medicine's crowning achievements. Particularly in the wake of the damning 1910 Flexner report, American physicians grasped such achievements to shore up their troubled professional authority. For these physicians, the X ray played a crucial role not only in medical diagnostics and therapeutics but also in enhancing the status of the medical profession. As a result, physicians grew increasingly reluctant to use the X ray, a symbolically and materially potent therapy, on a problem that hovered on the fringes of professional respectability. Seeking to preserve the "scientific" clout of radiation therapy, many physicians sought to restrict its application.[60]

Accordingly, physicians pressured one another to resist patients' requests for X-ray epilation. One physician urged his fellows to resist the natural temptation to comply with patients' willingness to employ x-radiation, since their condition was a "cosmetic defect" rather than a "serious disease."[61] As another practitioner summarized, "It is not customary to shoot at sparrows with cannon-balls. Why, if we treat a hairy surface of the face of a fair lady, for instance, resort to means as powerful as those we employ in carcinoma?"[62] Through such admonitions, X-ray experts slowly recast hypertrichosis as "a purely cosmetic defect," superfluous to the proper domain of truly scientific therapeutics.[63]

As professionalizing physicians backed away from X-ray epilation in an effort revive medicine's troubled authority, commercial practitioners were quick to fulfill demand for the therapy.[64] New commercial X-ray salons were opened not only by nonmedical beauty specialists but also by those physicians and scientists who had been squeezed out of reputable medical societies, professional publications, and other systems of collegial recognition due to their continued interest in X-ray hair removal. These practitioners easily presented clients with respectable diplomas and refereed publications testifying to their professional credentials. Their salons bore titles that alluded to the technique's scientific origins, such as "Hamomar Institute," "Kern Laboratories," "Hirsutic Laboratories," and the like. Commercial practitioners, in other words, emphasized the very same scientific aura physicians were attempting to define, protect, and control by disavowing this application of the ray. As professional scientists and physicians sought to prohibit X-ray epilation, they bolstered the technique's scientific prestige. The technique's popularity soared accordingly, and hundreds of X-ray salons opened across the country over the course of the next three decades.

The Triumph of Science over Superfluous Hair

While salon owners and managers promoted the X ray as the most "scientific" method of hair removal available, they never actually defined the term. Indeed, the ambiguity of "science" was perhaps its most marketable feature. Yet as loosely and broadly as science was defined in hair removal advertisements, it was quite systematically linked to a notion of progress. Whatever else science might have been, it was invariably and unceasingly advancing.[65] A brochure for the Virginia Laboratories of Baltimore emphasized the endless progress of science, noting that the thirty years of refinement of the X ray that had elapsed since Roentgen's discovery "can be said to be but a mere few minutes of time in the slow but certain development of knowledge." The advancement of science was further naturalized by likening it to the maturation of an awkward youngster. As the Baltimore salon put it, "great discoveries often go without much attention for a time, but inevitably they come into their own."[66]

Just as the epilation advertisements linked science to a notion of inexorable progress, that progress was in turn linked to the "strange power" of the X ray.[67] Science, accessible through the enigmatic new light, could now at last fulfill the dominant culture's "dream of faultless skin."[68] One 1933 brochure, paradoxically titled "Be Your True Self: We will tell you how," proclaimed: "Our MODERN SCIENTIFIC METHOD with filtered RAYS" is "not 'just another' hair remover," but the only one that can "banish" one's dark traces. As the brochure pledged, "It is only a step

Figure 3.3. Advertisement for Vi-Ro-Gen, of Pittsburgh, ca. 1935. (Reproduced with permission of the American Medical Association.)

from the Shadow into the Sunshine."[69] The enlightenment offered by scientific knowledge and the woman's own visible, physical "enlightenment" were further entangled in one salon's letter to a prospective client: "Thanks to the progress of science, no woman need now endure the torture and selfconsciousness [*sic*] which a physical blemish causes."[70] Or, more ominously, "Whatever you do— wherever you go—you need to have your skin CLEARED from this dark shadow."[71] The inevitable progress of science, the brochures suggest, eliminates all darkness in its path. Scientific advancement yields personal, physiological transformation (fig. 3.3).

As a notion of scientific advancement was linked to somatic enlightenment, so too were both narratives linked to dreams of upward class mobility. Through plen-

tiful photographs of plush consultation rooms and sleek treatment rooms, epilation promoters promised prospective clients a space unequaled in cleanliness and
luxury.[72] Simply by stepping into the salon, claimed one testimonial, a woman was
invited to experience the "padded carpets, the beautiful polychrome furniture, the
soft shaded lights and dainty draperies, and . . . the quiet, restful atmosphere of
. . . studious habits and refined tastes. . . . The epilation laboratory is a vision of
sanitary loveliness. The place is done in pure white enamel, not an atom of dust
can enter here."[73]

Many of these promotions made the economic importance of technological
self-fashioning far more explicit. Prefaced by a drawing of a pale, well-dressed
couple approaching a cap-wearing attendant, a Chicago Marveau Laboratories ad
demanded, "Can you afford to neglect your personal appearance any longer, in
this age when it counts so much in social and economic advancement?"[74] Similarly
underscoring the sexual (and hence, economic) benefits of feminine hairlessness,
the Dunsworth Laboratories of Indianapolis seconded this theme (fig. 3.4): "Freedom from Unwanted Hair Opens the Gates to Social Enjoyments that are Forever
Closed to Those so Afflicted."[75] Cleanliness, science, and class mobility merged in
the commodification of hair removal.

The salons knew their audience well. While it is impossible to determine the
racial identities of the thousands of anonymous American X-ray clients, we do
know that commercial epilation salons specifically distributed their promotions
to urban, non-English-speaking populations, a fact noted by health officials seeking to warn the public about the practice.[76] Medical and legal records indicate that
most epilation clients were working women employed in low- or middle-income
positions: telephone operators, secretaries, clerks, and so on. Some former clients
recalled receiving special group discounts at commercial salons for bringing in
large numbers of friends or coworkers.[77] Such discounts must have been attractive,
for even in the midst of the nation's worst depression, salons charged between five
dollars and thirty dollars for a series of ten to forty treatments.[78] That thousands
of working women struggled to save the vast sums necessary for these treatments
points to the larger resonance of the epilation advertisements: the hope that personal, physical transformation might bring passage to new economic opportunities — to a world of "refined tastes." In the salons' promotions, a mist-enveloped
"science" promised enlightenment at once somatic and social, with the X ray providing the purchasable avenue to both kinds of development.

Obviously, it is impossible to know exactly what role the advertisements' mul-

Figure 3.4. Advertisement for the Dermic Laboratories of
San Francisco and Los Angeles, ca. 1931. (Reproduced with
permission of the American Medical Association.)

tiple allusions to racial transformation and class mobility played in the lives of
the particular individuals who sought access to X-ray hair removal. Yet removing
the "dark shadow" that barred access to the world of "social enjoyments" appears
to have acquired a certain urgency during the interwar period, an era of increasingly restrictive U.S. immigration laws, widespread interest in eugenics, and increasingly desperate economic depression.[79] While women rarely address race explicitly in their letters about hair, they frequently connected their "affliction" to
broader economic and social concerns. In one young woman's letter to the American Medical Association (AMA), for example, anxiety about her financial malaise
and anxiety about her excessive hair flow into one another. Describing her frus-
tration at the expense of epilation treatments, the stenographer from Philadelphia
concluded, "I am ever so anxious to find a cure for this affliction. This has been
the cause of much unhappiness and actual sorrow to me."[80] Obviously troubled

by the "considerable sum" she had already exhausted on X-ray epilation, twenty-five-year-old Anne Steiman expressed similar concerns from Brooklyn in 1933:

> I am working for quite a small salary now and I have saved every penny I possibly could. denying myself luxuries that all girls love toward trying to get rid of this unwanted hair, and believe me that saving this money was quite an effort as things at home are quite bad.
>
> But, I cannot go on as I am now, as I am miserable through a freak of nature and I have more than once thought of putting an end to my misery.[81]

Steiman and other young women of the nation's urban working poor could scarcely afford to ignore the possibility of economic uplift seemingly carried by a "complexion clear and fair."[82]

The few existing traces of women's explicit responses to epilation advertisements suggest that consumers were in fact impressed by the promotions' themes of advancement. When writing to beauty specialists and medical experts for advice on the X ray, women often referred directly to advertisements that they had clipped and attached to their letters. One woman summarized the mood of these letters when writing to the AMA in 1931. Under the headline of a brochure from Philadelphia's Cosmique Laboratories, Katherine Moore asked simply, "Doesn't this sound pretty good?"[83]

The End of X-Ray Epilation

Clearly, the X ray did sound good, not only to Katherine Moore and to Anne Steiman but also to the women who would, nearly a decade later, sneak access to the technology in the old San Francisco rooming house. Assistant Special Agent Williams's confusion there between X-ray hair removal and back-alley abortion is illuminating, for it reminds us that however irrelevant or ridiculous the subject of hair removal may appear to readers today, for many women in the interwar period epilation was nothing less than a matter of life and death. While we can never know definitively why some individuals chose this technology from among the numerous others then available, two dominant themes in the cultural representations of the X ray—its suprasensuality and its association with scientific progress—suggest that the act of removing roots had more than one meaning for these women.

These meanings, of course, would change with the passage of time. By the late 1940s the practice of X-ray epilation had largely, though not completely, vanished.

The demise of X-ray epilation could be explained in various ways. On the broadest level, one might take into account shifting commitments to ideals of whiteness. Particularly after the atrocities of the Nazi eugenic program, white Americans were forced to reexamine their malignant veneration of racial purity.[84] The reevaluation of radiation risk in the aftermath of the bombing of Hiroshima and Nagasaki further tarnished the appeal of X-ray salons.[85]

While such major events abetted the demise of X-ray epilation, the end of the practice ultimately stemmed from persistent local activism. Opponents of X-ray epilation, including local and national Better Business Bureaus, local and state Departments of Health, Boards of Medical Examiners, women's magazine editors, and law enforcement agents, attacked the practice on multiple fronts throughout the 1930s. Some physicians lobbied for the revision of existing Medical Practices Acts, which had overlooked the unlicensed use of electrical devices in the treatment of superfluous hair, while others swayed X-ray manufacturers to stop production of commercial epilation equipment altogether.[86] Local Better Business Bureaus pressured daily papers to refuse advertisements for X-ray hair removal, while metropolitan health officials used radio announcements to warn illiterate consumers of the dangers of the practice.[87] The increasing prominence of severely disfigured or dying epilation clients and their retributive claims against epilation providers must have also lessened the popularity of the procedure.[88] By 1940 the practice had been driven out of the formal sector of the American economy, surviving primarily in surreptitious venues like the old rooming house at 126 Jackson.[89]

Fifty years after San Francisco police closed down operations on Jackson Street, several lessons emerge from the history of X-ray epilation. To begin with, we are reminded once again to shift attention from dramatic, singular events, such as the bombing of Hiroshima, toward more mundane, yet equally significant, uses of technology. As Ruth Schwartz Cowan argued persuasively more than twenty years ago, thinking only in "grandiose terms" about technological change often obstructs our view of the radical developments "going on right under our noses."[90] Turning to hair removal, a subject quite literally right under our noses, helps indicate a lacuna in mainstream technological history: the ongoing production of "whiteness." The making of race, this example demonstrates, is a symbolic and material process, one that merits further consideration by historians of technology.

As X-ray epilation demonstrates the role of race in histories of technology, so, too, does it demonstrate the centrality of technology in histories of race. Historians, literary critics, and philosophers have increasingly examined the construction or materialization of race and racial difference in early-twentieth-century narratives such as Nella Larson's 1929 *Passing* and George Schuyler's 1931 *Black No More*.[91] To date, however, little attention has been paid to the actual technologies used to produce "race" in this period of U.S. history. Even as they describe the mutability of race in interwar America, scholars continue to neglect the artifacts that render race visible or invisible.[92] The X ray was particularly central in this respect, used not only to eradicate some women's "dark shadows" but also to sterilize "undesirables" and to permanently lighten the skin of some African American men.[93] Further scrutiny of such technological practices can only expand our understanding of race's deadly effects, and thus increase our ability to counter them.

But perhaps the most haunting lesson of the North American Hiroshima Maidens lies in the persistent faith in scientific progress evident among survivors of X-ray epilation. Even while coming to terms with their scarred and twisted bodies, many X-ray epilation clients held on to their belief in the special, personal promise of science. Their faith is captured in the 1954 words of a former Tricho client:

> The treatments were supposed to have been guaranteed, but within the last few year's 'white spots' have appeared on my chin. This has been very heart breaking to me especially when on one's face.
>
> I have been wondering if ther might possibly be some new medical discovery which might help me. There are so many wonderful things happening these day's.[94]

Despite experiencing the dire physiological effects of attempts to reform bodies rather than social structures, some clients continued to seek personal salvation in the advancement of science. Their persistence, their hope in finding such salvation, seems the most bittersweet message of this story.

NOTES

For helpful comments, the author thanks members of the Department of Technology and Social Change at Linköping University and of the Program in Science, Technology, and Society at the Massachusetts Institute of Technology, participants at the 1996 SHOT annual meeting, and two anonymous referees. She extends particular gratitude to Deborah Fitzgerald and Leo Marx of MIT for their many helpful suggestions. Research for this article was supported by the Bakken Museum and Library of Electricity in Life, the Museum of Questionable Medical Devices, the I. Austin Kelly III/Richard M. Douglas Fund, Bates College, and the Dibner Fund for the History of Science and Technology.

1. J. W. Williams to C. B. Pinkham, memorandum, 2 March 1940, folder 0317-01, American Medical Association Historical Health Fraud Collection, Chicago (hereafter cited as AMA).

2. Ibid.

3. A. C. Cipollaro and M. B. Einhorn, "Use of X-Rays for Treatment of Hypertrichosis is Dangerous," *Journal of the American Medical Association* 135 (11 October 1947): 350.

4. H. Martin et al., "Radiation-Induced Skin Cancer of the Head and Neck," *Cancer* 25 (1970): 61–71.

5. Irving B. Rosen and Paul G. Walfish, "Sequelae of Radiation Facial Epilation (North American Hiroshima Maiden Syndrome)," *Surgery* 106 (December 1989): 946–50. The analogy between the twelve former epilation patients considered and the famous Hiroshima Maidens structures the physicians' presentation of the syndrome. The article opens, for example, not with a description of the twelve patients but with a description of the bombing of Hiroshima and its resultant "100,000 deaths" (946). It concludes by explaining how the epilation patients, "like their Japanese counterparts," managed to accept "their situation with fortitude and were able in most part to fashion a normal life" (950).

6. For further discussion of the original Hiroshima Maidens, see David Serlin, "Cut-Out Identities: Cosmetic Surgery and Cultural Imperialism," *Merge* 0 (1998): 46–50.

7. In this respect, X-ray hair removal devices might better be likened to cigarettes or tanning beds than to weapons of war. Judith A. McGaw's classic essay "No Passive Victims, No Separate Spheres: A Feminist Perspective on Technology's History" provides a cogent critique of "victim" approaches to the historiography of women and technology; see *In Context: History and the History of Technology,* ed. Stephen H. Cutcliffe and Robert C. Post (Bethlehem, Pa., 1989). For a more recent comment on feminist historiographies of technology, see Nina E. Lerman, Arwen Palmer Mohun, and Ruth Oldenziel, "Versatile Tools: Gender Analysis and the History of Technology," *Technology and Culture* 38 (1997): 1–8.

8. Cipollaro and Einhorn, 350. Tricho advertisements boasted that their method of hair removal had been used "by thousands of women long before the discovery was announced to the public," lending further indeterminacy to the number of individuals using X-ray epilation. See "The Tricho System," advertisement for George Hoppman's salon, Chicago, folder 0318-01, AMA.

9. While a few men used X-ray epilation for removal of hair from the face, ears, and neck, the vast majority of X-ray clients were women. Indeed, some men expressed concern that adopting this method of hair removal—a method marketed so strongly to women—might cause them to become "effeminate." See, e.g., Arthur Nelson to AMA, ca. 20 July 1934, folder 0317-01, AMA.

10. J. B. Loudon, "On Body Products," in *The Anthropology of the Body,* ed. John Blacking (London, 1977), 163. On the prevalence of hair removal in ancient Greece, for instance, see Martin Kilmer, "Genital Phobia and Depilation," *Journal of Hellenic Studies* 102 (1982): 104–12. For an account of contemporary white women's attitudes toward body hair, see Susan A. Basow, "The Hairless Ideal: Women and Their Body Hair," *Psychology of Women Quarterly* 15 (1991): 83–96.

11. Alfred F. Niemoeller, *Superfluous Hair and Its Removal* (New York, 1938), 14.

12. On the evolutionary significance of hair, see Londa Schiebinger, *Nature's Body: Gender and the Making of Modern Science* (Boston, 1993), and Cynthia Eagle Russett, *Sexual Science: The Victorian Construction of Womanhood* (Cambridge, Mass., 1989).

13. Attitudes concerning immigration, race, and hair are exemplified in "Racial Characteristics of Hair," *Scientific American Supplement,* 10 March 1917, 160. On ideals of white middle-class feminine beauty, see Lois Banner, *American Beauty* (Chicago, 1983); Kathy Peiss, *Hope in a Jar: The Making of America's Beauty Culture* (New York, 1998); Vincent Vinikas, *Soft Soap, Hard Sell: American Hygiene in an Age of Advertisement* (Ames, Iowa, 1992).

14. For one of the earliest uses of the term "hypertrichosis," see C. Krebs, "Case of Hypertrichosis (Homo Hirsutus)," trans. H. J. Garrigues, *Archives of Dermatology* 5 (1878): 161–62. A partial bibliography of period dermatological articles on hypertrichosis can be found in L. Brocq, "Cent Dix Hypertrichoses Traitées Par L'Électrolyse," *Annales de Dermatologie et Syphilogie* 8 (1897): 1083–84.

15. George Henry Fox, "On the Permanent Removal of Hair by Electrolysis," *Medical Record* 15 (1879): 270.

16. Christine Hope, "Caucasian Female Body Hair and American Culture," *Journal of American Culture* 5 (Spring 1982): 93–99.

17. See, e.g., Marie Fink to AMA, 17 August 1931, folder 0317-02, AMA; Mrs. B. Tellef to AMA, 6 August 1926, folder 0318-01, AMA; Mrs. Allen Stamper to *Hygeia,* 4 November 1929, folder 0317-03, AMA; Joseph Rohrer, *Rohrer's Illustrated Book on Scientific Modern Beauty Culture* (New York, 1924), 46.

18. Herman Goodman, "The Problem of Excess Hair," *Hygeia* 8 (May 1930): 433; Maurice Costello, "How to Remove Superfluous Hair," *Hygeia* 18 (July 1940): 586; William J. Young, "Hypertrichosis and Its Treatment," *Kentucky Medical Journal* 18 (June 1920): 217; Oscar L. Levin, "Superfluous Hair," *Good Housekeeping,* September 1928, 106. While physicians and beauty specialists all reported the intense suffering endured by hairy women, the presence of body hair did not necessarily lead women to depression. At least a few women viewed their bountiful body hair as neither excessive, superfluous, nor merely tolerable, but instead as a beautiful complement to the rest of their physique. See, e.g., Joseph Mitchell, "Profiles: Lady Olga," *New Yorker,* 3 August 1940: 20–28.

19. Paul E. Bechet, "The Etiology and Treatment of Hypertrichosis," *New York Medical Journal* 98 (16 August 1913): 313.

20. Ironically, some X-ray epilation clients—such as "Mrs. E.B.," aged 30—were so upset by the atrophy and other effects of radiation poisoning, particularly if their hypertrichosis had been only "mild" in the first place, that they attempted suicide. See F. J. Eichenlaub, "Some More Tricho Cases," *Journal of the American Medical Association* 94 (26 April 1930): 1341.

21. Niemoeller (n. 11 above), 13.

22. Historian Evelynn M. Hammonds discusses the reliance of race on the visual in "New Technologies of Race," in *Processed Lives: Gender and Technology in Everyday Life,* ed. Jennifer Terry and Melodie Calvert (New York, 1997), 109.

23. As Kobena Mercer has put it, "the legacy of . . . biologizing and totalizing racism is

traced as a presence in everyday comments made about our hair" (249). See his extended argument about the racial significance of human hair, "Black Hair/Style Politics," in *Out There: Marginalization and Contemporary Culture*, ed. Russell Ferguson et al. (Cambridge, Mass., 1990): 247–64.

24. Peter A. Browne, *Trichologia Mammalium; or, A Treatise on the Organization, Properties and Uses of Hair and Wool* (Philadelphia, 1853), 66.

25. William Stanton, *The Leopard's Spots: Scientific Attitudes toward Race in America, 1815–1859* (Chicago, 1960), 150–54.

26. See, e.g., Delmar Emil Bordeaux, *Cosmetic Electrolysis and the Removal of Superfluous Hair* (Rockford, Ill., 1942); Anna Hazelton Delavan, "Superfluous Hair," *Good Housekeeping*, March 1925, 96. On the valorization of whiteness in dominant middle-class U.S. beauty discourse more generally, see Peiss (n. 13 above), 34.

27. Ernest L. McEwen, "The Problem of Hypertrichosis," *Journal of Cutaneous Diseases Including Syphilis* 35 (1917): 830. See also Frank Crozer Knowles, "Hypertrichiasis in Childhood: The So-Called 'Dog-Faced Boy,' " *Pennsylvania Medical Journal* 24 (March 1921): 403.

28. "Beauty is your Heritage," advertisement for Marveau Laboratories, Chicago, folder 0317-02, AMA; "With the Advancement of Science Comes the Modern Way to Remove Superfluous Hair Permanently: TRICHO SYSTEM," advertisement for Tricho salon, Boston, folder 0317-17, AMA; Albert C. Geyser, "Facts and Fallacies about the Removal of Superfluous Hair," folder 0317-17, AMA; "Loveliness for the Most Discriminating Women," advertisement for Hair-X Salon, Philadelphia, folder 0317-02, AMA.

29. Condemnations of the technique in women's magazines were vivid and unambiguous. In 1928, for instance, "skin specialist" Dr. Oscar Levin described the perils of X-ray hair removal to the readers of *Good Housekeeping*: "the skin may become inflamed, scaly, wrinkled, streaked with prominent blood vessels . . . at times, later in life, warty and scaly growths may appear, which finally break down and ulcerate, and may even become cancerous . . . the weight of expert opinion is against the use of x-rays or any type of radiation for destroying superfluous hair." See Levin (n. 18 above), 190–91. On clients' resistance to authorities' condemnations, see H. L. J. Marshall to AMA, 30 May 1928, folder 0318-02, AMA.

30. Some women sought medical advice on the X-ray treatments of superfluous hair, and upon hearing condemnation of the practice unhesitatingly rejected it. Others, already well informed on the potential dangers of X-ray overexposure, were blatantly misinformed about the nature of the technology by X-ray providers themselves. But these were a small minority among epilation clients. See, e.g., the case report from G. V. Stryker in "The Tricho System Again," *Journal of the American Medical Association* 92 (16 March 1929): 919; Mary Mulholland to AMA, 24 June 1930, folder 0318-02, AMA.

31. Advertisement for the Velvet Mitten Company, Los Angeles, folder 0317-01, AMA; Albert Geyser, "Truth and Fallacy Concerning the Roentgen Ray in Hypertrichosis," *Scientific Therapy and Practical Research* (March 1926), reprint in folder 0317-17, AMA.

32. Young (n. 18 above), 217; Niemoeller (n. 11 above), 91.

33. Geyser, "Truth and Fallacy."

34. On Koremlu's toxicity, see *Journal of the American Medical Association* (30 July

1932): 407; A. J. Cramp, *Nostrums and Quackery and Pseudo-Medicine* (Chicago, 1936), 3: 35; and Gwen Kay, "Regulating Beauty: The Role of the Food and Drug Administration in the 1938 Food, Drug, and Cosmetics Act" (paper presented at the annual meeting of the History of Science Society, Atlanta, Ga., November 1996).

35. Cited by Niemoeller, 48.

36. Ernst Ludwig Franz Kroymayer, *The Cosmetic Treatment of Skin Complaints* (London, 1930), 69; McEwen (n. 27 above), 832.

37. As technical refinements were made to electrolysis equipment, the procedure was also practiced at home. See Niemoeller, chap. 21.

38. Adolph Brand, "Hypertrichosis," *New York Medical Journal* 97 (1913): 708.

39. Brocq (n. 14 above).

40. Ibid.; Young (n. 18 above), 219.

41. While historians typically resist explaining technological diffusion according to reified notions of "utility" (the idea that certain technologies succeed simply because they "work" better than others), the efficacy of X-ray hair removal was acknowledged by nearly all commentators in the early twentieth century. For a compelling critique of the notion of technological utility, see Michael Mulkay, "Knowledge and Utility: Implications for the Sociology of Knowledge," *Social Studies of Science* 9 (1979): 63–80.

42. AMA to Mrs. P. G. Range, 28 October 1927, folder 0314-03, AMA; AMA to Mr. A. E. Backman, 29 May 1930, folder 0314-03, AMA.

43. Mary Mulholland to AMA, 20 May 1930, folder 0318-02, AMA. This description of the Tricho machine may remind some readers of the shoe-fitting fluoroscopes employed in North American and European shoe stores in the mid-twentieth century. See "Shoe-Fitting Fluoroscopes," *Journal of the American Medical Association* 139 (9 April 1949): 1004–5; H. Kopp, "Radiation Damage Caused By Shoe-Fitting Fluoroscope," *British Medical Journal* 2 (1957): 1344–45.

44. Geyser, "Facts and Fallacies" (n. 28 above).

45. "Gone for Good," advertisement for Rudolph Tricho Institute, Detroit, folder 0318-01, AMA.

46. See, for example, "Tricho Method of Removing Superfluous Hair," advertisement for Mrs. L. P. Williams's Tricho salons, Connecticut and Massachusetts, folder 0318-01, AMA.

47. "Loveliness for the Most Discriminating Women" (n. 28 above).

48. "A Flawless Skin," *Boston Post,* 15 November 1928, copy in folder 0318-02, AMA.

49. "Gone for Good."

50. Banner (n. 13 above), 205.

51. Roentgen's discovery has been mythicized by Otto Glasser in his *Wilhelm Conrad Röntgen and the Early History of Röntgen Rays* (Springfield, Ill., 1934) and *W. C. Röntgen* (Springfield, Ill., 2nd ed., 1958). On the immediate public response to Roentgen's announcement, see Nancy Knight, "The New Light: X-Rays and Medical Futurism," in *Imagining Tomorrow: History, Technology, and the American Future,* ed. Joseph J. Corn (Cambridge, Mass., 1986); Ronald L. Eisenberg, *Radiology: An Illustrated History* (St. Louis, 1992); E. R. N. Grigg, *The Trail of the Invisible Light* (Springfield, Ill., 1965); Ruth Brecher and Edward

Brecher, *The Rays: A History of Radiology in the United States and Canada* (Baltimore, 1969). Curiously, most of these histories focus on the imaging and diagnostic capabilities of the X ray, to the neglect of the technology's myriad therapeutic applications.

52. John Daniel, "The X-rays," *Science*, 10 April 1896, 562–63.

53. Brecher and Brecher, *The Rays*, 82.

54. Eduard Schiff and Leopold Freund, "Beiträge zur Radiotherapie," *Weiner Medicinische Wochenschrift* 48 (28 May 1898): 1058–61.

55. Neville Wood, "Depilation by Roentgen Rays," *Lancet* 1 (27 January 1900): 231; T. Sjögren and E. Sederholm, "Beitrag zur therapeutischen Verwertung der Röntgenstrahlen," *Fortschritte auf dem Gebiete der Röntgenstrahlen* 4 (1901): 145–70; Edith R. Meek, "A Variety of Skin Lesions Treated by X-ray," *Boston Medical and Surgical Journal* 147 (7 August 1902): 152–53.

56. Mihran Krikor Kassabian, *Röntgen Rays and Electro-therapeutics* (Philadelphia, 1910), 80.

57. V.E.A. Pullin and W. J. Wiltshire, *X-Rays Past and Present* (London, 1927), 173–74.

58. A review of early awareness of X-ray sequelae may be found in Brecher and Brecher, *The Rays*, (n. 51 above), chap. 7.

59. See, for example, the long, autobiographical description of the grisly results of prolonged X-ray exposure in S. J. R., "Some Effects of the X-rays on the Hands," *Nature*, 29 October 1896, 621, or Elihu Thomson's early descriptions of his famous self-experiments with X-ray exposure: "Roentgen Rays Act Strongly on the Tissues," *Electrical Engineer* 22 (25 November 1896): 534, and "Roentgen Ray Burns," *Electrical Engineer* 23 (14 April 1897): 400.

60. For a general overview of the professionalization of medicine, see Paul Starr, *The Social Transformation of American Medicine* (New York, 1982). On the role of the X ray in professionalization in particular, see Joel Howell, *Technology in the Hospital* (Baltimore, 1995); Stanley Joel Reiser, *Medicine and the Reign of Technology* (Cambridge, 1978); and Bettyann Holtzmann Kevles, *Naked to the Bone: Medical Imaging in the Twentieth Century* (New Brunswick, N.J., 1997). Lisa Cartwright rightly critiques historians' repeated depiction of the X ray as the defining tool of modern professional medicine; see her *Screening the Body: Tracing Medicine's Visual Culture* (Minneapolis, 1995), esp. chap. 5.

61. George M. MacKee, "Hypertrichosis and the X-ray," *Journal of Cutaneous Diseases Including Syphilis* 35 (1917): 177.

62. Carl Beck, *Röntgen Ray Diagnosis and Therapy* (New York, 1904), 377.

63. William Allen Pusey and Eugene Wilson Caldwell, *Röntgen Rays in Therapeutics and Diagnosis* (Philadelphia, 1904), 360–61. For a relevant study of physicians' exclusion of "cosmetic" practices and practitioners during this era of medical professionalization, see Beth Haiken's award-winning work on the controversy surrounding the tension between "reconstructive" (hence medically legitimate) and "cosmetic" (hence illegitimate) plastic surgery in American medicine in the first two decades of the twentieth century: "Plastic Surgery and American Beauty at 1921," *Bulletin of the History of Medicine* 68 (1994): 429–53.

64. Although most professional physicians had disavowed X-ray epilation by 1918, it is important to note that this rejection was neither uniform nor total. As late as 1927 two

physicians in Chicago offered an epilating dose of X rays to a girl with a hairy chin and upper lip; see Mrs. P. G. Range to A. J. Cramp, 23 October 1927, folder 0314-03, AMA. Three years later, physician H. H. Hazen reported other colleagues still practicing X-ray epilation; see H. H. Hazen, "Injuries Resulting in Irradiation in Beauty Shops," *American Journal of Roentgenology and Radium Therapy* 23 (1930): 411. Some evidence hints at the medical use of cosmetic X-ray epilation as late as 1960. See Howard T. Behrman, "Diagnosis and Management of Hirsutism," *Journal of the American Medical Association* 172 (23 April 1960): 126.

65. See, e.g., "With the Advancement of Science" (n. 28 above).

66. "Permanent Freedom from Unwanted Hair," advertisement for the Virginia Laboratories, Baltimore, folder 0317-01, AMA. Baltimore's Virginia Laboratories employed the "Marton Method" and used Marton's text and illustrations in many of their advertisements.

67. On the "strange powers" of the ray, see "Gone for Good" (n. 45 above).

68. Ibid.

69. "Be Your True Self: We will tell you how," advertisement for Frances A. Post, Inc., Cleveland, folder 0317-01, AMA.

70. H. Gellert [Secretary of Hamomar Institute] to "Madam," 1933, folder 0317-01, AMA.

71. Ibid.

72. Miss H. Stearn to National Institute of Health, 5 July 1933, folder 0317-01, AMA; "BEAUTY: Woman's Most Precious Gift," advertisement for the Dermic Laboratories, San Francisco and Los Angeles, folder 0317-02, AMA.

73. M. J. Rush, "Hypertrichosis: The Marton Method, A Triumph of Chemistry," *Medical Practice* (March 1924): 956, reprint in folder 0317-01, AMA.

74. "Beauty is your Heritage" (n. 28 above).

75. "Permanent Freedom from Unwanted Hair" (n. 66 above).

76. Herman Goodman, "Correspondence," *Journal of the American Medical Association* 84 (9 May 1925) 1443; S. Dana Hubbard [City of New York Department of Public Health.] to A. J. Cramp, 8 January 1929, folder 0318-02, AMA; Dale Brown [Cleveland Better Business Bureau.] to A. J. Cramp, 29 July 1930, folder 0314-03; AMA. Of course, even if epilation salons recorded statistics on their clients' racial and ethnic backgrounds, these classifications would not necessarily coincide with clients' self-identifications, nor with contemporary racial and ethnic typologies; "race" is hardly stable.

77. Edward Oliver, "Dermatitis Due to 'Tricho Method,'" *Archives of Dermatology and Syphilology* 25 (1932): 948; D. E. H. Cleveland, "The Removal of Superfluous Hair by X-Rays," *Canadian Medical Association Journal* 59 (1948): 375.

78. B. O. Halling, internal memorandum, 22 October 1925, folder 0318-01, AMA; Mulholland to AMA, 20 May 1930, AMA; Mrs. B. Tellef to AMA, 6 August 1926, folder 0318-01, AMA. Many women received more than forty treatments; one young woman received fourteen hundred exposures over five years; see Dorothy R. Kirk to Arthur Cramp, 14 December 1928, folder 0315-14, AMA.

79. Historians have scrutinized early-twentieth-century eugenic discourses at length. My understanding of this period is most influenced by Donna Haraway's "Teddy Bear Patri-

archy: Taxidermy in the Garden of Eden, New York City, 1908–1936," in *Primate Visions: Gender, Race, and Nature in the World of Modern Science* (New York, 1989), 26–58.

80. Mary F. Amerise to AMA, 18 May 1925, folder 0314-03, AMA.

81. Anne Steiman to AMA, 5 September 1933, folder 0317-01, AMA (text appears as in original document).

82. "A Flawless Skin" (n. 48 above).

83. Katherine Moore to AMA, 27 August 1931, folder 0318-02, AMA.

84. While one would err by suggesting that the Holocaust marked the end of eugenic thinking, the specter of the Nazi program did (and still does) trouble discussions of racial purity in the United States and Canada. See Robert Proctor's discussion of changes in postwar eugenic discourse in *Racial Hygiene: Medicine under the Nazis* (Cambridge, Mass., 1989), 303–8.

85. As Anja Hiddinga has noted, widespread public discussion of radiation hazards began only after 1945. See Hiddinga, "X-ray Technology in Obstetrics: Measuring Pelves at the Yale School of Medicine," in *Medical Innovations in Historical Perspective,* ed. John Pickstone (London, 1992), 143. For an extended discussion on the significance of Hiroshima, see M. Susan Lindee, *Suffering Made Real: American Science and the Survivors at Hiroshima* (Chicago, 1994). For thorough discussions of the history of radiation standards in the United States, see Daniel Paul Serwer, "The Rise of Radiation Protection: Science, Medicine and Technology in Society, 1896–1935" (Ph.D. diss., Princeton University, 1976), and Gilbert F. Whittemore, "The National Committee on Radiation Protection, 1928–1960: From Professional Guidelines to Government Regulation" (Ph.D. diss., Harvard University, 1986).

86. On the revision of licensing laws, see C. B. Pinkham to Max C. Starkhoff, 3 April 1928, folder 0318-02, AMA; C. B. Pinkham to A. J. Cramp, 22 January 1929, folder 0318-02, AMA. On the distribution of X-ray equipment, see A. J. Cramp to Howard Fox, 15 November 1929, folder 0318-02, AMA; Rollins H. Stevens to Arthur Cramp, 8 November 1929, folder 0318-02, AMA.

87. Better Business Bureau of Rochester to A. J. Cramp, 15 May 1930, folder 0318-02, AMA; Howard Fox to Arthur J. Cramp, 13 November 1929, folder 0318-02, AMA.

88. The disintegration of the Tricho Sales Corporation in 1930, for example, was influenced by a collective lawsuit organized by a group of seven New York women, all former clients. See S. Dana Hubbard [New York City Department of Health] to A. J. Cramp, 8 January 1929, folder 0318-02, AMA. See also S. Dana Hubbard to C. B. Pinkham, 8 January 1929, folder 0318-02, AMA.

89. New franchises that publicly advertised the technique were opening in Canada as late as 1948. See Cleveland (n. 77 above), 374. It is impossible to determine the extent of back-alley X-ray epilation.

90. Ruth Schwartz Cowan, "The Industrial Revolution in the Home: Household Technology and Social Change in the 20th Century," *Technology and Culture* 17 (1976): 1. Conversations with Robert D. Friedel have encouraged me to uncover the histories of unglamorous, everyday technologies.

91. See, e.g., Valerie Smith, "Reading the Intersection of Race and Gender in Narratives

of Passing," *diacritics* 24 (Summer/Fall 1994): 43; Judith Butler, "Passing, Queering: Nella Larsen's Psychoanalytic Challenge," *Bodies That Matter* (New York, 1993), 167–85; Werner Sollers, *Neither Black nor White Yet Both* (New York, 1997). Also see Nella Larson, *Passing* (New York, reprint, 1969), and George S. Schuyler, *Black No More: Being an Account of the Strange and Wonderful Workings of Science in the Land of the Free, A.D. 1933–1940* (New York, reprint, 1971).

92. For one recent exception to this trend, see Tanya Hart, "From 'Heated Forks' to Pressing Combs: African American Women, Technology, and Progressive-Era Beauty Culture" (paper presented at the annual meeting of the Society for the History of Technology, Baltimore, Maryland, October 1998).

93. See, for example, in William J. Hammer Collection, 1874–1935, Archives Center, National Museum of American History, Washington D.C., "All Coons to Look White: College Professors Have Scheme to Solve Race Problem" (*New York City Morning Telegraph*, 22 January 1904), clipping in box 59, folder 2; "Radium and X-Ray Used to Beautify," *Boston Herald*, 8 May 1904, clipping in box 59, folder 3; "'Can the Ethiopian Change His Skin or the Leopard His Spots': Radium Light Turns Negro's Skin White," *Boston Globe*, 25 January 1904, box 60, folder 2; "Burning Out Birthmarks, Blemishes of the Skin and Even Turning a Negro White with the Magic Rays of Radium, the New Mystery of Science!" *New York American*, 10 January 1904, box 60, folder 2.

94. Helen L. Camp to AMA, 30 March 1954, folder 0318-02, AMA (text appears as in original document).

Situated Technology: Camouflage

RACHEL P. MAINES

In Herzig's essay, X-ray technology carried multiple meanings and had both personal and professional uses. Rachel Maines identifies another technology with medical and personal purposes: the technology of what we might call therapeutic and recreational orgasm. Not surprisingly, particular understandings of the purpose of such technology clearly shaped the choices people have made about its design and use. In this case, perhaps even more than that of the X ray, the labor-saving technology in question is camouflaged to the historical observer (and perhaps to some contemporary observers). Maines addresses this issue of camouflage directly, joining McGaw in challenging us to see technology at work in many situations, as well as gender ideologies shaping technologies (and our understandings of technologies) in often unexpected ways.

   In this article, and in her subsequent book, Maines finds late-nineteenth-century medical doctors apparently unaware of the sexual ramifications of what they called "electromechanical massage" (a vibrator) applied to female genitalia—they assumed that sex required penetration. A vibrator applied externally was not, therefore, sexual. As Maines explains in *The Technology of Orgasm: "Hysteria," the Vibrator, and Women's Sexual Satisfaction* (Baltimore, 1998), many doctors found another new device, the speculum, potentially immoral and corrupting of female virtue—precisely because it entered the vagina. What does Maines mean by "camouflaged technology"? Is the vibrator a "female technology" as McGaw labels brassieres and closets? A medical technology comparable to the X ray? Who are the producers and consumers in this story, and why was the technology recognized or camouflaged by the story's various actors?

"Socially Camouflaged Technologies: the Case of the Electromechanical Vibrator." © 1989 IEEE. Reprinted, with permission, from *IEEE Technology and Society Magazine* 8 (1989): 3–11.

Certain commodities are sold in the legal marketplace for which the expected use is either illegal or socially unacceptable. Marketing of these goods, therefore, requires camouflaging of the design purpose in a verbal and visual rhetoric that conveys to the knowledgeable consumer the item's selling points without actually endorsing its socially prohibited uses. I refer not to goods that are actually illegal in character, such as marijuana, but to their grey-market background technologies, such as cigarette rolling papers. Marketing efforts for goods of this type have similar characteristics over time, despite the dissimilarity of the advertised commodities. I shall discuss here an electromechanical technology that addresses formerly prohibited expressions of women's sexuality—the vibrator in its earliest incarnation between 1870 and 1930. Comparisons will be drawn between marketing strategies for this electromechanical technology, introduced between 1880 and 1903, and that of emmenagogues, distilling, burglary tools, and computer software copying, as well as the paradigm example of drug paraphernalia.

I shall argue here that electromechanical massage of the female genitalia achieved acceptance during the period in question by both professionals and consumers not only because it was less cumbersome, labor-intensive, and costly than predecessor technologies, but because it maintained the social camouflage of sexual massage treatment through its associations with modern professional instrumentation and with prevailing beliefs—about electricity as a healing agent.[1]

The case of the electromechanical vibrator, as a technology associated with women's sexuality, involves issues of acceptability rather than legality. The vibrator and its predecessor technologies, including the dildo, are associated with masturbation, a socially prohibited activity until well into the second half of the twentieth century.[2] Devices for mechanically assisted female masturbation, mainly vibrators and dildos, were marketed in the popular press from the late nineteenth century through the early thirties in similarly camouflaged advertising. Such advertisements temporarily disappeared from popular literature after the vibrator began to appear in stag films, which may have rendered the camouflage inadequate, and did not resurface until social change made it unnecessary to disguise the sexual uses of the device.[3]

For purposes of this discussion, a vibrator is a mechanical or electromechanical appliance imparting rapid and rhythmic pressure through a contoured working surface usually mounted at a right angle to the handle. These points of contact generally take the form of a set of interchangeable vibratodes configured to the anatomical areas they are intended to address. Vibrators are rarely employed internally in masturbation; they thus differ from dildos, which are generally straight-shafted and may or may not include a vibratory component. Vibrators are here distinguished also from massagers, the working surfaces of which are flat or dished.[4] It should be noted that this is a historian's distinction imposed on the primary sources; medical authors and appliance manufacturers apply a heterogeneous nomenclature to massage technologies. Vibrators and dildoes rarely appeared in household advertising between 1930 and 1955, massagers continued to be marketed, mainly through household magazines.[5]

The electromechanical vibrator, introduced as a medical instrument in the 1880s and as a home appliance between 1900 and 1903, represented the convergence of several older medical massage technologies, including manual, hydriatic, electrotherapeutic and mechanical methods. Internal and external gynecological massage with lubricated fingers had been a standard medical treatment for hysteria, disorders of menstruation and other female complaints at least since the time of Aretaeus Cappadox (circa 150 A.D.), and the evidence suggests that orgasmic response on the part of the patient may have been the intended therapeutic result.[6] Douche therapy, a method of directing a jet of pumped water at the pelvic area and vulva, was employed for similar purposes after hydrotherapy became popular in the eighteenth and nineteenth centuries.[7] The camouflage of the apparently sexual character of such therapy was accomplished through its medical respectability and through creative definitions both of the diseases for which massage was indicated and of the effects of treatment. In the case of the electromechanical vibrator, the use of electrical power contributed the cachet of modernity and linked the instrument to older technologies of electrotherapeutics, in which patients received low-voltage electricity through electrodes attached directly to the skin or mucous membranes, and to light-bath therapy, in which electric light was applied to the skin in a closed cabinet. The electrotherapeutic association was explicitly invoked in the original term for the vibrator's interchangeable applicators, which were known as "vibratodes." Electrical treatments were employed in hysteria as soon as they were introduced in the eighteenth century, and remained in use as late as the 1920s.

Hysteria as a disease paradigm, from its origins in the Egyptian medical corpus through its conceptual eradication by American Psychological Association fiat in 1952, was so vaguely and subjectively defined that it might encompass almost any set of ambiguous symptoms that troubled a woman or her family. As its name suggests, hysteria as well as its "sister" complaint chlorosis were until the twentieth century thought to have their etiology in the female reproductive tract generally, and more particularly in the organism's response to sexual deprivation.[8] This physiological condition seems to have achieved epidemic proportions among women and girls, at least in the modern period.[9] Sydenham, writing in the seventeenth century, observed that hysteria was the most common of all diseases except fevers.[10]

In the late nineteenth century, physicians noted with alarm that from half to three-quarters of all women showed signs of hysterical affliction. Among the many symptoms listed in medical descriptions of the syndrome are anxiety, sense of heaviness in the pelvis, edema (swelling) in the lower abdomen and genital areas, wandering of attention and associated tendencies to indulge in sexual fantasy, insomnia, irritability, and "excessive" vaginal lubrication.[11]

The therapeutic objective in such cases was to produce a "crisis" of the disease in the Hippocratic sense of this expression, corresponding to the point in infectious diseases at which the fever breaks. Manual massage of the vulva by physicians or midwives, with fragrant oils as lubricants, formed part of the standard treatment repertoire for hysteria, chlorosis, and related disorders from ancient times until the post-Freudian era. The crisis induced by this procedure was usually called the "hysterical paroxysm." Treatment for hysteria might comprise up to three-quarters of a physician's practice in the nineteenth century. Doctors who employed vulvular massage treatment in hysteria thus required fast, efficient, and effective means of producing the desired crisis. Portability of the technology was also a desideratum, as physicians treated many patients in their homes, and only manual massage under these conditions was possible until the introduction of the portable battery-powered vibrator for medical use in the late 1880s.

Patients reported experiencing symptomatic relief after such treatments, and such conditions as pelvic congestion and insomnia were noticeably ameliorated, especially if therapy continued on a regular basis. A few physicians, including Nathaniel Highmore in the seventeenth century and Auguste Tripier, a nineteenth-century electrotherapist, clearly recognized the hysterical paroxysm as sexual orgasm.[12] That many of their colleagues also perceived the sexual character of hysteria treatments is suggested by the fact that, in the case of married women, one of the therapeutic options was intercourse, and in the case of single women, marriage was routinely recommended.[13] "God-fearing physicians," as Zacuto expressed it in the seventeenth century, were expected to induce the paroxysm with their own fingers only when absolutely necessary, as in the case of very young single women, widows, and nuns.[14]

Many later physicians, however, such as the nineteenth-century hydrotherapist John Harvey Kellogg, seem not to have perceived the sexual character of patient response. Kellogg wrote extensively about hydrotherapy and electrotherapeutics in gynecology. In his "Electrotherapeutics in Chronic Maladies," published in *Modern Medicine* in 1904, he describes "strong contractions of the abdominal muscles" in a female patient undergoing treatment, and similar reactions such that "the office table was made to tremble quite violently with the movement."[15] In their analysis of the situation, these physicians may have been handicapped by their failure to recognize that penetration is a successful means of producing orgasm in only a minority of women; thus, treatments that did not involve significant vaginal penetration were not morally suspect. In effect, misperceptions of female sexuality formed part of the camouflage of the original manual technique that preceded the electromechanical vibrator. Insertion of the speculum, however, since it traveled the same path as the supposedly irresistible penis during intercourse, was widely criticized in the medical community for its purportedly immoral effect on patients.[16] That some questioned the ethics of the vulvular massage procedure is clear; Thomas Stretch Dowse quotes Graham as observing that "Massage of the pelvic organs should be intrusted to those alone who have 'clean hands and a pure heart.'"[17] One physician, however, in an article significantly titled "Signs of Masturbation in the Female," proposed the application of an electrical charge to the clitoris as a test of salacious propensities in women. Sensitivity of the organ to this type of electrical stimulation, in his view, indicated secret indulgence in what was known in the nineteenth century as "a bad habit."[18] Ironically, such women were often treated electrically for hysteria supposedly caused by masturbation.

However they construed the benefits, physicians regarded the genital massage procedure, which could take as long as an hour of skilled therapeutic activity, as something of a chore, and made early attempts to mechanize it. Hydrotherapy, in the form of what was known as the "pelvic douche" (massage of the lower pelvis with a jet of pumped water), provided similar relief to the patient with reduced demands on the therapist. Doctors of the eighteenth end nineteenth centuries frequently recommended douche therapy for their women patients who could afford spa visits. This market was limited, however, as both treatment and travel were

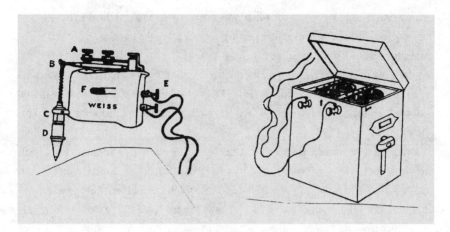

Figure 4.1. Joseph Mortimer Granville's "percuteur" of 1883, manufactured by the Weiss Instrument Company.

costly.[19] A very small minority of patients and doctors could afford to install hydrotherapeutic facilities in convenient locations; both doctor and patient usually had to travel to the spa. Electrically powered equipment, when it became available, thus had a decentralizing and cost-reducing effect on massage treatment.

In the 1860s, some spas and clinics introduced a coal-fired steam powered device invented by a Dr. George Taylor, called the "Manipulator," which massaged the lower pelvis while the patient either stood or lay on a table.[20] This too required a considerable expenditure either by the physician who purchased the equipment or by the patient who was required to travel to a spa for treatment. Thus, when the electromechanical vibrator was invented two decades later in England by Mortimer Granville and manufactured by Weiss, a ready market already existed in the medical community.[21] Ironically, Mortimer Granville considered the use of his instrument on women, especially hysterics, a morally indefensible act, and recommended the device only for use on the male skeletal muscles.[22] Although his original battery-powered model was heavy and unreliable, it was more portable than water-powered massage and less fatiguing to the operator than manual massage (Fig. 4.1).

Air-pressure models were introduced, but they required cumbersome tanks of compressed air, which needed frequent refilling. When line electricity became widely available, portable plug-in models made vibratory house calls more expeditious and cost effective for the enterprising physician. The difficulty of maintaining batteries in or out of the office was noted by several medical writers of the period predating the introduction of plug-in vibrators.[23] Batteries and small office generators were liable to fail at crucial moments during patient treatment, and required more engineering expertise for their maintenance than most physicians cared to acquire. Portable models using DC or AC line electricity were available with a wide range of vibratodes, such as the twelve-inch rectal probe supplied with one of the Gorman firm's vibrators.

Despite its inventor's reservations, the Weiss instrument and later devices on the same principle were widely used by physicians for pelvic disorders in women and girls. The social camouflage applied to the older manual technology was carefully maintained in connection with the new, at least until the 1920s. The marketing of medical vibrators to physicians and the discussion of them in such works as Covey's *Profitable Office Specialties* addressed two important professional considerations: the respectability of the devices as medical instruments (including their reassuringly clinical appearance) and their utility in the fast and efficient treatment of those chronic disorders, such as pelvic complaints in women, that provided a significant portion of a physician's income.[24] The importance of a prestige image for electromechanical instrumentation, and its role in the pricing of medical vibrators, is illustrated by a paragraph in the advertising brochure for the "Chattanooga," (Figs. 4.2 and 4.3), at $200 in 1904 the most costly of the physicians' office models: "The Physician can give with the 'Chattanooga' Vibrator a thorough massage treatment in three minutes that is extremely pleasant and beneficial, but this instrument is neither designed nor sold as a 'Massage Machine.' It is sold only to Physicians and constructed for the express purpose of exciting the various organs of the body into activity through their central nervous supply."[25]

I do not mean to suggest that gynecological treatments were the only uses of such devices, or that all physicians who purchased them used them for the production of orgasm in female patients, but the literature suggests that a substantial number were interested in the new technology's utility in the hysteroneurasthenic complaints. The interposition of an official-looking machine must have done much to restore clinical dignity to the massage procedure. The vibrator was introduced in 1899 as a home medical appliance and was by 1904 advertised in household magazines in suggestive terms we shall examine later on. It was important for physicians to be able to justify to patients the expense of $2–3 per treatment, as home vibrators were available for about $5.

The acceptance of the electromechanical vibrator by physicians at the turn of

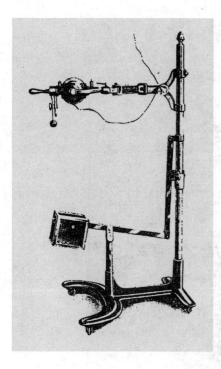

Figure 4.2. The Chattanooga, at $200 the most expensive medical vibrator available in 1904, could be wheeled over the operating table and its vibrating head rotated for the physician's convenience.

this century may also have been influenced by their earlier adoption of electro-therapeutics, with which vibratory treatment could be, and often was, combined.[26] Vibratory therapeutics were introduced from London and Paris, especially from the famous Hôpital Salpêtrière, which added to their respectability in the medical community.[27] It is worth noting as well that in this period electrical and other vibrations were a subject of great interest and considerable confusion, not only among doctors and the general public, but even among scientists like Tesla, who is reported to have fallen under their spell. "[T]he Earth," he wrote, "is responsive to electrical vibrations of definite pitch just as a tuning fork to certain waves of sound. These particular electrical vibrations, capable of powerfully exciting the Globe, lend themselves to innumerable uses of great importance."[28] In the same category of mystical reverence for vibration is Samuel Wallian's contemporaneous essay on "The Undulatory Theory in Therapeutics," in which he describes "modalities or manifestations of vibratory impulse" as the guiding principle of the universe. "Each change and gradation is not a transformation, as mollusk into mammal, or monkey into man, but an evidence of a variation in vibratory velocity. A

Figure 4.3. Chattanooga Vibrator parts.

certain rate begets a *vermis,* another and higher rate produces a *viper,* a *vertebrate,* a *vestryman.*"[29]

In 1900, according to Monell, more than a dozen medical vibratory devices for physicians had been available for examination at the Paris Exposition. Of these, few were able to compete in the long term with electromechanical models. Mary L. H. Arnold Snow, writing for a medical readership in 1904, discusses in some depth more than twenty types, of which more than half are electromechanical. These models, some priced to the medical trade as low as $15, delivered vibrations from one to 7,000 pulses a minute. Some were floor-standing machines on rollers; others could be suspended from the ceiling like the modern impact wrench.[30] The more expensive models were adapted to either AC or DC currents. A few, such as those of the British firm Schall and Son, could even be ordered with motors custom-wound to a physician's specifications. Portable and battery-powered electromechanical vibrators were generally less expensive than floor models, which both looked more imposing as instruments and were less likely to transmit fatiguing vibrations to the doctor's hands.

Patients were treated in health spa complexes, in doctor's offices or their own homes with portable equipment. Designs consonant with prevailing notions of what a medical instrument should look like inspired consumer confidence in the physician and his apparatus, justified treatment costs, and, in the case of hysteria treatments, camouflaged the sexual character of the therapy. Hand- or foot-powered models, however, were tiring to the operator; water-powered ones became too expensive to operate when municipalities began metering water in the early twentieth century. Gasoline engines and batteries were cumbersome and difficult to maintain, as noted above. No fuel or air-tank handling by the user was required for line electricity, in contrast with compressed air, steam and petroleum as power sources. In the years after 1900, as line electricity became the norm in urban communities, the electromechanical vibrator emerged as the dominant technology for medical massage.

Some physicians contributed to this trend by endorsing the vibrator in works like that of Monell, who had studied vibratory massage in medical practice in the United States and Europe at the turn of the twentieth century. He praises its usefulness in female complaints: "[P]elvic massage (in gynecology) has its brilliant advocates and they report wonderful results, but when practitioners must supply the skilled technic with their own fingers the method has no value to the majority. But special applicators (motor-driven) give practical value and office convenience to what otherwise is impractical."[31] Other medical writers suggested combining vibratory treatment of the pelvis with hydro- and electrotherapy, a refinement made possible by the ready adaptability of the new electromechanical technology.

At the same period, mechanical and electromechanical vibrators were introduced as home medical appliances. One of the earliest was the Vibratile, a battery-operated massage device advertised in 1899. Like the vibrators sold to doctors, home appliances could be handpowered, water-driven, battery or street-current apparatus in a relatively wide range of prices from $1.50 to $28.75. This last named was the price of a Sears, Roebuck model of 1918, which could be purchased as an attachment for a separate electrical motor, drawing current through a lamp socket, which also powered a fan, buffer, grinder, mixer and sewing machine. The complete set was marketed in the catalogue under the headline "Aids that Every

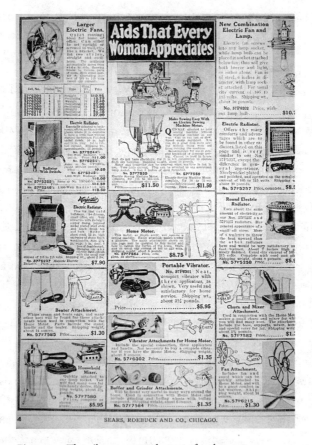

Figure 4.4. The vibratory attachments for the 1918
Sears Roebuck home motor were only one of
many electromechanical possibilities.

Woman Appreciates." (Fig. 4.4). Vibrators were mainly marketed to women, although men were sometimes exhorted to purchase the devices as gifts for their wives, or to become door-to-door sales representatives for the manufacturer.[32]

The electromechanical vibrator was preceded in the home market by a variety of electrotherapeutic appliances which continued to be advertised through the twenties, often in the same publications as vibratory massage devices. Montgomery Ward, Sears Roebuck, and the Canadian mail order department store T. Eaton and Company all sold medical batteries by direct-mail by the end of the nineteenth century. These were simply batteries with electrodes that administered a mild shock. Some, like Butler's Electro-Massage Machine, produced their own electricity with friction motors. Contemporaneous and later appliances sometimes had special features, such as Dr. H. Sanche's Oxydonor, which produced ozone in addition to the current when one electrode was placed in water. "Electric" massage rollers, combs and brushes with a supposedly permanent charge, retailed at this time for prices between one and five dollars. Publications like the *Home Needlework Magazine* and *Men and Women* advertised these devices, as well as related technologies, including correspondence courses in manual massage.

Vibrators with water motors, a popular power source, as noted above, before the introduction of metered water, were advertised in such journals as *Modern Women,* which emphasized the cost savings over treatments by physicians and further emphasized the advantage of privacy offered by home treatment. Such devices were marketed through the teens in *Hearst's* and its successors, and in *Woman's Home Companion.*[33] Electromechanical vibrators were sold in the upper-middle-class market, in magazines typically retailing for between ten and fifteen cents an issue. As in the case of medical vibrators, models adapted to both AC and DC current were more expensive than those for use with DC only; all were fitted with screw-in plugs through the twenties.[34]

All types of vibrators were advertised as benefiting health and beauty by stimulating the circulation and soothing the nerves. The makers of the electromechanical American vibrator, for example, recommended their product as an "alleviating, curative and beautifying agent. . . . It will increase deficient circulation — develop the muscles — remove wrinkles and facial blemishes, and beautify the complexion."[35] Advertisements directed to male purchasers similarly emphasized the machine's advantages for improving a woman's appearance and disposition. And ad in a 1921 issue of *Hearst's* urges the considerate husband to "Give 'her' a Star for Christmas" on the grounds that it would be "A Gift That Will Keep Her Young and Pretty." The same device was listed in another advertisement with several other electrical appliances, and labeled "Such Delightful Companions!"[36] A husband, these advertisements seem to suggest, who presented his wife with these progressive and apparently respectable medical aids might leave for work in the morning secure in the knowledge that his spouse's day would be pleasantly and productively invested in self-treatment. Like other electrical appliance advertising of the time, electromechanical vibrator ads emphasized the role of the device in making

a woman's home a veritable Utopia of modern technology, and its utility in reducing the number of occasions, such as visiting her physician, on which she would
be required to leave her domestic paradise.[37]

Advertisements for vibrators often shared magazine pages with books on sexual
matters, such as Howard's popular *Sex Problems in Worry and Work* and Walling's
Sexology, handguns, cures for alcoholism, and, occasionally, even personals, from
both men and women, in which matrimony was the declared objective. Sexuality
is never explicit in vibrator advertising; the tone is vague but provocative, as in
the Swedish Vibrator advertisement in *Modern Priscilla* of 1913, offering "a machine that gives 30,000 thrilling, invigorating, penetrating, revitalizing vibrations
per minute . . . Irresistible desire to own it, once you feel the living pulsing touch
of its rhythmic vibratory motion." Illustrations in these layouts typically include
voluptuously proportioned women in various states of *dishabille.* The White Cross
vibrator, made by a Chicago firm that manufactured a variety of small electrical appliances, was also advertised in *Modern Priscilla,* where the maker assured
readers that "It makes you fairly tingle with the joy of living."[38] It is worth noting
that the name "White Cross" was drawn from that of an international organization devoted to what was known in the early twentieth century as "social hygiene,"
the discovery and eradication of masturbation and prostitution wherever they appeared. The Chicago maker of White Cross appliances, in no known way affiliated
with the organization, evidently hoped to trade on the name's association with
decency and moral purity.[39] A 1916 advertisement from the White Cross manufacturer in *American Magazine* nevertheless makes the closest approach to explicit
sexual claims when it promises that "All the keen relish, the pleasures of youth,
will throb within you."[40] The utility of the product for female masturbation was
thus consistently camouflaged.

Electromechanical vibrator advertising almost never appeared in magazines
selling for less than 5 cents an issue (10 to 20 cents is the median range) or more
than 25 cents. Readers of the former were unlikely to have access to electrical current; readers of the latter, including, for example, *Vanity Fair,* were more likely
to respond to advertising for spas and private manual massage. While at least a
dozen and probably more than twenty U.S. firms manufactured electromechanical vibrators before 1930, sales of these appliances were not reported in the electrical trade press. A listing from the February *1927 NELA Bulletin* is typical; no
massage equipment of any kind appears on an otherwise comprehensive list that
includes violet-ray appliances.[41] A 1925 article in *Electrical World,* under the title

"How Many Appliances Are in Use?" lists only irons, washing machines, cleaners, ranges, water heaters, percolators, toasters, waffle irons, kitchen units, and ironers.[42] *Scientific American* listed in 1907 only the corn popper, chafing dish, milk warmer, shaving cup, percolator, and iron in a list of domestic electrical appliances.[43] References to vibrators were extremely rare even in popular discussions of electrical appliances.[44] The U.S. Bureau of the Census, which found 66 establishments manufacturing electrotherapeutic apparatus in 1908, does not disaggregate by instrument type either in this category or in "electrical household goods." The 1919 volume, showing the electromedical market at a figure well over $2 million, also omits detailed itemization. Vibrators appear by name in the 1949 *Census of Manufactures,* but it is unclear whether the listing for them, aggregated with statistics for curling irons and hair dryers, includes those sold as medical instruments to physicians.[45] This dearth of data renders sales tracking of the electromechanical vibrator extremely difficult. The omissions from engineering literature are worth noting, as the electromechanical vibrator was one of the first electrical appliances for personal care, partly because it was seen as a safe method of self-treatment.[46]

The marketing strategy for the early electromechanical vibrator was similar to that employed for contemporaneous and even modern technologies for which social camouflage is considered necessary. Technologically, the devices so marketed differ from modern vibrators sold for explicitly sexual purposes only in their greater overall weight, accounted for by the use of metal housings in the former and plastic in the latter. The basic set of vibratodes is identical, as is the mechanical action. The social context of the machine, however, has undergone profound change. Liberalized attitudes toward masturbation in both sexes and increasing understanding of women's sexuality have made social camouflage superfluous.

In the case of the vibrator, the issue is one of acceptability, but there are many examples of similarly marketed technology of which the expected use was actually illegal. One of these, which shares with the vibrator a focus on women's sexuality, was that of "emmenagogues" or abortifacient drugs sold through the mail and sometimes even off the shelf in the first few decades of this century. Emmenagogues, called in pre-FDA advertising copy "cycle restorers," were intended to bring on the menses in women who were "late." Induced abortion by any means was of course illegal, but late menses are not reliable indicators of pregnancy. Thus, women who purchased and took "cycle restorers" might or might not be in violation of antiabortion laws; they themselves might not be certain without a medical examination. The advertising of these commodities makes free use of this ambiguity in texts like the following from *Good Stories* of 1933: "Late? End Delay-Worry. American Periodic Relief Compound double strength tablets combine Safety with Quick Action. Relieve most Stubborn cases. No Pain. New discovery. Easily taken. Solves women's most perplexing problem. RELIEVES WHEN ALL OTHERS FAIL. Don't be discouraged, end worry at once. Send $1.00 for Standard size package and full directions. Mailed same day, special delivery in plain wrapper. American Periodic Relief Compound Tablets, extra strength for stubborn cases, $2.00. Generous Size Package. New Book free."[47] The rhetoric here does not mention the possibility of pregnancy, but the product's selling points would clearly suggest this to the informed consumer through the mentions of safety, absence of pain, and stubborn cases. The readers of the pulp tabloid *Good Stories* clearly did not require an explanation of "women's most perplexing problem."

Distilling technology raises similar issues of legality. During the Prohibition period, the classified section of a 1920 *Ainslee's* sold one and four gallon copper stills by mail, advising the customer that the apparatus was "Ideal for distilling water for drinking purposes, automobile batteries and industrial uses."[48] Modern advertisements for distilling equipment contain similar camouflage rhetoric, directing attention away from the likelihood that most consumers intend to employ the device in the production of beverages considerably stronger than water.[49]

Although changes in sexual mores have liberated the vibrator, social camouflage remains necessary for stills and many other modern commodities, including drug paraphernalia. The peering Prep Kit, for example, is advertised at nearly $50 as a superlative device for grinding and preparing fine powders, "such as vitamin pills or spices."[50] Burglary tools are marketed in some popular (if lowbrow) magazines with the admonition that they are to be used only to break into one's own home or automobile, in the event of having locked oneself out. The camouflage rhetoric seems to suggest that all prudent drivers and homeowners carry such tools on their persons at all times. Most recently, we have seen the appearance of computer software for breaking copy protection, advertised in terms that explicitly prohibit its use for piracy, although surely no software publisher is so naive as to believe that all purchasers intend to break copy protection only to make backup copies of legitimately purchased programs and data.[51] As in vibrator advertising, the product's advantages are revealed to knowledgeable consumers in language that disclaims the manufacturers' responsibility for illegal or immoral uses of the product.

The marketing of socially camouflaged technologies is directed to consumers who already understand the design purpose of the product, but whose legally and/or culturally unacceptable intentions in purchasing it cannot be formally recognized by the seller. The marketing rhetoric must extol the product's advantages for achieving the purchaser's goals—in the case of the vibrator, the production of orgasm—by indirection and innuendo, particularly with reference to the overall results, i.e., relaxation and relief from tension. The same pattern emerges in the advertisement of emmenagogues: according to the manufacturer, it is "Worry and Delay" that are ended, not pregnancy. In the case of software copyright protection programs, drug paraphernalia and distilling equipment, the expected input and/or output are simply misrepresented, so that an expensive finely calibrated scale with its own fitted carrying case may be pictured in use in the weighing of jelly beans. As social values and legal restrictions shift, the social camouflaging of technologies may be expected to change in response, or to be dispensed with altogether, as in the case of the vibrator.

NOTES

1. This research was made possible in part by a grant from the Bakken (Museum and Library of Electricity in Life), August 1985. Various versions of this paper have benefited from comments and criticism from John Senior at the Bakken, Joel Tarr of Carnegie Mellon University, Shere Hite of Hite Research, Karen Reeds of Rutgers University Press, my former students at Clarkson University, and participants in the Social and Economic History Seminar, Queens University (Canada), the Hannah Lecture series in the History of Medicine at the University of Ottawa, and the 1986 Meeting of the Society for the History of Technology with the Society for the Social Study of Science. Anonymous Referees of *IEEE Technology and Society Magazine* and other journals have also provided valuable guidance in structuring the presentation of my research results.

2. Jayme A. Sokolow, *Eros and Modernization: Sylvester Graham, Health Reform and the Origins of Victorian Sexuality in America* (Rutherford, N.J., 1983), pp. 77–99; John S. Hailer and Robin Hailer, *The Physician and Sexuality in Victorian America* (Urbana, Ill., 1973), pp. 184–216; Donald E. Greydanus, "Masturbation: Historic Perspective," *New York State Journal of Medicine* 80, no. 12 (November 1980): 1892–96; Thomas Szasz, *The Manufacture of Madness* (New York, 1977), pp. 180–206; E. H. Hare, "Masturbatory Insanity: The History of an Idea," *Journal of Mental Sciences* 18 (1962): 1–25; Vern Bullough, "Technology for the Prevention of 'Les Maladies Produites par la Masturbation,'" *Technology and Culture* 28, no. 4 (October 1987): 828–832.

3. On the vibrator in stag films, see Roger Blake, *Sex Gadgets* (Cleveland, 1968), pp. 33–

46. An early postwar reference to the vibrator as an unabashedly sexual instrument is Albert Ellis, *If This Be Sexual Heresy* (New York, 1963), p. 136.

4. Vibrators and dildos are illustrated in Paul Tabori, *The Humor and Technology of Sex* (New York, 1969); the dildo is discussed in a clinical context in William H. Masters, *Human Sexual Response* (Boston, 1966). Vibrators of the period to which I refer in this essay are illustrated in Sam J. Gorman, *Electro Therapeutic Apparatus* (10th ed. Chicago, c.1912); Wappler Electric Manufacturing Co. Inc., *Wappler Cautery and Light Apparatus and Accessories* (2nd ed. New York, 1914), pp. 7 and 42–43; Manhattan Electrical Supply Co., *Catalogue Twenty-Six: Something Electrical for Everybody* (New York, n.d.); and Mary Lydia Snow and Arnold Hastings, *Mechanical Vibration and Its Therapeutic Application* (New York, 1904 and 1912). For modern vibrators, see Helen Singer Kaplan, "The Vibrator: A Misunderstood Machine," *Redbook* (May 1984): 34, and Mimi Swarz, "For the Woman Who Has Almost Everything," *Esquire* (July 1980): 56–63.

5. See, for examples of such advertising, which in fact included a persistent abdominal emphasis, "Amazing New Electric Vibrating Massage Pillow," Niresk Industries (Chicago, Ill.) advertisement in *Workbasket* (October 1958): 95; "Don't be Fat," body massager, Spot Reducers advertisement in *Workbasket* (September 1958): 90; and "Uvral Pneumatic Massage Pulsator," *Electrical Age for Women* 2, no. 7 (January 1932): 275–76.

6. This therapy is extensively documented but rarely noted by historians. For only a few examples of medical discussions of vulvular massage in the hysteroneurasthenic disorders, see Aretaeus Cappadox, *The Extant Works of Aretaeus the Cappadocian,* ed. and trans. by Francis Adams (London, 1856), pp. 44–45, 285–87, and 449–51; Petrus Alemarianus Forestus (Pieter van Foreest), *Observationem et Curationem Medicinalium ac Chirurgicarum Opera Omnia* 3, book 28 (Rothomagi, 1653), pp. 277–340; Galen of Pergamon, *De Locis Affectis,* trans. by Rudolph Siegel (Basel and New York, 1976), book VI, chapter II: 39; and Sigismond A. Weber, *Traitement par l'Electricité et le Massage* (Paris, 1889), pp. 73–80. Of modern scholars, only Audrey Eccles discusses this therapy in detail in her *Obstetrics and Gynecology in Tudor and Stuart England* (London and Canberra, 1982), pp. 76–83.

7. Simon Baruch, *The Principles and Practice of Hydrotherapy: A Guide to the Application of Water in Disease* (New York, 1897), pp. 101, 211, 248, and 365; William H. Dieffenbach, *Hydrotherapy* (New York, 1909), pp. 238–45; Mood Health Publishing Company, *20th Century Therapeutic Appliances* (Battle Creek, Mich., 1909), pp. 20–21; William Snowdon Hedlev, *The Hydro-Electric Methods in Medicine* (London, 1892); Guy Hinsdale, *Hydrotherapy* (Philadelphia and London, 1910), p. 224; John Harvey Kellogg, *Rational Hydrotherapy* (Philadelphia, 1901); J. A. Irwin, *Hydrotherapy at Saratoga* (New York, 1892), pp. 85–134 and 246–48; Curran Pope, *Practical Hydrotherapy: A Manual for Students and Practitioners* (Cincinnati, 1909), pp. 181–92 and 506–38; Russell Thacher Trail, *The Hydropathic Encyclopedia* (New York, 1852), pp. 273–95. Women were reportedly in the majority as patients at spas, and some were owned by women entrepreneurs and/or physicians. See T. Whyman, "Visitors to Margate in the 1841 Census Returns," *Local Population Studies* 8, (1972): 23. Since at least the time of Jerome, baths and watering places have had a reputation for encouraging unacceptable expressions of sexuality. For female masturbation with water, see J. Aphrodite, pseud., *To Turn You On: Sex Fantasies for Women* (Secaucus, N.J., 1975), pp.

83–91, and E. Halpert, "On a Particular Form of Masturbation in Women: Masturbation with Water," *Journal of the American Psychoanalytic Association* 21 (1973): 526.

8. A bibliography of nineteenth-century American works on women and sexuality in relation to hysteria is available in Nancy Sahli, *Women and Sexuality in America: A Bibliography* (Boston, 1984). See also Edward Shorter, "Paralysis: the Rise and Fail of the 'Hysterical' Symptom," *Journal of Social History* 19, no. 4 (Summer 1986): 549–82; Roberta Satow, "Where Has All the Hysteria Gone?" *Psychoanalytic Review* 66 (1979–80): 463–73; Désiré Magloire Bourneville and P. Regnard, *Iconographie Photographique de la Salpetrière* vol. 2 (Paris, 1878), pp. 97–219; Jean-Martin Carcot, *Clinical Lectures on Certain Diseases of the Nervous System,* trans. E. P. Hurd (Detroit, 1888), p. 141; Havelock Ellis, *Studies in the Psychology of Sex* (New York, 1940), 1: 270; Alan Krohn, *Hysteria: The Elusive Neurosis* (New York, 1978), pp. 46–51; William J. McGrath, *Freud's Discovery of Psychoanalysis: The Politics of Hysteria* (Ithaca, N.Y., 1986), pp. 152–72; Ilza Veith, *Hysteria: The History of a Disease* (Chicago, 1965); Franz Wittels, *Freud and His Time* (New York, 1931), pp. 215–42; and Dewey Ziegler and Paul Norman, "On The Natural History of Hysteria in Women," *Diseases of the Nervous System* 15 (1967): 301–6.

9. Carol Bauer, "The Little Health of Ladies: An Anatomy of Female Invalidism in the Nineteenth Century," *Journal of the American Medical Woman's Association* 36, no. 10 (October 1981): 300–306; Barbara Ehrenreich and D. English, *Complaints and Disorders: The Sexual Politics of Sickness* (Old Westbury, N.Y., 1973), pp. 15–44; Russell Thacher Trail, *The Health and Diseases of Women* (Battle Creek, Mich., 1873), pp. 7–8.

10. Thomas Sydenham, "Epistolary Dissertation on Hysteria," in *The Works of Thomas Sydenham* vol. 2, trans. R.G. Latham (London, 1848), pp. 56 and 85; Joseph Frank Payne, *Thomas Sydenham* (New York, 1901), p. 143.

11. Only a minority of writers on hysteria associated the affliction with paralysis until Freud made this part of the canonical disease paradigm in the twentieth century.

12. Franz Josef Gall, *Anatomie et Phisiologie du Système Nerveaux en Général* vol. 3 (Paris, 1810–19), p. 86; Auguste Elisabeth Philogene Tripier, *Leçons Cliniques sur les Maladies de Femmes* (Paris, 1883), pp. 347–51; Nathaniel Highmore, *De Passione Hysterica et Affectione Hypochondriaca* (Oxon., 1660), pp. 20–35; Ellis, *Studies in the Psychology of Sex,* 1: 225. See also Pierre Briquet, *Traitè Clinique et Thérapeutique de l'Hystèrie* (Paris 1859), pp. 137–38, 570, and 613.

13. William Cullen, *First Lines in the Practice of Physic* (Edinburgh, 1791), pp. 43–47; Robert Burton, *The Anatomy of Melancoly,* ed. Dell Floyd and Paul Jordan Smith (New York, 1927), pp. 353–55; Gregor Horst, *Dissertatronem . . . inauguratem De Mania. . . . Gissae: typis Viduae Friederici Kargeri* (1677), pp. 9–18; A.F.A. King, "Hysteria," *American Journal of Obstetrics* 24, no. 5 (May 18, 1891): 513–32; *Medieval Woman's Guide to Health,* trans. Beryl Rowland (Kent, Ohio, 1981), pp. 2, 63, and 87; Philippe Pinel, *A Treatise on Insanity,* trans. D. Davis (Facsimile edition of the London 1806 edition; New York, 1962), pp. 229–30; Wilhelm Reich, *Genitality in the Theory and Therapy of Neurosis,* trans. by Philip Schmitz (1927; reprint, New York, 1980), pp. 54–55 and 93.

14. Abraham Zacuto, *Praxis Medico Admiranda* (London, 1637), p. 267. Zacuto is at pains to point out that some physicians regard vulvular massage as indecent: "Nom autem

ex hoc occasione, liceat Medico timenti Deum, sopitis pariter cunctis sensibus, & una abo-
lita respiratione in foeminis quasi animam agentibus, see in maximo vitae periculo constitutes, veneficium illud semen, foras ab utero, titillationibus, & frictionbius portion obscoe-
narium elidere, different eloquenter."

15. John Harvey Kellogg, "Electrotherapeutics in Chronic Maladies," *Modern Medicine*
(October–November 1904): 4. Kellogg's background is described in detail in Richard W.
Schwartz, *John Harvey Kellogg, MD* (Nashville, 1970).

16. Women who regularly undergo the discomfort of gynecological examination with
this instrument are justifiably amused by its nineteenth-century mythology. For an example of conservative views on the speculum, see Wilhelm Griesinger, *Mental Pathology and
Therapeutics,* trans. C. Lockhart Robinson and James Rutherford (London, 1867), p. 202.
On the inefficiency of penetration as a means to female orgasm, the standard modern work
is of course Shere Hite, *The Hite Report on Female Sexuality* (New York, 1976), but the phenomenon was widely noted by progressive physicians and others before the seventies. Most
of these latter, however, regarded the failure of penetration to fully arouse about three-
quarters of the female population as either a pathology on the women's part or as evidence
of a natural diffidence in the female. Hite is the first to point out that the experience of
the majority constitutes a norm, not a deviation. For examples of various male views on
this subject, see Marc H. Hollender, "The Medical Profession and Sex in 1900," *American
Journal of Obstetrics and Gynecology* 108, no. 1 (1970): 139–48; Carl Degler, "What Ought to
Be and What Was," *American Historical Review* 79 (1974): 1467–90, and *At Odds: Women
and the Family in America from the Revolution to the Present* (New York, 1980), pp. 249–78;
and Gilles de la Tourette, *Traité Clinique et Thèrapeutique de l'Hystérie Paroxystique* vol. 1
(Paris, 1895), p. 46. Feminine views are seldom recorded before this century; a few examples
are those reported by Katherine B. Davis, summarized in Robert L. Dickson and Henry
Pierson, "The Average Sex Life of American Women," *Journal of the American Medical Association* 85 (1925): 113–17; Sofie Lazarsfeld, *Woman's Experience of the Male* (London, 1967),
pp. 112, 181, 271, and 308. It has also been noted that few women have difficulty achieving
orgasm in masturbation, and that the median time to orgasm in masturbation is substantially the same in both sexes: Alfred Charles Kinsley, *Sexual Behavior in the Human Female*
(Philadelphia, 1953), p. 163.

17. Thomas Stretch Dowse, *Lectures on Massage and Electricity in the Treatment of Disease* (Bristol, 1903), p. 181.

18. E. H. Smith, "Signs of Masturbation in the Female," *Pacific Medical Journal* (February 1903).

19. For examples of spa expenses in the Unites States, see Samuel A. Cloyes, *The Healer:
The Story of Dr. Samantha S. Nivison and Dryden Springs, 1820–1915* (Ithaca, N.Y., 1969),
p. 24; Estrelita Karsh, "Taking the Waters at Stafford Springs," *Harvard Library Bulletin* 28,
no. 3 (July 1980): 264–81; Marilyn MacMillan, "An Eldorado of Ease and Elegance: Taking
the Waters at White Sulphur Springs," *Montana* 35 (Spring 1985) 36–49; and Harold Meeks,
"Smelly, Stagnant and Successful: Vermont's Mineral Springs," *Vermont History* 47, no. 1
(1979): 5–20.

20. Taylor wrote indefatigably on the subject of physical therapies for pelvic disorders,

and devoted considerable effort to the invention of mechanisms for this purpose. See George Henry Taylor, *Diseases of Women* (Philadelphia and New York, 1871); *Health for Women* (New York, 1883) and eleven subsequent editions; "Improvements in Medical Rubbing Apparatus," U.S. Patent 175,202 dated March 21, 1876; *Mechanical Aids in the Treatment of Chronic Forms of Disease* (New York, 1893); *Pelvic and Hernial Therapeutics* (New York, 1885); and "Movement Cure," U.S. Patent 263,625 dated August 29, 1882.

21. An example of the early Weiss model is available for study at the Bakken (Library and Museum), Minneapolis, Minn., accession number 82.100.

22. Joseph Mortimer Granville, *Nerve-Vibration and Excitation as Agents in the Treatment of Functional Disorders and Organic Disease* (London, 1883), p. 57; his American colleague Noble Murray Eberhart advises against vibrating pregnant women "about the generative organs" for fear of producing contractions. See his *A Brief Guide to Vibratory Technique* 4th ed. (Chicago, 1915), p. 59. For examples of enthusiastic endorsements of the new technology, see Franklin Benjamin Gottschalk, *Practical Electra-Therapeutics* (Hammond, Ind., 1904), and *Static Electricity, X-Ray and Electro-Vibration: Their Therapeutic Application* (Chicago, 1903); International Correspondence Schools, *A System of Electrotherapeutics* 4 (Scranton, Penn., 1903); Anthony Matijaca, *Principles of Electro-Medicine, Electro-Surgery and Radiology* (Tangerine, Fla., Butler, N.J., and New York, 1917); Samuel Howard Monell, *A System of Instruction in X-Ray Methods and Medical Uses of Light, Hot Air, Vibration and High Frequency Currents* (New York, 1902); Maurice Riescher Pilgrim, *Mechanical Vibratory Stimulation; its Theory and Application in the Treatment of Disease* (New York, 1903); May Cushman Rice, *Electricity in Gynecology* (Chicago, 1909); Alphonse David Rockwell, *The Medical and Surgical Uses of Electricity,* new ed. (New York, 1903); Snow and Hastings, *Mechanical Vibration* (n. 4 above); and Melanchthon R. Waggoner, *The Note Book of an Electra-Therapis* (Chicago, 1923); Samuel Spencer Wallian, *Rhythmotherapy* (Chicago, 1906), and "Undulatory Theory in Therapeutics," *Medical Brief* (May and June, 1905): 231.

23. See, for example, A. Lapthorn Smith, "Disorders of Menstruation," in *An International System of Electro-Therapeutics,* ed. Horatio Bigelow (Philadelphia, 1894), p. G163.

24. Alfred Dale Covey, *Profitable Office Specialities* (Detroit, 1912), pp. 16, 18, and 79–95; Edward Trevert Bubier, *Electra-Therapeutic Hand Book* (New York, 1900); J. J. Duck Co., *Anything Electrical: Catalog No. 6* (Toledo, Ohio, 1916), p. 162; Golden Manufacturing Co., *Vibration: Nature's Great Underlying Force for Health, Strength and Beauty* (Detroit, Mich., 1914); Sam J. Gorman Co. *Physician's Vibragenitant* (Chicago, n.d.); Keystone Electric Co., *Illustrated Catalogue and Price List of Electro-Therapeutic Appliances . . . etc.* (Philadelphia, 1903), pp. 63–66; Schall and Son, Ltd., *Electro-Medical Instruments and their Management. . . 17th ed.* (London and Glasgow, 1925); Vibrator Instrument Co., *Treatise on Vibration and Mechanical Stimulation* (Chattanooga, Tenn., 1902); Vibrator Instrument Co., Clinical Dept., *A Course on Mechanical Vibratory Stimulation* (New York, 1903); "Vibratory Therapeutics," *Scientific American* 67 (October 22, 1892): 265. Most of these manufacturers were quite respectable instrument firms; see Audrey B. Davis, *Medicine and Its Technology: An Introduction to the History of Medical Instrumentation* (Westport, Conn., 1981), p. 22.

25. Vibrator Instrument Co., *Chattanooga Vibrator* (Chattanooga, Tenn., 1904), pp. 3, 26.

26. Auguste Vigouroux, *Etude sur la Rèsistance Electrique chez les Melancoliques* (Paris, 1890); Richard J. Cowen, *Electricity in Gynecology* (London, 1900); George J. Engelmann, "The Use of Electricity to Gynecological Practice," *Gynecological Transactions* 11 (1886); David V. Reynolds, "A Brief History of Electrotherapeutics" in *Neuroelectric Research,* ed. D. V. Reynolds and A. Sjoberg (Springfield, Ill., 1971), pp. 5–12; John V. Shoemaker, "Electricity in the Treatment of Disease," *Scientific American Supplement* 63 (January 5, 1907): 25923–24.

27. "Vibratory Therapeutics," *Scientific American* 67 (October 22, 1897): 265.

28. John J. O'Neill, *Prodigal Genius: The Life of Nikola Tesla* (New York, 1944), p. 210.

29. *Medical Brief* (May 1905): 417. See also the theory of light vibrations employed in the Master Electric Company's advertising brochure *The Master Violet Ray* (Chicago, n.d.).

30. Monell, *A System of Instruction* (n. 22 above), 595; Snow and Hastings, *Mechanical Vibration* (n. 4 above).

31. Monell, *A System of Instruction,* 591.

32. See, for example, "Wanted, Agents and Salesman . . ." Swedish Vibrator Company, *Modern Priscilla* (April 1913): 60.

33. "Agents! Drop Dead Ones!" Blackstone Water Power Vacuum Massage Machine, *Hearst's* (April 1916): 327; "HydroMassage" Warner Motor Company, *Modern Women* 11, no. 1 (December 1906): 190.

34. Wall receptacles are a relatively late introduction. See Fred E. Schroeder, "More 'Small Things Forgotten': Domestic Electrical Plugs and Receptacles, 1881–1931," *Technology and Culture* 27, no. 3 (July 1986): 525–43.

35. "Massage is as old as the hills . . . ," American Vibrator Company, *Woman's Home Companion,* April 1906, p. 42.

36. "Such Delightful Companions!" Star Electrical Necessities, 1922, reproduced in Edgar R. Jones, *Those Were the Good Old Days* (New York, 1959), unpaged; "A Gift that will Keep her Young and Pretty," Star Home Electric Massage, *Hearst's International,* December 1921, p. 82.

37. See, for example, the Ediswan advertisement in *Electrical Age for Women* 2, no. 7 (January 1932): 274, and review on page 275 of the same publication.

38. "Vibration is Life," Lindstrom-Smith Co., *Modern Priscilla,* December 1910, p. 27.

39. David J. Pivar, *Purity Crusade: Sexual Morality and Social Control, 1868–1900* (Westport, Conn., 1973), pp. 110–17.

40. See also *American Magazine* 75, no. 2 (December 1912), 75, no. 3 (January 1913), and 75, no. 7 (May 1913): 127; *Needlecraft,* September 1912, p. 23; *Home Needlework Magazine,* October 1908, p. 479, and October 1915, p. 45; *Hearst's,* January 1916, p. 67, February 1916, p. 154, April 1916, p. 329, and June 1916, p. 473; and *National Home Journal,* September 1908, p. 15.

41. J. E. Davidson, "Electrical Appliance Sales During 1926," *NELA Bulletin* 14, no. 14 (February 1927): 119–20.

42. *Electrical World,* December 5, 1925, p. 1164. See also George A. Hughes, "How the Domestic Electrical Appliances are Serving the Country," *Electrical Review* 72 (June 15, 1918): 983; E. A. Edkins, "Prevalent Trends of Domestic Appliance Market," *Electrical World,*

March 30, 1918, p. 670–71; and "Surveys Retail Sale of Electrical Appliances," *Printer's Ink* 159 (May 19, 1932): 35.

43. "Electrical Devices for the Household," *Scientific American* 96 (January 26, 1907): 95.

44. The vibrator is not included in extensive lists of appliances in Helen Lamborn, "Electricity for Domestic Uses," *Harper's Bazaar* 44 (April 1910): 285, and H. S. Knowlton, "Extending the Uses of Electricity," *Cassier's Magazine* 30 (June 1906): 99–105.

45. U.S. Bureau of the Census, *Census of Manufactures,* 1908, 1919, and 1947, pp. 216–17, 203, and 734 and 748 respectively.

46. On the early history of appliances, see Earl Lifshey, *The Houseware Story* (Chicago, 1973). For the safety issue, see "Electromedical Apparatus for Domestic Uses," *Electrical Review,* October 22, 1926, p. 682.

47. *Good Stories,* October 1933, p. 2. See also similar advertisement in the same issue for Dr. Roger's Relief Compound, p. 12.

48. "Water Stills," *Ainslee's Magazine,* October 1920, p. 164.

49. See, for example, Damark International, Inc., *Catalog B-330* (Minneapolis, 1988), p. 7, which emphasizes the "Alambiccus Distiller's" usefulness for distilling herbal extracts.

50. *Mellow Mail Catalogue* (New York, 1984), pp. 32–39.

51. Steven Levy, *Hackers: Heroes of the Computer Revolution* (Garden City, N.Y., 1984), p. 377.

PART II

Entwined Categories: Technology Constructs Gender

What is gender?
How has gender been technologically constructed?
How do gendered meanings get assigned to specific technologies/ technological practices?

As noted in the introduction, the relation between gender and technology is a reciprocal one. Ideas about society, including gender, shape the ways we make, do, and design things; these things, in turn, become part of how we identify, structure, represent, and perform gender. Technologies from mag wheels to makeup, blacksmithing to baking have gendered meanings (in particular contexts). In the previous section, we suggested focusing on the technologies and the ways gender influenced their design, manufacture, and use, and also the ways gender influenced how such technologies have been seen by (or remained invisible to) both contemporaries and historians.

The following essays offer a chance to shift perspective and explore the ways people use and understand the material world to construct, maintain, alter, or negotiate gender boundaries. In examining these issues, it is crucial to recognize the industrial and North American context under examination here. In contrast, consider the planter George Washington, whose masculinity was defined by powdered wig and hosiery covering a well-turned calf; or consider the agricultural practices of many North American native peoples of the same era, for whom farming was considered women's work, inappropriate for men. Social rules for gender and technology vary with time, place, and culture. In this volume, the gender systems as well as the technologies belong to an industrial society, and for the most part to the industrializing United States. These "industrial genders" were entwined with the economic, technological, and organizational changes we typically label with words like urbanization, industrialization, consumer society, mass production, secularization, modernization. How were gender expectations preserved or changed as the economy shifted from an agricultural base to a manufacturing one? When did new technologies challenge older gender constructions, and how did people respond? How, in turn, did ideas about gender shape the technological choices people made during the process?

FIVE

Industrial Genders: Constructing Boundaries

NINA E. LERMAN

Nina Lerman takes us back to the 1850s, the era of the Colt revolver, the sewing machine, the gold rush, and the telegraph. In
this article, she explores technological knowledge in a context
of rapidly changing technologies, asking how particular kinds of
knowledge and particular technological activities were assigned
to different groups of Americans. Focusing on children in a range
of different institutions in a single industrial city (Philadelphia), she
explores the ways the material world, and knowledge of its workings, was mustered to maintain or challenge the various social
boundaries of the nineteenth-century city.

How does Lerman define technology? How does she understand the relationship between gender and technology? Between
race and class and technology? In the institutional examples presented here, do you find a reciprocal relationship—how did the
technologies shape gender ideas and practices?

"'Preparing for the Duties and Practical Business of Life': Technological Knowl-
edge and Social Structure in Mid-19th-Century Philadelphia," *Technology and
Culture* 38, no. 1 (1997): 31–59. Reprinted by permission of the Society for the
History of Technology.

Industrialization is commonly described as a social as well as a technological transformation. Nonetheless, the tight links between the history of technology and social history suggested by such a phrase have remained largely elusive; the "social" and the "technological" are most often treated in opposition to one another.[1] But scholars discuss both technology and social categories like gender in terms of their "social construction," suggesting another, more parallel or even interdependent relationship. Indeed, phrases such as "social shaping of technology" invite transgression of traditional boundaries. Exploring the "social and technological transformation" of American industrialization in this spirit, I find that technological knowledge (a topic long since introduced in these pages) provides an ideal approach: the study of technological knowledge places *people making and doing things* — from cooking to carpentry — at the focus of our scrutiny, and demands integration of the technical and the social.[2]

Industrialization as a process of rapid technological, economic, and social change means, by definition, the great upheaval and overhaul of technological knowledge as well as of social structures and ideas. By focusing on the *acquisition* of technological knowledge, or technical education, this article begins to explore the multilayered interconnections between social and technological change, and to study closely the fabric of the industrial transformation. Examining the technical education adults provide children — "preparing for the duties and practical business of life," as one board of managers described girls' sewing and boys' shoemaking[3] — allows us to explore important questions about technological knowledge: What kinds of choices are made by whom in the acquisition and development of technological knowledge? Who learns how to do what? — and further, who is allowed to learn how to do what? What knowledge is ubiquitous, and what is unusual? What does it mean, in a particular context, to say that a particular person possesses a particular skill? These questions draw our attention to the reciprocal shapings of technological knowledge and social ideologies like gender, race, and class.

This article examines the particular case of technical education in mid-nineteenth-century Philadelphia, a city caught up in the whirlwind boom and bustle of changing technologies, changing populations, and the changing social strategies of urban industrialization. Mid-century Philadelphia was a city in transition—an old social and technological order was, for many, being replaced with a larger-scale, more bureaucratic and industrial vision. The population of urban Philadelphia had more than tripled since 1820; when the whole of Philadelphia County was consolidated into one municipal jurisdiction in 1854, its twenty-four wards housed more than 400,000 people, more than a quarter of whom had been born outside of the United States.[4] Manufacturing in Philadelphia was diverse and generally proprietary rather than corporate: the 1850 manufacturing census found 4,542 businesses worth a total of $32 million—an average of about $7,000 each—dotting the city.[5] The technological enthusiasm of the day, too, manifested itself in diversity: the Franklin Institute's Twenty-Second Exhibition of American Manufactures, held in 1852, boasted nearly 900 items spread over several galleries and arranged in twenty-three categories ranging from "Models and Machinery" to "Carpets, Oil Cloths, Silks, and Umbrellas"; from "Lamps and Gas Fixtures" and "Chemicals, Perfumery, Paints and Colors" to "Clothing, Needlework, and Miscellaneous Articles," this last among the most numerous.[6]

By this time, too, a broad range of institutions offered young Philadelphians training in technological knowledge suited to their gender, their ethnic background, and their expected economic position. Philadelphia's public school system was by mid-century capped by a "people's college," Central High School, open by competitive examination to high-scoring grammar school boys. It also included a Girls' Normal School training future schoolteachers. A "practical" curriculum similar to Central's, including modern languages and science, was also available to fatherless boys at the Girard College for Orphans. Stephen Girard, whose bequest founded the school, was known not only for his wealth and eccentricity but also for his humble beginnings and self-made success. A traditional classical education was of course available at several old Philadelphia academies, as it was at the extraordinary Institute for Colored Youth, a Quaker-endowed school run largely by the African-American community it served. More typical benevolent institutions included the Pennsylvania School for the Deaf, entering its fourth decade of teaching English to deaf children from Pennsylvania and neighboring states, and the Philadelphia House of Refuge, an institution founded a quarter century earlier as an alternative to adult jails for juvenile delinquents. The House of Refuge added a Colored Department in 1850. Other institutions had a more explicitly technical focus: the Pennsylvania School of Design for Women offered training in industrial textile design to young women, and the apprenticeship program at Baldwin

Locomotive Works trained young men to work in the "high technology" of the antebellum era.

Whatever its primary focus, technical education in any institution was intended to prepare children for their roles as adults. Curricula varied widely between institutions, yet all the institutions discussed here shared a common concern with the future productive value of children's lives. As such, the records and reports of these institutions provide a wealth of information — if not about how to cane chairs or design textiles, at least about who should learn, how long it might take, what institutional resources might be provided in the process, and what a "successful" pupil might hope for in adult life. Using such records to make comparisons across the most conspicuous social boundaries illuminates the shifting social and technological dimensions of mid-century industrialization, and highlights the broad range of available technological knowledge.

Most of the published scholarship on technological knowledge focuses on engineering and machine shop knowledge. While authors studying engineering recognize that it is only one kind of technological knowledge, as a group they have been most concerned with the relationship of engineering to science, and of technological knowledge to scientific knowledge. Recent studies of craft and artisanal knowledge, meanwhile, have focused primarily on metalwork.[7] But technological knowledge includes materials and activities not generally associated with machine tools or with engineering, as Eugene Ferguson intimates when he uses textiles to illustrate the connections between what we see and what we know about it. "When we see a piece of cloth lying on a chair, we can say without touching it that it is probably soft (or harsh), pliable (or stiff), and a good (or poor) insulator against cold. A tailor who has used that kind of cloth can add a whole new order of facts and insights regarding its attributes."[8] Rarely, of course, do we describe the tailor's activities as "construction" or "building" or "engineering," despite his familiarity with patterns, measurements, fitting (if not filing), and design. When we read Ferguson's description, however, we recognize that a tailor, like an engineer, makes use of his "mind's eye." We also know that his needle by itself does not make suits any more than a rivet gun makes a bridge truss; that tailoring, like construction in other materials, requires technological knowledge.

And yet we are more likely to refer to tailoring as "skill" or "craft" rather than "technological knowledge." Why do we maintain a tradition of labeling them differently? If we return to nineteenth-century America and discuss "sewing" rather than "tailoring," we find still another set of distinctions: sewing was the technique by which tailors built things, but sewing was also generally considered women's work. In fact, sewing was seen as women's work — and was considered quite ordinary — *except* when performed by tailors, or sailmakers, or cordwainers. In these instances we refer to sewing as a "skilled craft." Nineteenth-century seamstresses, on the other hand, are often referred to as "unskilled" workers.[9] Their knowledge, even more than tailors', has rarely been considered "technological," even as we comfortably define needle, thread, and fabric as "technological" artifacts.

To suggest that the gender differences between tailors and seamstresses — rather than their work — might account for these differences in perception of what people do will hardly be surprising, but it underscores the idea that gender plays a role in defining which activities can readily be labeled "technological." Further, it raises the twin question of how technological activities shape the structures and meanings of gender (or other) difference. What remains, then, is to disentangle the content of technological knowledge from its social meanings (defined not only by gender but also by ethnicity, economic position, etc.). For these meanings have shaped access to different kinds of technological knowledge, historically; they have also shaped our present-day choices about what topics we should study.

I argue that technological knowledge in America has been and is perceived hierarchically, and which kinds one acquires can both depend on and help define one's social circumstance. In making this argument I am not advocating rigid determinism, but suggesting instead that the relationship between social hierarchies and perceived hierarchies of technological knowledge is a reciprocal one.[10] The content and symbolic meanings of words like "sewing" or even "skill" are different but not unrelated, because knowing about a particular technology carries social meaning at particular places and times.[11]

Philadelphia educators confronting the turbulence of mid-century provided technical education adapted to social boundaries as they understood them. They did not apply the word "gender" to children although they taught boys and girls differently; neither did they consider their teachings a form of "technological knowledge," even as they related much of their work to the needs of "industry." In their own terms, they contended with changing expectations of gender, race, and class roles in their industrializing society. For example, the contradictions between women's prescribed place in the home and their increasing employment outside it limited educational choices for girls and shaped the justifications proponents could offer. For the middle class, the steady growth of school learning and the increasing use of paper and written forms of technological knowledge (drafting, mathematics, bookkeeping) meant that demarcations between technical education and literary education were not always clearly defined. And for schools catering to the "lower sort," the moral aspects of industrious behavior, along with the value of steady work habits to employers paying hourly wages, blurred institutional distinctions between technical education and strict socialization: technical education began with what educators called "habits of industry," or the willingness to work hard at whatever task might present itself. Finally, managers of institutions, parents, and youths had also to contend with the changing labor market in devising and choosing educational programs.

For the historian, this complex web of relationships between economic valuation, social categories, and ideas about technological knowledge means that distinctions of gender, race, and class in technical education cannot long be considered in isolation from each other. The following discussion begins with boundaries based on gender, but moves quickly toward an integrated analysis of race and class difference as well.

"Home" and "Work"

Throughout the nineteenth century, gender was perhaps the most easily identifiable determinant of the technical education a child would receive. At the Pennsylvania School for the Deaf, boys were taught crafts like tailoring and shoemaking; girls learned housewifery.[12] At the Philadelphia House of Refuge, some boys made razor strops or wire umbrella "furniture" and others caned chairs; all girls learned housewifery.[13] In general, at nineteenth-century residential institutions, being male predicted little about the content of one's technical education, though for boys (as for girls) the value of hard work was consistently taught. Being female, however, meant learning and performing "housewifery," usually for the residents of the institution as a whole. Even institutions training women for work outside the home used the rhetoric of domesticity in descriptions of their programs. In the prevailing language of annual reports and institutional publications, women might work for a while as teachers or mill hands, but "preparation for life" meant preparation for housewifery.

"Housewifery" in the nineteenth century denoted all the tasks of running a household, such as cooking, cleaning, sewing, and laundry. Depending on the nature of the household and one's position in it, however, "housewifery" could mean performing such tasks for one's husband and children, managing servants who

TABLE 5.1. *House of Refuge, Girls' Work*

	1850		1853		1854	
	White	Colored	White	Colored	White	Colored
Total number of items sewed	3,481	1,952	4,447	1,595	5,630	2,161
Average number of girls	52	~30	48	34	47	36
Boys per girl	3.5	~2.6	3.7	2.4	4.1	2.1
Items per boy	18.9	24.4	24.8	19.7	29.0	28.8
Items per girl	66.9	65.1	92.6	46.9	119.8	60.0

Source: Data drawn from House of Refuge, *Annual Report [for 1850]* (n. 13 above), pp. 23, 27; HR, *Annual Report [for 1853]*, pp. 21, 26–27; HR, *Annual Report [for 1854]*, pp. 20, 27.

performed them, or serving in someone else's family. Increasingly, the tasks of housewifery included management of consumption rather than production, but this transformation was neither swift nor thorough even in later periods. For girls confined in the House of Refuge, for example, "housewifery" meant not only the maintenance work of laundry, cleaning, and help with kitchen and dishes, but also significant production.

Thus, in 1850, when the new Colored Department opened, the staff on the colored girls' side faced the colossal task of clothing all the children in the new Department, and doing so with the help of a small and inexperienced group of girls.[14] Nonetheless, an average of thirty girls sewed all the essentials of institutional domesticity, including 324 jackets, 208 shirts, 132 frocks, 150 "comfortables," 134 bed ticks, 215 sheets, and 86 bedspreads. Similarly, in 1854 when the white department moved to enlarged quarters, forty-seven girls sewed 450 pillow ticks, 450 bed ticks, and 614 pillowcases in addition to 637 jackets, 625 shirts, and 716 pairs of pants. Jackets, shirts, and pants, of course, were worn by the boys—who outnumbered the girls by about four to one that year. As in any ordinary household, the quantity of sewing accomplished by the girls varied more closely with the number of people wearing clothes than with the number of people sewing them (see Table 5.1).[15]

All of these figures were carefully tabulated in the *Annual Reports* of the Refuge. The quantities of boys' work—chairs caned, shoes stitched—were consistently reported with monetary values specified. For girls' production, no such value was ever calculated. Indeed, as Jeanne Boydston has argued, by mid-century domestic production—let alone the extensive maintenance work of cleaning and mending—was becoming increasingly invisible as an economic activity. "Work" increasingly referred to earning wages; tasks not earning wages, then, seemed less and less often to be real "work."[16] In residential institutions, of course, housework could not easily be ignored by male managers. They understood its necessity, and noticed when an extreme shortage of females made its performance difficult. During such times, they might even engage in a monetary exchange and hire someone to help.[17] Nonetheless, housewifery was portrayed as a noneconomic activity; girls would someday be provided for "at home." Boys, on the other hand, were taught assuming a need to earn a livelihood; they would have to "work."

For discussing technological knowledge, however, the nineteenth century's idealized depiction of "home" and "work" is insufficient. Hierarchies of technological knowledge, bound as they were to notions about preparing children for their places as adults, were far more complex than any two-part analysis could suggest. Rather, nineteenth-century educators and reformers employed multiple social distinctions both within and among the various institutions. Just as the term "housewifery" had different meanings—some women managed servants, and some women served mistresses—"work" was cast differently for different boys. Surviving records of educational institutions allow for more complex comparisons.

By the mid-1850s, for example, both "colored" and "white" inmates of the House of Refuge lived in new facilities. The architecture of these buildings cast social boundaries quite literally in stone: the thick stone outer wall of the Refuge, twenty-five to thirty feet high, was replicated between the otherwise adjacent Colored and White Departments. Further, within each enclosure, a tall brick wall, abutting the outer walls and the large central building, divided the boys' yard from the girls'; similar segregation took place on the inside, as well. Thus, the House of Refuge housed four distinct social groups, divided by both race and gender. Its records and reports, in general, were similarly divided, allowing comparisons not only within the institution but also between single Departments and other contemporary institutions.

Daily activities within each Department at the Refuge were determined by gender. There was no significant difference between the work colored and white girls, or colored and white boys, performed. But the House of Refuge managers had guardianship of its charges beyond their stay within the House walls. After children of either race had been "reformed" in the Refuge for a year or two, they were often encouraged to sign indentures of apprenticeship to a carefully chosen master. Despite the ostensibly parallel treatment of all male and all female inmates, racial differences among boys emerged at this stage: colored and white boys were apprenticed very differently. In 1853 and 1854, for example, about 60 percent of each group of apprentices were bound to farmers. The other 40 percent of colored boys were apprenticed to waiters and barbers; one boy was apprenticed to a bricklayer.

The nonfarming 40 percent of white boys, on the other hand, were apprenticed to forty-three different trades, including boot and shoe makers, carpenters, machinists, and blacksmiths. The forty-three white boys' trades did not include barber, waiter, or even bricklayer. The trades available to colored apprentices, particularly those already labeled "delinquent," were few indeed.[18]

Apprenticeship, historically, had multiple and overlapping meanings. It provided a means of transferring technological knowledge—and thus economic opportunity—from generation to generation; it also doubled, as at the Refuge, as a kind of foster care.[19] Despite the reported decline of traditional craft apprenticeship by mid-century, the practice remained current in much of Philadelphia's manufacturing community. Sons of proprietors often completed formal schooling, and then, at age fourteen (or later if they could be spared and could get in to the selective Central High), they began an apprenticeship with their fathers, leading eventually to partnership in the firm.[20] At Matthias Baldwin's renowned locomotive works, boys learned to be machinists as apprentices. And when managers of the Girard College for Orphans sought to bind out their charges in conformity with the terms of the school's bequest, they found a ready market for reputable, well-schooled white male orphan apprentices.

Traditional craft apprenticeship was an exchange of rights and responsibilities: a master cared for and educated the child, and the child served faithfully as an increasingly knowledgeable assistant. The knowledge acquired by a male apprentice in particular was generally assumed to prepare him for appropriate adult work—to provide means of economic independence. While nineteenth-century indentures often specified cash payments in place of bed, board, or clothing, the real economic incentive came when a young man no longer served as apprentice, but used his knowledge to hire himself out as a journeyman or to set up shop for himself.

Baldwin's apprenticeship program was an updated version of the craft apprenticeship tradition. By the 1840s, Baldwin could describe the company's "usual terms" as including a five-week trial period without wages, followed by a five-year term in which a portion of the boy's wages were held back, and the resulting $130 presented to the boy on completion of the full term. The boy was "to be put under the care of a friend who will be responsible for his conduct out of the shop."[21] In 1853 Baldwin's partner, Matthew Baird, took over the program and reinstituted formal indentures. Records show that the company both attempted to track down its own runaways, and immediately fired any boy found to be under indenture elsewhere. Rules were strict, but apprenticeship at Baldwin was a good opportunity, and both Baldwin employees and railway mechanics sought to place their sons with the company. Considering that apprentices' monetary wages began at $2.25 per week and rose only to $3.25, the value of the apprenticeship period resided largely in the training acquired, and in the connections to the firm made by both the boys and their fathers.[22] Apprenticeship for white boys was servitude, certainly—but for well-placed white boys, like those at Baldwin, the apprenticeship years were an investment. Journeymen at Baldwin were paid $10 to $15 per week, a fairly respectable salary.

The viability of white male apprenticeship at mid-century extended beyond the realm governed by paternal cooperation. The Girard College for Orphans, required under the terms of its bequest to apprentice its charges to trades, found in 1854 that "a much larger number of applications [seeking boys] has been received than the college could possibly supply."[23] Like the House of Refuge, Girard College was an institutional "parent" seeking placements for youths. Unlike House of Refuge delinquents, Girard boys were, as orphans went, a highly privileged set. Because the will specified that the school was for "poor male white children," with preference given to boys born in Philadelphia, and then in Pennsylvania,[24] Girard College orphans were native-born, if not of American parentage. By contrast, more than one-third of the white children (boys and girls) at the House of Refuge in 1853 were born outside of the United States, slightly higher than the general population. Less than one-quarter had American-born parents; nearly half of known parents came from Ireland.[25]

A comparison of apprenticeships at Girard and the House of Refuge is therefore instructive (see Table 5.2). Fewer than 10 percent of the Girard boys were apprenticed to farmers, as opposed to 60 percent at the House of Refuge, and half of those were farmers involved in some kind of manufacture: tanning, milling, or papermaking. While the two most numerous crafts for white House of Refuge boys were boot- and shoemaking and carpentry, more than one-fourth of Girard College apprentices learned to be either printers or druggists. Other Girard College crafts not represented among Refuge indentures included brass founder, saddler, silver chaser, silver plater, stereotyper, and coachmaker.[26]

Education for Work

The distribution of trades to which Girard boys were apprenticed was much wider than that of House of Refuge boys, and the trades themselves were gen-

TABLE 5.2. *House of Refuge and Girard College Apprenticeships, 1854*

Trade	House of Refuge, White Department		Girard College for Orphans		Total
	N	percent	N	percent	
Farmer	86	58.9	4	4.5	94 total
Farmer and tanner			2	2.2	Included above
Farmer and miller			1	1.1	Included above
Farmer and papermaker			1	1.1	Included above
Boot and shoe maker	11	7.5	2	2.2	13
Printer	1	0.7	12	13.5	13
Druggist/drug store			12	13.5	12
Carpenter	7	4.8	2	2.2	9
Machinist	5	3.4	1	1.1	6
Mariner	2	1.4	3	3.4	5
Merchant	2	1.4	3	3.4	5
Wool manufacturer	4	2.7			4
Blacksmith	3	2.1			3
Coach maker			3	3.4	3
Cooper	1	0.7	2	2.2	3
Painter and glazier			3	3.4	3
Tin smith	1	0.7	2	2.2	3
Baker	2	1.4			2
Brass founder			2	2.2	2
Hatter	1	0.7	1	1.1	2
Manufacturer of piano keys			2	2.2	2
Plasterer	1	0.7	1	1.1	2
Saddler			2	2.2	2
Silver chaser			2	2.2	2
Silver plater			2	2.2	2
Stereotyper			2	2.2	2
Store keeper	2	1.4			2
Turner in ivory			2	2.2	2
Weaver	2	1.4			2
Other trades	15	10.3	20	22.5	35
Total indentured	146	100.0	89	100.0	235
Total trades	31		44		64

Data drawn from House of Refuge, *Annual Report [for 1854]* (Philadelphia: 1855), p. 19; Louis Romano, *Manual and Industrial Education at Girard College, 1831–1965: An Era in American Educational Experimentation* (New York: Arno Press, 1980), Appendix G, pp. 337–340.

erally perceived as higher status. Besides being more "American," Girard boys quite simply were orphans, and not delinquents. No moral imperative compelled the Girard directors to keep boys away from the city as the Refuge managers attempted to do, and the full range of urban trades was potentially available. But in addition, Girard College provided a very different kind of training before the boys arrived at their masters' workplaces. Girard College students entered between ages six and ten, and attended school full time. The most advanced portion of the Girard curriculum included geometry, algebra, natural philosophy, history, geography, grammar, trigonometry, French, Spanish, writing, bookkeeping, and architectural, mechanical, and perspective drawing.[27] A Girard College boy was well prepared to learn the vagaries of printing or pharmacy.

House of Refuge boys, on the other hand, mostly young teens, attended a few hours of school in the morning and evening, before and after a full day in the workshops making razor strops and caning chairs. In the classes of the upper division, white boys' textbooks might include a "Rhetorical Reader," a "Geography and Atlas," a "Geography of Pennsylvania," and a "History of the United States." There is, in one *Annual Report,* mention of a text in "Elementary Astronomy." They were provided with basic schooling, not a high school–style curriculum; even the "ordinary branches" of education were not, after all, "the paramount object" of their detention in the Refuge. In addition, as many as two-fifths of white boys could not write their names on admission; nearly one-fifth did not know the alphabet.[28] Even without the stigma of delinquency, their relative lack of schooling cannot have enhanced their attractiveness as potential apprentices.

Limited as the white boys' schooling was, however, it easily surpassed the schoolwork in the other departments, both before admission and during incarceration. While about 60 percent of white boys and only slightly fewer white girls (55 percent) could read at least easy material on admission, the percentage dropped to about 36 percent for colored inmates of either sex. The percentage of children who could write more than just their names on admission fell from 40 percent of white boys to only 20 percent of white girls and colored boys, and less than 10 percent of colored girls. The schooling provided within the Refuge perpetuated such differences: white girls' texts were similar to white boys', but with no mention of astronomy and with "Sacred History" in place of the "rhetorical reader"; in the Colored Department, the only "text books" listed other than elementary readers and arithmetic texts were "Sacred History" and the "Old and New Testaments," for boys and girls alike. School reports for white boys placed more emphasis on the ability to "write on paper," as well.[29]

By this time, the prestige of "white collar work" was becoming something to seek, and the use of paper in technical undertakings had acquired clear status connotations. In an increasingly wage-oriented context, managerial money-related skills—bookkeeping, for example—became more highly valued. Similarly, mechanical and architectural drawings had the rhetorical power of both paper and quantitative precision. Stuart Blumin has emphasized the emerging distinction between "manual" and "non-manual" work, but the use of one's hands is not what defines the issue.[30] Instead, the label "manual" makes rhetorical claims. In literal terms, bookkeeping and clerking involved steady "manual" work, making them hard to differentiate effectively from etching or watchmaking. Rather, emerging middle-class hierarchies of technological knowledge valued abstractions of technology, and their various notations on paper, more than specific craft skills.

Such abstractions and notations, not surprisingly, were taught selectively. Gender distinctions, for example, seemed natural. At mid-century, women were generally excluded both from clerking activities and from learning the precise graphical forms of drafting and mechanical drawing. At Central High, a carefully constructed curriculum included graded lessons in linear drawing, perspective, and architectural and mechanical drawing; at the Girls' Normal, one instructor was responsible for both drawing and penmanship.[31] The Central High curriculum, intended to prepare students for "commerce, manufactures and the useful arts," emphasized modern languages and an extensive course in scientific studies. A telescope and observatory topped the school. Subjects taught at the Girls' Normal, on the other hand, were math, history, grammar, reading, drawing and writing, and music.[32] Subjects such as geography or science had yet to be introduced; women were not yet presumed to need them for elementary school teaching.

Teaching at the lower levels had become a woman's trade: in the 1850s, fewer than 10 percent of Philadelphia's teachers were male.[33] Teaching was also, however, a white trade, even when the students were colored. Not until the 1860s were the first African-American teacher and principal appointed in the public schools for colored children—having outdone their white competitors on the city teachers' examination.[34] Both, not coincidentally, were graduates of Philadelphia's notable Institute for Colored Youth (ICY).

Like Girard, ICY had been established by bequest, but with the original purpose of "instructing the descendants of the African Race in school learning, in the various branches of the mechnik [*sic*] arts and trades and in Agriculture." By 1852, however, students pursued almost entirely "school learning"; while earlier incarnations of the ICY had focused on manual labor and craft apprenticeship, the focus at mid-century was on "literary subjects."[35] The high school–level ICY included a library and scientific lecture series open to the public. In 1853, a preparatory department was added for younger students; in 1856, under the school's second principal, trigonometry, advanced algebra, Latin, and Greek were added for high school students of both sexes. Although the initial impulse of the Quakers administering the bequest had been to create a farm school for orphans, when African-

American leaders were allowed to shape the institution they created a tuition-free high school of stellar academic repute. Impressive by any standard, perhaps the ICY's most notable feature to members of the surrounding community was the fact that all of its teachers, as well as all of its students, were of the "African Race."

Not until several decades later would ICY educators press the board for the introduction of industrial training, intending to expand its mission beyond the training of teachers. In the 1850s, acquiring craft or industrial knowledge was not much of an issue; getting paid for the use of such knowledge was a more formidable barrier. For females, textile mill work was an unlikely option; African-American women worked as domestics and laundresses. For African-American males, even access to craft skills would not necessarily enable them to support themselves. The Friends' census of 1856 found that 38 percent of skilled colored craftsmen — a small group to begin with — were not working at their trade, "on account of the unrelenting prejudice against them."[36] Despite their generalized rhetoric, even white reformers at the House of Refuge knew that their colored charges were being apprenticed to a limited range of trades; by 1860 they had stopped listing the trades to which colored boys were apprenticed, counting only, as for girls, how many had been indentured.[37] Hopes for mobility among Philadelphia's African-American residents had to be differently structured than among European-Americans: a colored laborer might wish his son a barber, but a barber would be optimistic indeed to wish his son a printer or a machinist. For students at ICY, knowledge of ancient Greek and natural or mental philosophy was probably just as "practical" as mechanical drawing or textile design. On the other hand, the absence of a drawing curriculum, or any mention of experimental science courses,[38] meant that ICY graduates lacked the kind of technological knowledge expected by the upper echelons of Philadelphia's manufacturing community, even had access been possible.

In contrast, adding more specific training at Girard College had become an appealing option by the end of the 1850s, both to produce healthy, well-rounded boys and to enhance their attractiveness in the apprenticeship market, which was in turn the boys' path to "independent" self-sufficiency. "A workshop, in which the orphans might learn to use their hands while they are learning to use their heads, is almost as necessary as a gymnasium. The greater part of the pupils skill[ed] in the handling of tools . . . would make them more desirable and efficient as apprentices, and consequently increase the demand for them. It is believed that many of the older pupils would prefer the shop to the play-ground as a place of recreation, and that the practical knowledge which they would acquire, would be such as Mr. Girard intended when he said they should learn 'facts and things.'"[39] A workshop, however, sounded suspiciously like manual labor to some directors, and was therefore a tricky topic. Limited attempts to provide alternatives to school for some of the older boys were made in the next few years, but not until 1863 did the managers resolve the problem by hiring a "Professor of Industrial Science including Polytechnics." The curriculum mixed study of physics, chemistry, and anatomy with "applied mechanics" such as typesetting, typecasting, printing, and carpentry, and "applied chemistry" such as photography, daguerreotyping, electroplating, and telegraphy.[40] President Richards Vaux of the Board of Directors reported that year that "those, therefore, who predicted that Girard College was to be leveled down to a manual-labor school have been mistaken."[41] But even this solution proved short lived; Vaux stepped down as President in 1865, the professor resigned in 1866, and the issue remained inconclusively resolved until the early 1880s.[42]

Adulthood and Independence

The question of the workshop shows the Girard College Board of Directors coming to terms with the various paths to adulthood available to the boys in their charge. Industrialization meant shifting expectations. Girard's will, written in the late 1820s, required the school to apprentice boys to trades; it was less specific about technical education within the institution. The issue of "adulthood" was also evident when the Directors confronted the question of what it meant to provide a "home" for orphans. Should boys who fell on hard times be allowed to return? The topic generated a full range of inherently contradictory opinions: "families" should take back their sons, but Girard boys should not be taught dependency. "Dependency" had, by mid-century, acquired a new and stark definition: any receipt of charity was dependency, and—for males—any support not earned by one's own work was charity. As one official put it in response to the suggestion that temporary accommodations be provided in case of need, "A palace for orphan loungers would indeed be a novelty. No American boy, 14 or 18 years of age, with a good education, should be considered an orphan. Let the great lesson of independence and self-reliance be fully taught:—*Every orphan bound out from the College should carve his own way through life, and on no account be received back into the College*."[43] For a white boy worth his salt—an "American" boy—"independence" should mean that the existence or nonexistence of his parents was immaterial.

For white females, however, issues of dependence and independence were construed differently. The theoretical dictates of gender ideologies might not have articulated the fact, but by the 1850s wage work outside the home was part of the economic life of a growing number of women. Women's traditional means of earning cash had included taking in laundry or needlework, domestic service, and prostitution. At mid-century, young white women also commonly earned money in textile mills and by teaching. By definition, the pay for women's work was low—with the exception of prostitution, the income from which was a continuing lure. Women's pay was based on the idea that women were supported by men, even though many of the women seeking paid work were doing so because they supported themselves, and often their children. Low women's pay enhanced the perceived status of working men, but only while men could retain their better-paid jobs; as a result, male workers regularly voiced their resentment of women's invasion of their workplaces.[44]

This set of issues is illustrated, in all its confusion, by the early career of the Philadelphia School of Design for Women. Founded in 1848 by Sarah Peter, who had been exposed to the plights of a different class of women in some of her charity work, the School of Design was intended to provide a better means of support for "respectable" women who needed paid work.[45] By the time Peter sought the support of the managers of the Franklin Institute for the Promotion of the Mechanic Arts in 1850, she had developed a convincing argument for her strategy, connecting women's need for work with Philadelphia's need for local textile designers. She detailed both women's "want of a wider scope in which to exercise their abilities for the maintenance of themselves and their children," and the continued reliance of American textile manufacturers on foreign textile designs. But the location of the work made it ultimately acceptable: "I selected this department of industry, not only because it presents a wide field, as yet unoccupied by our countrymen, but also because these arts can be practiced *at home,* without materially interfering with the routine of domestic duty, which is the peculiar province of women."[46] The Committee on Instruction recommended the Institute's backing of the School, echoing Peter's arguments about the need for paid women's employment in the city, as well as her concern with American dependence on foreign design. Their concern with the location of women's work, however, went beyond Peter's preservation of the domestic routine. They found it "very desirable that whatever mode may be devised for the employment of female industry, should be of such a nature as shall allow it to be exercised at their own homes, or at least without crowding them together in workshops, and especially without forcing them into contact with the opposite sex,—practices which are too frequently destructive to female delicacy, (a quality not less valuable to the community than beautiful in itself,) even when they do not lead to actual habits of immorality."[47] Not only should women work in a domestic setting because their duty demanded it; male workplace settings could endanger respectable femininity. Women's industrial work was twice limited.

Through the Institute, the Philadelphia School of Design for Women gained an advantageous connection to Philadelphia's manufacturing community. Peter's combination of charitable, patriotic, and economic purposes—respectable women's employment and the prospect of American sources of textile designs— proved strong enough to raise the money the school needed. These two justifications are linked repeatedly in surviving documentation; a printed fund-raising letter, for example, told recipients that they were being asked to give $100 or more "from a belief . . . either in your philanthropy or interest as a merchant, manufacturer, or artizan [*sic*]." The letter went on to identify the purposes of the institution: "1st.—To increase the number of occupations for Women, and to afford a just and proper encouragement to female talents and industry. 2nd.—To furnish Patterns to our manufacturers, and thereby to promote the manufacturing interests of our Country." It went on to announce that designs for paperhangings and textiles had already been executed, "for all of which blocks are carved to order."[48]

Within a few years, disputes over the administration of the school led to its incorporation as an independent entity. The Constitution of PSDW, printed in its first *Annual Report* in the fall of 1854, declared the school's purpose in relatively neutral terms as "the instruction of women in decorative art, and the various practical applications thereof to industrial pursuits."[49] The report made no reference to female underemployment, American dependency on foreign design, or the place of domestic duties in women's lives. It did, however, define two different charges for tuition: "professional pupils" paid $4 per quarter, while "pupils who study for accomplishment" paid $9. A printed announcement for the current term, included in the report, explained that "when Pupils are sufficiently advanced to execute orders in Drawing or Coloring, Designing original Patterns, Lithography or Wood Engraving, they will receive for the products of their industry three-fourths of the fees for such orders—one-fourth being retained by the Institution." The Rules of

Government, most of which specified schedule and behavior, also added that when executing orders, "the pupils in the industrial classes will be permitted to remain at the school during ordinary working hours."[50]

The school seems to have operated as both an institution of women's art education, providing "accomplishment," and a source of both training and income for "professional"—wage-earning—students. The existence of "industrial classes" signals a clear understanding of the purpose of that part of the curriculum, and the option to stay beyond the time of formal instruction each day belies the early claims about industrial design as a domestic activity. Of course it is no surprise that production of woodblocks for textile printing, let alone lithography, was not performed in Victorian parlors. Peter's original conception of the school, despite her conformist marketing strategies, was based on an argument at odds with prevailing mid-century gender ideologies. The male businessmen whose support she needed generally assumed that women had homes—presumably provided by men—in which to confine their activities. Peter, on the other hand, knew that some women needed "family wages," although she never borrowed the terminology; she sought to make available a kind of technological knowledge that would carry the economic value of "skilled work."

Justifying the inclusion of female workers in industrial pursuits, however, required a complex balance of arguments, in which females remained respectable and the role of male workers was not threatened. In addition to assertions about domestic viability and the noticeable absence of American men in the field, proponents could cite women's alleged natural taste and discrimination of form and color. The Franklin Institute managers had done so in their initial report on the project: "Heretofore little, if any, attention has been paid to the cultivation of this peculiar faculty, and even in our most elaborate systems of female instruction it appears to be considered as very subordinate in importance to other branches less fitted to the peculiar capacities of their minds."[51] Artistic design could thus be construed as an extension of the talents women brought to their creation of comfortable domestic havens at home. In the mid-nineteenth century, women were assumed possessed of talents in dealing with the material world—but not of any need for economic independence.

Ideas about adulthood and independence for different groups are also evident in the *Annual Reports* of the House of Refuge. Published excerpts of letters from the apprentices' masters make plain that the managers' reform tactics carried entirely different meanings for different groups of children. A white boy, even one involuntarily confined in a reform school and then offered little choice about signing an indenture binding him to a farmer until his twenty-first birthday, could decide, if he chose, to "pass" as a member of his new community, taking on the identity the managers recommended for him. The excerpted letters about white boys include phrases such as "appears to regard me as a father and friend," "appears to be attached to the family," "is as one of our own family," "shares the same advantages that my own children do," and "all my men are fond of him, so much so that he is quite a pet."[52] One letter sums up the possibility explicitly: "He appears to enjoy himself both at home and at school; and the scholars and neighbors appear to respect him and enjoy his company about as much as if he were a native of the immediate vicinity."[53] For a colored child indentured to a white master, such assimilation would never be possible. Most of the comments about colored boys, not surprisingly, note their industriousness and the quality of their work, but say little of a personal nature.

In theory, white girls, like white boys, could become part of and "attached" to the family, but the published excerpts from letters do not reflect such sentiments. Instead, teenaged girls performing housework in someone else's house could easily be regarded as servants rather than as city children in need of families. As one writer put it, "were she now free, I would give one dollar per week rather than part with her." In this role, a girl's race was less important; excerpts from letters about girls of both races characterize them as "good girls," and comment on their morality and their work. A letter from the excerpts on colored girls related: "My mother speaks of her in flattering terms, and values her as an excellent servant."[54]

Even though white boys, generally, had access to adult "independence," their opportunities were certainly not equal. More privileged white boys were apprenticed differently from poor ones, who were in turn apprenticed differently from colored boys. Rhetorically, white House of Refuge boys were set on a path to independence when their indentures were signed, or even when they were returned to their "friends," schooled and well behaved. But the practical differences in opportunity available to the eleven House of Refuge boot- and shoemakers — most of whom could write and do simple arithmetic — as opposed to the twelve Girard College printers or the twelve Girard College druggists — who had studied French and Spanish as well as bookkeeping, trigonometry, and natural philosophy — were probably substantial. And for the eighty-six House of Refuge farmers, independence carried an entirely different meaning from that granted to the "desirable young men to be trained to business" graduating with a diploma from Central

High.[55] Independence depended on many factors, not least of which was techno-logical knowledge. And many specific kinds of technological knowledge had such clear associations with "manual" labor that the highest mobility seemed granted by a "practical" education that included no craft skills beyond bookkeeping and drafting.

Despite the many distinctions between white boys, it was nonetheless true that white girls, colored girls, and colored boys shared an entrenched dependency white boys might escape. This economic status also mirrored more concrete po-litical structures: neither women nor African-American men had the vote in Penn-sylvania in this period.[56] "Dependency's" other face, of course, was the work of caretaking. For respectable white women, caretaking was described in terms of delicate feminine virtues, rather than economic value. Such caretaking in fact in-volved extensive technological knowledge, but middle-class femininity was a con-struction that did not involve "work."[57] Caretaking in other people's families, ser-vice for wages, was work and admittedly involved skill, but it was considered far less respectable and was also increasingly the province of immigrant and colored women. Colored men, too, often made careers in caretaking roles, as waiters, bar-bers, and porters. Social ideologies and ideas about appropriate kinds of techno-logical knowledge were often so intertwined as to be virtually indistinguishable.

These tangled ideologies, in turn, shaped ideas about work and about inter-actions with technology. When the notion of a "woman's sphere" defines what women's responsibilities are, it also defines what men's are not.[58] The relegation of certain kinds of work to women and nonwhite men, and white men's distanc-ing of themselves from those tasks, gave white men license to take such tasks for granted. In fact, the skillful performance of the tasks of wives and maids, of waiters and porters rendered their work invisible — as economic activity, as technological activity. By mid-century, these divisions cut deeply: it was accidental neither that the white boys' case histories at the House of Refuge were the only ones devoid of the notation "has lived at service," nor that their school reports emphasized the ability to write on paper. Even among these underprivileged children, reflections of hierarchies of technological knowledge were evident in differences based on gender and race.

But to say certain tasks and technologies were rendered "invisible" is not at all to suggest they were impotent. The ideologies that defined which tasks did not be-long to men also defined the traits of the tasks that did, and vice versa. That girls were taught needlework and not metalwork illustrates that gender assumptions —

like other social constructions—shaped *access* to these hierarchies. That needle-work was a skill all women should possess, however, suggests that ideas about gender are entwined in perceptions of technological knowledge; these ideas in turn shape the hierarchies themselves.

Notions of desirable men's work reflected gender characteristics in both positive and negative terms. Manly work was performed away from home, the female domain. Manly work supported one's family but was not service, a distinction which helps explain the relative prestige of manufacturing work even when the worker could expect little control of the process. Women's work, much of it clearly production, was not viewed as such, or valued in monetary terms. Manly work was also demonstrably abstract, precise, or quantified, in contrast to soft, intuitive, or even imprecise female thinking. Traits valued in and taught to women—caretaking, softness, intuition, emotional understanding—came to be dissociated from the upper reaches of modern hierarchies of technological knowledge. Such traits came to be seen as "un-technological."

Social History and the History of Technology

The trends evident in these institutions stretch well beyond the 1850s and beyond Philadelphia. Few stories of industrialization confine themselves to a single decade; neither do such stories confine themselves to traditional academic categories. Examining the social and the technological together offers new perspectives in both realms.

Within social and labor history, discussions addressing the question of the relationship of gender to class, or race to class, have mostly been framed in terms of the development of American class structure and class consciousness. In women's history, discussions of gender and race or gender and class have been organized around women's changing roles in society. Beginning as I do here with a broad swath of activities, with a cross section of people "making and doing" things rather than a focus on gender or race or class specifically, allows for comparative analysis. Such analysis provides a crucial vantage point for assessing interconnections not only between technology and social structure but also between the various social dimensions we already expect in a particular historical context.

It is not my intention here to lay out a new social theory, but rather to suggest directions for further exploration. In working through the larger study from which this article is drawn, I have come to conceptualize gender, race, and class

each as an axis or dimension—we might call them "dimensions of difference."[59]
Such dimensions function on several levels: as scholars studying gender have suggested, they involve self-definition and identity, structures and institutions, and
symbols or representations. But these dimensions function as dimensions of social
power, as well.[60] In a given context, some dimensions will be the salient ones and
others will recede: in an all-male workplace like Baldwin Locomotive, class and
age and labels like "skill" may be more obvious than gender; in a working-class
neighborhood like the one near the House of Refuge, gender and ethnicity will
perhaps prevail. In each case we must look carefully both for changing boundaries,
and for assumptions so common—in our sources or in our own times—they seem
natural.

Focusing on technological knowledge highlights the interrelatedness of these
dimensions. For example, these sources provide clear evidence, here spoken largely
in a white middle- or upper-class male voice, of gender and racial solidarity that
transcended class: white boys could and should be expected to be independent.[61]
Wages—especially wages associated with "skilled work" and with independence—
were gendered male, and they were strongly associated with racial whiteness, too.[62]
In a capitalist economy, money is powerful; in an individualistic and slaveholding
society, independence is powerful; we can even argue that in an industrializing
city, factory technologies are powerful—and it makes sense that different kinds of
power reinforce each other. These features of industrializing America do not admit
of separate "social" and "technological" and "economic" explanations. Neither
does this view justify consideration of class, for example, without race and gender:
in the United States, gender and racial solidarity are related to class distinctions
and to class mobility. Further, these social boundaries of gender and race and class
are intertwined with technological knowledge and thus with technological change.

Ideas and ideologies about technology demand no less scrutiny than the traditional categories examined by social history. To treat technology as a social product
is to recognize that technology in our society has reflected, reinforced, and been
built into the social boundaries we construct and reconstruct. Historians of technology have long known that technological knowledge could provide a path of
upward social mobility, most often for white males. On the other hand it does not
necessarily do so: for example, we know that mechanical know-how is not the same
as management of capital, patents, partnerships, or employees; that great skill with
needle and thread or even yarn and loom might not provide a comfortable life for
one's children.[63]

Once we think in terms of multiple axes along which power can be appropriated, and ask of specific contexts what differences and what kinds of power are the important ones, we must also ask how one acquires power. In the present case, in what ways did changing technology and technological knowledge make these structures malleable, or in what ways reinforce them? Both effects are evident here: in institutions providing limited training based carefully on the status of the child, and in institutions like Girard College or the Philadelphia School of Design for Women whose founders made deliberate attempts to reshape the social order in specific ways.

Further, we must explore relationships between structures and ideologies, and ask how these hierarchies are reflected in the sources we study. The multiple alignments of power discussed here suggest that symbols, metaphors, and labels borrowed from these various categories can be mixed as needed: disempowerment helps explain why "unskilled" workers are by definition poorly paid, why women's work is by definition "unskilled," and why we find gendered language cropping up frequently and unselfconsciously in other kinds of power struggles.

Finally, recognizing that there are, particularly in a culture that values technology so highly, hierarchies of technology and technological knowledge, as well as social hierarchies, means we must ask ourselves fundamental questions about the ways we choose and then describe what we study. Hierarchies such as the ones discussed here influence us as historians twice over: nineteenth-century hierarchies influenced the nature and survival of the sources we use, and twentieth-century hierarchies shape the questions we ask. If we really seek to understand industrialization as a process of "technology and culture," we must fight the exclusion of any part of "making and doing things" from its place in the realm of the "technological"—even as we recognize that, being a human product, technology is no stranger in the realm of the "social."

NOTES

Nina Lerman thanks Lindy Biggs, Christian Gelzer, Mark T. Hamel, Gabrielle Hecht, Michael B. Katz, Sally Gregory Kohlstedt, Helen Longino, Judith McGaw, Arwen Mohun, Ruth Oldenziel, Philip Scranton, John Staudenmaier, and the *Technology and Culture* referees for their comments on the several incarnations of this paper.

1. Studies integrating social and technological change include Merritt Roe Smith, *Harper's Ferry Armory and the New Technology: The Challenge of Change* (Ithaca, N.Y., 1977);

Anthony Wallace, *Rockdale: The Growth of an American Village in the Industrial Revolution* (New York, 1972); Judith McGaw, *Most Wonderful Machine: Mechanization and Social Change in Berkshire Paper-Making, 1801–1885* (Princeton, N.J., 1987).

2. As Walter Vincenti has put it, "emphasis on knowledge . . . brings history of technology into symbiotic relation, not only with intellectual history and philosophy, but with social history and sociology as well." Walter Vincenti, *What Engineers Know and How They Know It: Analytical Studies from Aeronautical History* (Baltimore, 1990), p. 5, citing Hugh Aitken, Edwin Layton, and Rachel Laudan. Philosopher Helen Longino has made a related argument about science: "What I urge is a contextualism which understands the cognitive processes of scientific inquiry not as opposed to the social, but as themselves social." Helen E. Longino, "Essential Tensions—Phase Two: Feminist, Philosophical, and Social Studies of Science," in *The Social Dimensions of Science*, ed. Ernan McMullin (Notre Dame, Ind., 1992).

"Making and doing things" and related phrases have been used to define technology for decades now; see Melvin Kranzberg, "At the Start," *Technology and Culture* 1 (1959): 1–10; Brooke Hindle, "The Exhilaration of Early American Technology: An Essay" in *Technology in Early America: Needs and Opportunities for Study* (Chapel Hill, N.C.,1966), reprinted in Judith McGaw, ed., *Early American Technology: Making and Doing Things from the Colonial Era to 1850* (Chapel Hill, N.C., 1994).

3. Pennsylvania Institution for the Deaf and Dumb (PIDD), *Annual Report [for 1849]* (Philadelphia, 1850), p. 21.

4. Population in 1820 from Susan Klepp, *Philadelphia in Transition: A Demographic History of the City and Its Occupational Groups, 1720–1830* (New York, 1989), pp. 336–38; 1850 population from Theodore Hershberg et al., "A Tale of Three Cities: Blacks, Immigrants and Opportunity in Philadelphia, 1850–1880, 1930, 1970," in *Philadelphia: Work, Space, Family, and Group Experience in the Nineteenth Century*, ed. Theodore Hershberg (New York, 1981), p. 465. Philadelphia's foreign-born population was 29 percent of the total in 1850; 17.6 percent of inhabitants were born in Ireland.

5. Philip Scranton, *Proprietary Capitalism: The Textile Manufacture at Philadelphia, 1800–1885* (Cambridge, 1983), pp. 46–47.

6. *Catalogue of the Twenty-Second Exhibition of American Manufactures, held in the City of Philadelphia, by the Franklin Institute of the State of Pennsylvania for the Promotion of the Mechanic Arts. From the 19th to the 30th day of October, 1852* (Philadelphia, 1852).

7. John Staudenmaier grouped technological knowledge with discussions of science and technology in his 1985 survey of the field, although he pointed to a growing attention to artisanal rather than engineering knowledge among historians of technology. John Staudenmaier, S.J., "Science, Technology, and the Characteristics of Technological Knowledge," chap. 3 in *Technology's Storytellers: Reweaving the Human Fabric* (Cambridge, Mass., 1985). On engineering, see Eugene Ferguson, *Engineering and the Mind's Eye* (Cambridge, Mass., 1992); Vincenti (n. 2 above); Edwin Layton, "Through the Looking Glass, or, News from Lake Mirror Image," *Technology and Culture* 29 (July 1987): 594–607. On artisanal knowledge, see Robert Gordon, "Who Turned the Mechanical Ideal into Mechanical Reality?," *Technology and Culture* 29 (1988): 744–78; and Douglas Harper, *Working Knowledge: Skill*

and Community in a Small Shop (Chicago, 1987). Useful papers by Steven Lubar, Charles and Janet Keller, and Douglas Harper (all focused on aspects of metalwork) were presented in a session on "Knowing, Thinking, and Doing" at the Critical Problems Conference, Society for the History of Technology, Madison, Wis., 1991. Lubar's paper has now appeared in the volume of selected proceedings of that conference: Steven Lubar, "Representation and Power," *Technology and Culture* 36 (1995): S54–S82. Other older explorations of technological knowledge include Brooke Hindle, *Emulation and Invention* (New York, 1981), Eugene Ferguson, "The Mind's Eye: Nonverbal Thought in Technology," *Science* 197 (1977): 827–836; and Layton, "Technology as Knowledge," *Technology and Culture* 15 (1974).

Feminist approaches to scientific knowledge, which emphasize the relationship of knowledge to power, have also influenced me here; see, for example, the work of such diverse philosophers as Helen Longino, Genevieve Lloyd, or Sandra Harding. Alice Kessler-Harris summarizes some of these arguments in a brief discussion of the concept of work: "all knowledge is socially situated and . . . the claims to knowledge of the dominant group are conditioned by its desire to preserve power." This approach, she writes, "puts us in a position of problematizing all knowledge as the product of particular social situations." Alice Kessler-Harris, "Treating the Male as 'Other': Redefining the Parameters of Labor History," *Labor History* 34 (1993): 190–204, quotes p. 194.

8. Ferguson, *Engineering and the Mind's Eye*, p. 42.

9. Christine Stansell provides an excellent discussion of the work and status of nineteenth-century seamstresses in "Wage Work," part 3 in *City of Women: Sex and Class in New York, 1789–1860* (Urbana, Ill., 1987). Discussions of skill appear in studies from a range of disciplines: see, for example, Howard Gardner, introduction to *Frames of Mind: The Theory of Multiple Intelligences* (New York, 1985); David Pye, *The Nature and Art of Workmanship* (Cambridge, 1968); Charles More, *Skill and the English Working Class, 1870–1914* (New York, 1980), chap. 1; Harper; and Steven Shapin, "The Invisible Technician," *American Scientist* 77 (1989): 554–56; as well as the works cited in n. 7 above. The best work in the history of technology remains Gordon. Ava Baron provides useful discussions of gender and deskilling; see for example her articles "Questions of Gender: De-skilling and Demasculinization in the U.S. Printing Industry," *Gender and History* 1 (1989): 178–99, and "An 'Other' Side of Gender Antagonism at Work: Men, Boys, and the Remasculinization of Printers' Work, 1830–1920," in *Work Engendered: Toward a New History of American Labor*, ed. Ava Baron (Ithaca, N.Y., 1991). See also her introduction to *Work Engendered*, especially pp. 14 and 36, and Ruth Oldenziel, "Gender and the Meanings of Technology: Engineering in the U.S., 1880–1945" (Ph.D. diss., Yale University, 1992).

10. On the contrary, if there is a link here to arguments about determinism, it is the following: If we neglect technological knowledge in favor of artifacts, we deemphasize the idea that technology is a human activity, which means we can hardly blame other historians for stubbornly locating agency in hardware. See also Staudenmaier, esp. pp. 174–81.

11. That these hierarchies look obvious enough from the vantage point of the late twentieth century says only that they became entrenched, and long lived. Precisely because the nineteenth century saw major transformations in the economic, social, and technological definitions of both household and workplace, particularly in the urban United States, I

do not assume that the configurations, relationships, and hierarchies we take for granted now were inevitable then. Instead, I seek to explore the processes by which social and technological hierarchies were constructed and maintained—hierarchies which include engineering knowledge as well as needlework knowledge, metalworking knowledge as well as breadmaking knowledge.

12. PIDD, *Annual Report* for 1849 (n. 3 above), p. 21; see also subsequent annual reports.

13. Philadelphia House of Refuge (HR), *Annual Reports* for 1848–54 (Philadelphia, 1849–55).

14. HR, *Annual Reports* 1848–1854. Although early records show admission of at least one "colored" inmate, its services seem quickly to have been limited to white youths. See Negley Teeters, "The Early Days of the Philadelphia House of Refuge," *Pennsylvania History* (1960): 165–87; Mary Glazier, "The Origins of Juvenile Justice Policy in Pennsylvania" (Ph.D. diss., University of Pennsylvania, 1985); HR, "Address to the Citizens . . ." (1826), p. 12; HR, "Minutes of Visiting Committee," HR-A201 11/10/29 and Superintendent's Daily Log, HR-B-1 11/10/29, House of Refuge/Glen Mills School papers, Historical Society of Pennsylvania. (All citations of HR-A, HR-B, etc. refer to documents in this collection. The author thanks principal Randy Ireson for permission to cite this material.)

15. "Comfortables" here most likely refers to bedclothes, "comforters" in modern parlance. A comfortable could also be a scarf or muffler, or a knit wristband. Comfortables are listed among items of bedding in the report.

16. For an extended discussion, see Jeanne Boydston, *Home and Work: Housework, Wages, and the Ideology of Labor in the Early Republic* (New York, 1990).

17. For example, the House of Refuge Colored Department had some clothes made by the Northern Liberties House of Industry and temporarily hired one of the older white girls to help the staff; on tailoring, see HR Executive Committee Minutes, HR-A-3, 1/31/50, 2/7/50, 5/2/50; on the employment of white girls, HR-A-3 12/20/49, 1/3/50, 2/7/50, HR-L-11 2/20/50. The Pennsylvania Institution for the Instruction of the Blind, whose girls did not perform all of the housewifery for the institution, had hired help; see Pennsylvania Institution for the Instruction of the Blind (PIIB), *Annual Report* for 1855 (Philadelphia, 1856), pp. 8, 19; see also PIIB *Annual Reports* throughout the decade.

18. In 1860, 59.2 percent of the 103 white boys indentured were apprenticed to farmers; the others were to work at twenty-six different trades. The Superintendent's report for the Colored Department that year notes that nineteen boys were indentured but does not specify to what trades. Discharges by indenture in 1860, excluding those returned to court, almshouse, or died: 51.0 percent of white boys, 42.2 percent of colored boys, 47.9 percent of white girls, 73.1 percent of colored girls. HR, *Annual Report* for 1860, pp. 40–41. The discussion in this article focuses on the early part of the decade.

19. On apprenticeship, see Ian Quimby, *Apprenticeship in Colonial Philadelphia* (New York, 1985; reprint of M.A. thesis, University of Delaware, 1963); Christine Marie Daniels, "Alternative Workers in a Slave Economy: Kent County, Maryland, 1675–1810" (Ph.D. diss., Johns Hopkins University, 1990); W. J. Rorabaugh, *The Craft Apprentice: From Franklin to the Machine Age* (New York, 1986); and Nina E. Lerman, "From 'Useful Knowledge' to

'Habits of Industry': Gender, Race, and Class in 19th-Century Technical Education" (Ph.D. diss., University of Pennsylvania, 1993), chap. 1.

20. Philip Scranton, "Learning Manufacture: Education and Shop-Floor Schooling in the Family Firm," *Technology and Culture* 27 (1986): 40–62.

21. He was also prohibited from joining "any Military or Fire Company during his term of service." This discussion is based on John K. Brown, "The Baldwin Locomotive Works, 1831–1915: A Case Study in the Capital Equipment Sector" (Ph.D. diss., University of Virginia, 1992), pp. 205–14, quotes on p. 206.

22. Male textile workers in Kensington earned about $5/week ($20/month), women about $2/week ($8/month) in 1850. Scranton, *Proprietary Capitalism* (n. 5 above), p. 193.

23. Henry W. Arey, Secretary of Girard College and Superintendent of Binding Out, in the *Seventh Annual Report of the Board of Directors of the Girard College for Orphans* (Philadelphia, 1854), quoted in Louis Romano, *Manual and Industrial Education at Girard College, 1831–1965: An Era in American Educational Experimentation* (New York, 1980), p. 123.

24. Third and fourth priorities were New York City and New Orleans. Will of Stephen Girard, quoted in Romano, pp. 287, 296; see also Cheesman A. Herrick, "Admission and Discharge," chap. 14 in *History of Girard College* (Philadelphia, 1927). On Girard himself, see both Romano and Herrick.

25. The exact figure was 46.1 percent. HR, *Annual Report* for 1853, p. 20. Parentage was probably determined by father, but the *Annual Report* does not address the question. It is also worth noting that the managers found these data significant enough to print in their annual reports. Irish immigrants were often considered a separate "race" at mid-century; even so, Irish children lived on the "white" side of the wall at the House of Refuge.

26. One House of Refuge boy was also apprenticed to each of the trades of boot fitter, pump maker, house painter, coach painter, victualler, cartwright, book gilder, cupper, bleeder, and barber, attorney-at-law, marble cutter, slate quarryman, grocer, saddle-tree maker, stereotype founder, and gas fitter. Girard boys were apprenticed one each to an architect, a barber, a cabinet maker, a chair maker, a chemist, a confectioner, a conveyancer, an engraver, a fresco painter, a horticulturist, a lithographer, a manufacturer of Britannia ware, a manufacturer of tailor's trimmings, an ornamental carver, a philosophical [scientific?] instrument maker, a silversmith, a tanner, a watchmaker, a wheelwright, and a wood engraver.

27. Romano, p. 94 n. 26 and Appendix F, p. 335.

28. HR, *Annual Reports* for 1853, pp. 32–33, 37–39; 1850, p. 32; 1854, school reports.

29. HR, *Annual Report* for 1853, school reports. These reports remained similar throughout the decade. In 1854, 160 of 205 (78.0 percent) white boys wrote on paper.

30. Stuart Blumin, *Emergence of the Middle Class: Social Experience in the American City, 1760–1900* (Cambridge, 1989).

31. On boys, see David Labaree, *The Making of an American High School: The Credentials Market and the Central High School of Philadelphia, 1838–1939* (New Haven, Conn., 1988); on girls, see John Trevor Custis, *The Public Schools of Philadelphia—Historical, Biographical, Statistical* (Philadelphia, 1897). Also useful on drawing is Peter Marzio, *The Art*

Crusade: an Analysis of Nineteenth Century American Drawing Manuals, Chiefly 1820–1860 (Washington, D.C., 1976).

32. Labaree, pp. 24–25, fig. 2.1; Custis, p. 154.

33. Labaree, p. 98.

34. Linda Marie Perkins, "Quaker Beneficence and Black Control: The Institute for Colored Youth, 1852–1903," in *New Perspectives on Black Educational History,* ed. Vincent P. Franklin and James D. Anderson (Boston, 1978), pp. 19–43. Graduates teaching in the public schools are discussed on pp. 23–24.

35. This discussion is based largely on Perkins, although skeletal versions of the story appear in later annual reports of the ICY. Will of Richard Humphreys quoted in Perkins, p. 19. The phenomenon of African-American control seems to have been largely dependent on one particular manager, Alfred Cope. Perkins found that after Cope's death in 1875, interactions between Institute leaders became a matter of continuous negotiation.

36. Benjamin C. Bacon, *Statistics of the Colored People of Philadelphia* (Philadelphia, 1859), quoted in Theodore Hershberg, "Free Blacks in Antebellum Philadelphia: A Study of Ex-Slaves, Freeborn, and Socioeconomic Decline," in Hershberg, *Philadelphia,* (n. 4 above). Hershberg found that in 1838, the five occupations of laborer, porter, waiter, seaman, and carter accounted for 70 percent of the male work force. More than 80 percent of day-working women were in domestic service of some kind (52 percent were "washers"), pp. 376, 382.

37. HR, *Annual Report* for 1860, pp. 40–41. See also n. 18 above.

38. Institute for Colored Youth (ICY), *Annual Report* for 1860 (Philadelphia, 1861), pp. 8–10.

39. Girard College, *Twelfth Annual Report* (1859), quoted in Romano (n. 23 above), p. 138.

40. Romano, p. 158. Apparently some Girard boys also learned shoemaking, which Romano reports "had proved to be self-supporting."

41. Quoted in Herrick (n. 24 above), p. 232.

42. See Herrick, chap. 8; Romano, pp. 140–64. By the 1880s "manual training" had become part of progressive educational reform and could be incorporated in a high school curriculum. See Romano, also Lerman (n. 19 above), chap. 4.

43. "Report of the Committee on Girard College for the month of October, 1858," *Journal of the Common Council of the City of Philadelphia,* quoted in Romano, p. 135. Emphasis in original.

44. For a discussion of women's work, both in and outside the home, see Boydston (n. 16 above); on opportunities for earnings see Stansell (n. 9 above).

45. Nina de Angeli Walls, "Art and Industry in Philadelphia: Origins of the Philadelphia School of Design for Women, 1848 to 1876," *Pennsylvania Magazine of History and Biography* 117 (1993): 177–99; Letter of Sarah Peter, "At a stated meeting . . . held April 19, 1850," *Proceedings of the Franklin Institute of the State of Pennsylvania for the Promotion of the Mechanic Arts, Relative to the establishment of a school of design for women,* Library Company of Philadephia.

46. Letter of Sarah Peter, p. 1.

47. Letter of Sarah Peter, p. 4.

48. Philadelphia School of Design for Women (PSDW), "School of Design, No. 70 Walnut Street, Philadelphia . . . ," n.d., n.p.

49. PSDW, *First Annual Report* (Philadelphia, 1854), p. 10. The school became independent in 1853; see Walls, p. 10.

50. PSDW, *First Annual Report,* pp. 16, 17.

51. Letter of Sarah Peter (n. 45 above), p. 4. Note that French textile designers argued that these same attributes were characteristically male; see Carol E. Harrison, "A Noble Emulation: Bourgeois Voluntary Associations in Eastern France, 1830–1870" (D.Phil. thesis, Oxford University, 1993).

52. Letters from masters printed in House of Refuge *Annual Report* for 1854, pp. 45–48.

53. House of Refuge *Annual Report* for 1853, p. 47.

54. House of Refuge *Annual Report* for 1854, pp. 48, 52. By mid-century the more egalitarian notion of "help" had been largely superseded by the job of "domestic," also. See Faye E. Dudden, *Serving Women* (Middletown, Conn., 1983).

55. Philadelphia Board of Public Education, *Annual Report* for 1850, p. 118, quoted in Labaree (n. 31 above), p. 20.

56. *The Peoples of Philadelphia: A History of Ethnic Groups and Lower-Class Life, 1790–1940,* ed. Allen Davis and Mark Haller (Philadelphia, 1973). The franchise was denied in 1838 and reinstated in 1870.

57. On gender ideologies and work, see, among others, Boydston (n. 16 above); Stansell (n. 9 above); Baron (n. 9 above); Ruth Schwartz Cowan, *More Work for Mother: The Ironies of Household Technologies from the Open Hearth to the Microwave* (New York, 1983); Roszika Parker, *The Subversive Stitch: Embroidery and the Making of the Feminine* (London, 1984); Clyde Griffen, "Reconstructing Masculinity from the Evangelical Revival to the Waning of Progressivism: A Speculative Synthesis," in *Meanings for Manhood: Constructions of Masculinity in Victorian America,* ed. Mark C. Carnes and Clyde Griffen (Chicago, 1990). For further discussion, see "The Shoulders We Stand On" in this book.

58. On the relation of gender divisions to male and female identity in relation to technology, see Judith McGaw, "No Passive Victims, No Separate Spheres: A Feminist Perspective on Technology's History" in Cutcliffe and Post (n. 7 above), pp. 172–91. See also Boydston; Joan Wallach Scott, *Gender and the Politics of History* (New York, 1988) especially the essay "Gender: A Useful Category of Historical Analysis"; Baron; Carroll Smith-Rosenberg, *Disorderly Conduct: Visions of Gender in Victorian America* (New York, 1985).

59. They are hardly the only dimensions of difference; we can think similarly about age, place, sexual orientation, literacy, or skill, among others. Ava Baron uses the phrase "axes of difference" once in the introduction to her edited volume *Work Engendered* (n. 9 above), p. 35.

60. Scott, "Gender: A Useful Category." See also Sandra Harding, *The Science Question in Feminism* (Ithaca, N.Y., 1986). For further discussion see the introduction to this volume.

61. See also Kessler-Harris (n. 7 above). Kessler-Harris's comments about labor history in several instances parallel mine about the history of technology (here "economic" and "class" stand in, not at all coincidentally, for "technological" and "technology"): "The ten-

dency of labor history to separate historians of women from those of labor and to exclude women's activities from economic purpose—and therefore from a direct relationship to class and class formation—suggests the remarkably male terms in which class is still defined" (p. 192). She also suggests that her conclusions about gender might apply to analysis of race (p. 193).

62. See also David Roediger, *Wages of Whiteness: Race and the Making of the American Working Class* (London, 1991), pp. 11–13. Roediger borrows here from W.E.B. DuBois, whose "wages" came not simply in monetary form, but in other privileges as well. "Wages of skill" and "wages of independence" might be treated similarly.

63. For a clear discussion of patent management, see Carolyn Cooper, *Shaping Invention: Thomas Blanchard's Machinery and Patent Management in Nineteenth-Century America* (New York, 1991).

SIX

Industrial Genders: Home/Factory

ARWEN P. MOHUN

Increased attention to gender analysis has demanded a broader awareness of the boundaries and categories by which people sort their experiences. Not only male/female, but also producer/consumer, home/work, clean/dirty, and so on are all negotiable boundaries, and usually interrelated ones. They often appear obvious and natural until they are examined closely.

Arwen Mohun's research explores the construction of such boundaries and the ways gender can be used in the process of that exploration. Laundry, typically a "female technology" (as McGaw would put it), was by the late nineteenth and early twentieth century becoming an industrial process: one could gather "raw materials" in the form of dirty clothes and take them to a steam laundry, where skilled workers would process them using machines of various sorts, thus producing clean clothing to be returned to the owners. Men owned factories and other such mechanized businesses; women did laundry. What, then, might one mean by a "laundryman"? How might a laundryman establish his masculinity—and, given the competition from Chinese laundries, his whiteness—amid female employees, female customers, and a set of traditionally female activities? What kinds of social and cultural boundary work does Mohun identify in this story? What role does technology play in the process?

"Laundrymen Construct their World: Gender and the Transformation of a Domestic Task to an Industrial Process," *Technology and Culture* 38, no. 1 (1997): 97–120. Reprinted by permission of the Society for the History of Technology.

Laundry is women's work. For some centuries, this is what most Europeans and Americans assumed. Whether in manor houses or army camps, by streamsides or in tenement rooms, laundrywork has traditionally been one of the most powerfully gendered of all domestic tasks. Men, on the other hand, avoided doing laundry for reasons beyond the fact it was hot, difficult work. The man who submerged his arms into a washtub or picked up an iron (except in a tailor's shop) risked unsexing himself. A man who needed his washing done sought out a woman to do it for him—either wife, or servant, or washing woman.[1] Women were also the guardians of knowledge about the processes involved. They taught their daughters how to make soap and to season an iron, how to brew starch and remove stains. The tools and the knowledge and the task itself were indisputably theirs.[2]

In the mid-nineteenth century, "steam" or "power" laundries began to appear in American and British cities, offering consumers an alternative to the rigors of washday or the uncertainties of employing a washerwoman. These enterprises distinguished themselves from their domestic counterparts by adapting industrial methods and technologies to the problem of cleaning clothing, sheets, towels, and other items made of cloth. Rather than scaling up primitive washing machines then available for home use, early laundry proprietors bought or modified textile machinery that had been designed for bleaching and dying cloth.[3] In common with other industrial employers, they also reorganized the work process, dividing up the traditional parts of the laundress's job—washing, drying, starching, and ironing—between multiple workers and, eventually, specialized machines.

While women continued to constitute a majority of wage workers in these laundries and to do the wash at home, men controlled the industrialization of laundrywork. They created, propagated, and controlled the technology known collectively as a "steam laundry."[4]

These first commercial laundries are hardly more than shadows in the historical record, evidenced primarily by their presence in city directories and fire insurance maps. In the 1880s, however, large numbers of new entrepreneurs entered into the laundry business. This period of growth was accompanied by the first efforts of laundryowners and managers, "laundrymen" as they now called themselves, to create a public identity for themselves. Through trade journals and associations, laundrymen created local, national, and trans-Atlantic (Anglo-American) communities where technical information could be shared and common interests articulated and promulgated. This community was explicitly male. The few women who owned or managed laundries were denied a public voice or visible role in shaping what was now referred to as the "laundry industry."

Laundrymen were practical people struggling to make a living in a business plagued with problems. They could not, for example, control the amount or the quality of the "raw material" that came into the laundry nor the cultural dimensions of the ways customers judged the finish and cleanliness of the finished product. As with all technologically dependent businesses, success in the laundry industry depended in part on the judicious selection and knowledgeable use of appropriate machinery. Laundrymen constructed the technological part of their world by balancing economics, labor questions, and the constraints and possibilities of available technology.

Laundrymen also felt compelled to construct their industry with regard to its cultural meanings. Beginning in the late nineteenth century and lasting well into the twentieth, the gendered connotations attached to laundry technology and to its users are a pervasive theme in the laundrymen's discourse about themselves and their female employees, customers, and competitors. Confronted with the persistent notion that laundrywork was women's work, they presented their own complex arguments suggesting that while individual parts of the process might better be carried out by women, laundries as technological systems were essentially masculine; that they required masculine ways of thinking about and organizing technology in order to function properly; that, in fact, women unsexed themselves by claiming more than a small degree of authority in this realm.

This article explores the ways in which laundrymen used trade journals, advertisements, and industry histories to make gendered claims about commercial laundry technology and their relationship to it and to women workers and consumers. It argues that the creators of these documents recast laundrywork into a larger set of preexisting cultural idioms and symbols that linked masculinity and machinery. They claimed, for example, that commercial laundries were modern, mechanical, and scientific. Since it was neither desirable nor possible for them to eliminate women from employment or consumption, they also adopted a set of rhetorical strategies that cast women in specific roles defined by gendered ideologies such as domesticity and paternalism.

The examples that follow come from both British and American sources. Comparison highlights the culturally different symbolic frameworks into which British and American laundrymen tried to fit the particulars of their industry. Most importantly, this example suggests that Britons and Americans may not always have expressed masculinity through technology in the same ways. Because the existing secondary literature on masculinity is so small and the literature on masculinity and technology smaller still, much research remains to be done to establish the full meanings of these differences and the degree to which they are typical of men of similar classes and aspirations.[5] This article can be seen, in part, as an effort to add to that understanding.

Comparison is also a useful tool in understanding how trans-Atlantic differences in work force composition affected laundrymen's claims to authority. The work force of American laundries was far less homogeneous than its British counterparts'. Before 1920, it was composed primarily of large numbers of immigrant women. In the interwar period, African Americans gradually replaced their immigrant sisters in larger American cities like New York and Chicago.[6] For American laundrymen, race and ethnicity added an extra locus to the matrix of class and gender which shaped social relations within laundries.

More generally, this story helps illuminate how and why gendered meanings come to be attached to a wide variety of technologies. Laundrymen were far from alone in using trade journals and industry histories to create meanings. The "how" part of their story suggests the necessity of rethinking many of the sources historians of technology frequently use. Explaining why people as ordinary and prosaic as the laundrymen put much time and effort into creating meaning helps reveal the inseparability of constructing material and cultural worlds. Gendering technology is a pervasive and important cultural process with significant economic and social consequences.

Laundrymen Construct an Industry

Laundrymen most likely made their gendered claims in all kinds of ways: on the shop floor, in interactions with customers, in letters and conversations. That ephemeral discourse has largely disappeared from the world and with it the evidence for a kind of day-to-day negotiation over the relational meanings of gender. Much of what remains is trade literature.

Laundry trade journals began to appear in the late 1870s in both Britain and the United States. The sudden appearance of these materials probably had two causes. Limited secondary literature on trade journals suggests they began to appear for a wide variety of industries in the late nineteenth century.[7] Their appearance also coincided with exponential growth in the industry and the formation of the first laundry trade associations. The most prominent journals were the *National Laundry Journal* (United States), the *American Laundry Journal,* and *The Laundry Journal* (Great Britain).[8] These were long lived and, in several cases, tied to trade associations. By the 1890s, journal publishers and others had also begun to spin off advice books, collections of articles, books of designs for advertisements, and other specialized trade literature.

Trade journals have long been an important source for historians of technology searching for information about the technical details and social relationships of specific industries. They have less often been used as sources for cultural questions about technology and even more rarely for thinking about gender and technology.[9] While some of the content of these trade journals served the straightforward goal of sharing information about how to solve technical problems or use new products, much of it had a more complex purpose. A significant portion of these texts is made up of jokes, anecdotes, remembrances, and descriptions of social events such as conventions and news of members.

The editors of laundry journals, particularly of the British journals, sometimes liked to claim that these publications were for everyone in the laundry trade. Not surprisingly, much of the text was, in fact, shaped as an exchange between men. Laundrymen were unselfconscious about claiming to speak for everyone in the trade while generally speaking only to each other. In these journals, authorial voice was perceived and utilized as the voice without gender—the norm—just as whiteness in American society is often seen as being without race.[10]

The degree to which this was made explicit varied (one frequent device was a letter between colleagues or advice from an older man to a younger man). Laundrymen most often talked to each other and about women—women workers, women customers, and women laundresses with whom they competed. This discourse helped to create a community in which knowledge about technology and technological processes became masculinized and the mechanisms for sharing that knowledge became closed to women (although, of course, women could, and apparently occasionally still did read the journal).[11] In a world in which economic and social status depended on the respect of other men, this discourse was also a public assertion that despite the gendered connotations of the technology, laundrymen were manly men by the standards of their culture. Unlike the hen-pecked

Figure 6.1. "Don't Get the Clothes Too Blue!" stereopticon. (Warshaw Collection of Business Americana, Archives Center, National Museum of American History, neg. #96-3281. Reproduced with permission from the Smithsonian Institution.)

males humorists sometimes depicted doing the family wash, they had not given up their masculinity (fig. 6.1).[12]

A comparison between British and American journals and of changes within journals between the 1890s and the 1930s reveals the complexities of this voice: for while it is almost always male, the masculinities expressed change over time. In part, this had to do with changing editors and authors. More importantly, these authors reshaped their rhetorical strategies to fit changing cultural and technological contexts. At first reading, these journals seem full of familiar and unchanging stereotypes of men and women and their relationship to technology. A closer examination reveals that while these writers claimed that masculinity and femininity were unchanging, the characteristics, behaviors, and relationships they described changed over time.

In the first number of the *National Laundry Journal* for 1902, the editor, Charles Dowst, published an article in the form of a letter addressed to himself by an older laundryman. Though not much more than a page long, the article is a complex document with a number of functions, including memorializing deceased and retired laundrymen, naming and identifying the place of individuals in the community, and mentoring, with Dowst acting as the surrogate for the reader. It also utilizes several different strategies to make the point that this is a masculine business.

Much of the language borrows from what Warren Susman has called the "culture of character"—reminiscent of Horatio Alger and the rhetoric of "luck and pluck."[13] "My boy," says the ancient laundryman, "don't for a moment deceive yourself into believing that luck has all to do with success in life . . . the little band who gathered in Chicago and organized our glorious National Association were all hard, tireless workers in the battle for success. Those who climbed the highest, commenced the lowest; their struggles were most desperate. Gray matter may have been the powerful adjunct in reaching the fulfillment of their dreams, yet back of all was the indomitable iron will of resolve."[14]

The ancient laundrymen's speech also made a virtue of the liabilities with which many of the *National Laundry Journal*'s readers entered the industry. Many were small-time businessmen who had either failed in their first business or who were going into business for themselves for the first time. Unless they had worked in a laundry, they most likely knew little or nothing about the process of laundering clothes. Significantly, most American laundrymen's descriptions of how they learned the technical aspects of the business lack outside agency. As they told it, they learned by experimentation or from other laundrymen. Even when it is clear that they absorbed the business from hanging around someone else's laundry, female workers never appear in their stories.[15]

In the textual world of the ancient laundryman, it is men who solve the technical problems of laundrywork: "to the rescue came T. S. Wiles of Troy, elucidator of all laundry problems, the Alma Mater, the educator of aboriginal laundrymen. In his brain was solved the wicked and knotty question of grey streaks in cuffs, it was he who conquered the 'Scattergood Phantom' . . . he alone of all toilers in the thorny vineyard of tribulations attained the perfection of laundry work."[16]

The technical problem that T. S. Wiles so heroically solved involved discoloration from bleach or bluing. It is a problem that is also described in advice books for housewives from the same period, but it is given a very different, masculinized meaning here.

T. S. Wiles was not alone in the pantheon. Throughout the pages of the journal, the names of men who have solved these small problems are celebrated. The process of giving credit shapes gender divisions: men are creators, women either reproduce what men thought up (if they are workers) or consume the fruits of the laundrymen's genius (if they are consumers).

On the rare occasions when women who own or manage laundries do appear in the pages of the American journals, either they are oppositional figures or they are bound by conventional stereotypes. In the same volume of the *National Laundry Journal* as the letter to Dowst from the ancient laundryman is a rare letter from a woman laundryowner. The first four words of her letter identify her sex: "I am a woman who about six months ago took it into my head to start into business for myself." According to the letter, she chose a steam laundry not because it seemed appropriate for a woman or because she could use knowledge she already had but because she thought "it would be different from everything else." The rest of the letter details her inability to find a reliable man to run the laundry for her.[17] In one sense, her letter helps confirm the laundrymen's hegemony. She is the exception that proves the rule. She enters the business without informed intention. Her search for a male manager also makes a distinction between having the capital to buy a laundry and having the knowledge to run it. The term "laundryman" is ambiguous in that it conflates and muddies those two categories.

Where female laundryworkers appear in the American trade literature, either they are portrayed collectively—usually in articles about strikes—or they are infantilized and portrayed as technologically ignorant. It is difficult to find examples in laundry journals of a laundryman giving credit to a female worker for innovations or improvements in laundry processes. A poem entitled "The New Machine," published in the *National Laundry Journal* in 1904 for the amusement and reassurance of the laundrymen, suggests some of these gender politics:

> Strange how a laundry girl acts when e'er a new machine
> Is given her to operate instead of a "has-been."
> She kicks and pouts because it doesn't work the same way
> And here are some of her expressions used throughout the day:

The laundry girl goes on to disparage the owner, the manager, and the engineer. The narrator then explains:

> Of course, the only remedy, we know is the old way
> Of treating all the ladies—that's let her have her say,
> Then go to her next morning, tell her we have changed our mind,
> And that her old machine was better than the new, we find.

The laundry girl has by then discovered the virtues of the new machine and is not willing to give it up. The laundryman was, of course, right about the machine and about feminine psychology:

We walk away a-smiling, for we knew it very well,

That, soon as she got used to it, she'd make an awful yell

If we should think of changing it. But that's the usual scene

When we get one of our laundry girls an up-to-date machine.[18]

American laundryowners also employed the strategy of rendering their workers invisible. In their depiction, machines carried out the processes. Invisibility was a useful strategy not only in bolstering the authority of laundrymen but in hiding workers from consumers who worried about black or ethnic hands touching their private things. Laundrymen seemed to believe that one of the appeals of machinery for consumers was that it had no race or class.[19]

While the editorial voice of British journals is also male, women, both workers and managers (very rarely consumers) are a much more pervasive presence than in American journals. Many of the masculine idioms employed by Americans made less sense in British culture. British laundrymen more often derived their authority from paternalism and class — ways of social organization that necessarily involved women. Moreover, they often constructed the idea of laundrywork as men's work not in contrast to a general domestic model but rather in specific opposition to washerwomen, who were much more powerful cultural figures in Britain than in the United States.[20]

British laundrymen evoked the domestic connotations of laundrywork and of domestic laundry technology far more often than Americans. As a later section of this article will show, they sometimes took on the tools of the traditional washerwoman as their own symbols. They also alternatively denigrated and sentimentalized the person of the traditional laundress, her craft knowledge, and her tools. A 1907 article in *The Laundry Journal* suggests the complex class and gender connotations of traditional laundrywork: "Six days at the wash tub is now the most popular remedy for riotous suffragettes. It is favoured by persons in authority as well as the man in the street."[21]

Direct claims to being able to fix or operate machinery more competently than women are largely lacking in the British journals. While British laundrymen implied that they understood the principles of machinery, they seldom demonstrated their masculinity by portraying themselves as rolling up their sleeves and getting their hands dirty. British laundry journals often described a figure called the "laundry engineer" who had no explicit American counterpart. This was a man hired to set up and tend the mechanical parts of the laundry. Like the machinery and supply salesman, the male drivers, and the washhouse man, he was a potential but not accepted member of the brotherhood of laundrymen.[22]

Perhaps because British laundrymen were less likely to consider hands-on knowledge of machinery an important measure of their masculinity, they also voiced more willingness to acknowledge the technological capabilities of their female employees than did their American counterparts. They also made more of a distinction between abstract knowledge about how a machine worked (a masculine domain) and the ability to use machinery. At least one editor was hopeful about the ability of women to learn to operate new machinery. He argued that "surely if girls can work the telegraph, they might easily become familiar with the working of the many and less delicate and complicated machines which are used in the laundry."[23]

One of the mostly startling characteristics of British laundries is the widespread employment of women managers. While American laundries sometimes employed forewomen who managed a single department, they did not have the kind of broad authority of their British counterparts. The use of women managers in British laundries had a practical basis—they knew more about laundrywork and they cost less than their male counterparts. They asserted authority over the relatively homogeneous work force by virtue of their higher class status.

The presence of female managers would seem to undermine the authority of the British laundrymen. Partly the problem was solved by making a sharper distinction between managing personnel problems (clearly women's work) and solving technical problems. Female managers might teach employees how to use machinery, but they were not credited with making decisions about purchases or laundry set-up. Journal articles that mention women managers often use the strategy of praising their abilities by emphasizing how they stemmed from gendered qualities. Lady managers are good with people. They are valued for qualities such as the ability to be firm but kind.[24] In one article, a prominent laundryowner named Hugh Trenchard made the "frank confession" that "if I had trouble with the workers I lock myself up and leave my manageress to deal with it."[25] One could easily substitute "servants" and "wife" into Trenchard's confession. Asked to toast "the ladies" at an 1899 dinner of the National Laundry Protective Association, a Mr. Roberts recontextualized qualities that might otherwise be seen as manly as ill-fitting by saying that he had "always looked upon laundry ladies as so desperately serious and businesslike that he hardly knew whether to tackle the subject in a serious or jocular style."[26] This domestic language contrasts with the masculine analogies sometimes used by American journals to describe male managers. An article from the 1920s compared the role of the male manager to that of an engineer on the famous train, the Twentieth Century Limited. The "superintendent is the man at the throttle" stated the article's author, "a man of decision, with the force and character to have his orders carried out without question or debate." Should he successfully carry out his duties, "his passengers (the public) will take more frequent rides on his particular train."[27]

British laundrymen often used the domestic connotations of laundrywork in a different way from their American counterparts in justifying their authority over female workers. Claims about technological knowledge have less importance to them than having social relations inside the laundry mirror those outside. Because almost all British laundryworkers were native-born white women, British laundrymen could cast these women in a familial framework, with themselves as pater familias. For instance, they sometimes portrayed laundries as a good place for women to work. Like the Victorian home, laundries could provide appropriately womanly work under the ultimate authority of a suitable man.

In an 1887 article in the British journal the *Laundry Record,* the (male) author advised that "waste girls" who were having difficulty finding husbands should go into laundry work as "it is particularly suited to the feminine genius." Moreover, not only can women use their innate abilities but they will benefit in turn because "unlike other female occupations, it [laundrywork] has no tendency to unfit a girl for the cares and responsibilities attaching to the dignity of wife and mother."[28] The intention of this particular article (and others like it) was to attract a better class of girl, as laundrymen were convinced that the unladylike behavior of their employees sullied the reputation of the industry.[29] Nowhere does one get a sense of the heat and monotony and danger of these jobs. Many of these articles romanticized the laundress and her work. Girls learning laundrywork in a state-run boarding school are described as unwilling to leave their work in the "school board fairyland" even for dinner.[30]

Even in Britain, the rhetoric of paternalism tended to break down in the face of state-sponsored demands that laundryowners actually look after the welfare of their workers. Then the worker's knowledge became a valuable commodity that she could easily use to protect herself against misuse by her employer. In response to a proposal that the Factory Acts be extended to laundry workers, one British laundryman argued that the Acts were to protect the young and the weak. "Laundry workers are, if young—as few are—by no means weak." He went on to explain that a skilled laundry worker could always find alternate employment "should her employer, or his prices, or her fellow workers, or indeed anything displease her."[31]

In American journals, romantic or paternalistic images of laundry workers are largely absent. Paternalism was largely an unappealing strategy for laundry-men who might not want to make their immigrant and (by the 1920s) African-American workers part of the family in any capacity. The argument that laundry-work is culturally appropriate work for future wives and mothers was never made. Reformers who wanted to argue the opposite case made it most successfully on the grounds that laundrywork is destructive of the reproductive physiology of women — not their femininity (which was considered questionable anyway in the case of immigrant and nonwhite women).[32]

In the first decades of the twentieth century, laundrymen became more and more preoccupied with female consumers. They recognized that the majority of housewives still did their family's laundry at home or handed it over to servants or washerwomen.[33] Laundrymen had good reason to believe that sending the laun-dry out was an appealing alternative to the backbreaking labor of stirring a copper, ironing sheets, or turning a wringer. Even women who could afford a washer-woman still had to endure wet laundry in their kitchens or the risks of sending it off to tenement rooms to be done.

Laundrymen sensed an enormous potential for new business. Where once they had been content with whatever business found its way to them, more and more they began to pursue what was known in the trade as the "family bundle." By the 1920s, this segment of the laundry market had become the subject of much dis-cussion in the trade journals and the target of a variety of advertising schemes.

As it turned out, pursuing the family bundle was a complex matter. To get the bundle, laundrymen had to negotiate the gendered meanings of laundrywork with their female customers. They needed to claim the superiority of their machinery-intensive approach while not undermining women's claims to authority as care-takers of the health, well-being, and cleanliness of their families. It is indicative of the continued identification of laundry as gendered female that laundrymen were often surprisingly acquiescent to women customer's opinions and expertise on the subject.

Laundrymen made some effort to try to understand what women wanted be-cause they came to believe that, directly or indirectly, women consumers had the power to define standards for the trade by defining what cleanliness meant and by making technological choices. Acknowledging middle-class consumer's expertise but not that of working-class laundresses (who surely knew more about how to get clothes clean than their middle-class sisters) allowed laundrymen to split women along class lines.

In the 1920s, large numbers of trade journal articles began appearing with titles like "The Customer's Point of View," "The Housewife and the Modern Laundry," and "Housewife Fires Broadside at Laundry." Toward this end, they also invited club women or home economists to offer their views.[34] These articles not only acknowledged women's expertise as consumers but reinforced the dichotomy of domestic laundrywork as feminine and commercial laundrywork as male. Consistent with the gender ideology of their age, these articles often portrayed middle-class women as mysterious, capricious, and demanding. One article summed up the gender divide with a terse bit of advice to laundrymen, "women's ways are not men's ways."[35]

Advertisers were not above capitalizing on the anxiety about gender relations evidenced in these discussions to sell their products to worried laundrymen. For instance, a 1924 advertisement captioned "Mrs. Housewife is the Judge" was placed in *Laundry Age* to sell a product called Erusto Salts to laundryowners. The ad is illustrated with a woman dressed in judge's robes yelling at a forlorn-looking laundryman half her size. The caption explains that "the Court knows a whole lot about the Family Wash but not so very much about the Modern Laundry. She thinks a great many things about the Modern Laundry that are not true. That's why you are on the Defense. The Court must be convinced before she will hand down a favorable decision and award you the Family Wash" (fig. 6.2).[36]

Laundrymen believed that women consumers cared about how the process was carried out as well as what the end product looked like. They continually reiterated that what set laundries apart from other alternatives was their modern, industrial character—that the processes were carried out by factory methods using science and engineering to set standards and design controls.

By the 1920s, the Laundryman's National Association (LNA) ran an advertising and publicity campaign that opened up laundries for tours by women customers. In an advertising campaign sponsored by the LNA, women's magazines such as *Good Housekeeping* featured the adventures of "Alice in Launderland." The text and pictures of a 1928 ad suggest the complexities of constructing all the gendered meanings of laundrywork to preserve the laundrymen's authority without offending the female customer (fig. 6.3).

Alice, fashionably dressed in a cloche hat and fur-trimmed coat, is being shown through a laundry by a man described as her "genial" guide. In the background are workers sorting laundry. "Here's our 'raw material' as it comes from the home,"

May 1, 1924   LAUNDRY AGE   73

Mrs. Housewife is the Judge

The case is that of the Family Wash vs. the Modern Laundry.

The Court knows a whole lot about the Family Wash but not so very much about the Modern Laundry. She thinks a great many things about the Modern Laundry that are not true. That's why you are on the Defense. The Court must be convinced before she will hand down a favorable decision and award you the Family Wash.

dry smell," and with even the most dainty fabrics unharmed by chemical action.

Such evidence has great weight with the Court. Use it! Try a keg of ERUSTO. We pay the costs if it isn't satisfactory. Tear off the coupon and mail it.

Sterling Products Company
Dept. L, Easton, Pa.

Figure 6.2. "Mrs. Housewife is the Judge," *Laundry Age,* May 1, 1924, p. 73. (Courtesy of the Library of Congress.)

says the guide, using a factory analogy. It is the method, the technology, that is emphasized. Nothing is said about the women workers in the picture. Further text picks up other of the laundryman's themes: "Whether you supervise the laundress at home, or send clothes out to questionable quarters, you will find that modern laundries offer freedom from work and worry." And finally, a small box at the bottom of the page offers: "Go with Alice into Launderland — This delightful journey booklet may be had from any modern laundry displaying on its trucks this picture of Alice in Launderland."[37]

Laundrymen Construct Their Past

Beginning in the 1920s, laundrymen also increasingly employed another type of rhetorical strategy in constructing and justifying their identity. In printed sources meant both for public and trade consumption, they told a series of creation stories about the origins of the industry. In a culture fascinated with inventors and invention and deeply (if not rigorously) historical in its understanding of the world, creation stories had a powerful resonance.

The appearance of these creation myths can be ascribed in part to the widespread fascination with advertisement and publicity that took hold of both British and American culture in the interwar period. Creating a public image for business became an increasingly important role for trade associations. The peculiar form of laundry histories in this period can also be explained by changes within the community of laundryowners. The deaths of many of the older laundrymen who had entered the industry in the 1880s freed a younger generation from the

Figure 6.3. "Let the Laundry Do It!" Alice in Launderland advertisement, *Good Housekeeping,* December 1928, p. 183. (Copyright International Fabricare Institute, reprinted with permission.)

obligation to memorialize specific individuals and from a collective memory that often suggested the real beginnings of the industry commenced from their efforts. These newer stories were more generalized, shaped to fit broader cultural ideas about origins. The two examples described below were published in industry histories designed not just for an audience of laundrymen but also to publicize the industry to the general public.

In 1950, Fred de Armond, a longtime associate editor of *Laundry Age* and publicist for the industry, published an official history of the industry which included an origin story which had been repeated in various forms since the 1920s.[38] De Armond states that the technology that made the industry possible was born when an inventive miner in the California Gold Rush hooked up a jury-rigged washing machine to a donkey engine. He had decided that "an easier way to get the yellow metal was by doing washing for other womanless forty-niners" than by digging for it himself.[39]

No good evidence exists to support the veracity of the story, and, true or not, it is clear from a number of other sources that the first prototypes of the technology predated the miner's innovation by at least half a century.[40] De Armond's story (and its similar predecessors) draws on two of the most powerful sets of images in American mythohistory: its setting is the Turnerian frontier, its protagonist is the Yankee inventor. Each device is significantly gendered. In Frederick Jackson Turner's model, the frontier is initially a world without women. Women come as civilizers who bring with them the accoutrements of civilization including, in this case, clean linen. In their absence, the inventive miner comes up with a technological solution to what is construed as a social problem. The introduction of a technological solution mitigates the humiliating circumstance of men having to do women's work themselves. In some sense, they're no longer doing it, the machine is doing it.

Because of laundry's domestic linkages, de Armond's story also lends itself to some of the gendered conventions of inventor narratives: It is a tale of Yankee ingenuity in which an amateur inventor takes a piece of familiar household technology and alters it subtly but with enormous ramifications.

For British laundrymen, both the frontier and the Yankee inventor were culturally meaningless (except as intriguing Americanisms). Instead, twentieth-century British stories derive their power and authority by drawing on symbols of status and tradition. In an official industry history, Antcliffe Prince describes the arrangement of symbols in the coat of arms of the Company of Launderers of the City of

Figure 6.4. Coat of Arms of the Company of Launderers of
the City of London. (Ancliffe Prince, *The Craft of Laundering: A History of Fabric Cleaning* [London, 1970], p. 25.)

London (fig. 6.4), a business organization of owners and managers of power laundries organized in 1960.[41] In the center of the coat of arms, says Prince, is a picture
of one of the earliest designs for a washing machine in the British patent record
(1782) "between three antique flat irons fired proper."[42] I know of no evidence that
the "Sidgier Machine," as it was known among laundrymen, was ever put to practical use. Its design suggests that, in any case, it would have been of little use in a
commercial laundry. Its authority rests in its place in the patent record; in the fact
that an individual inventor's name can be attached to it.

The mythical roots of the British laundry industry in Prince's version are hardly
womanless. Traditional laundresses (washerwomen) are pervasive, conflicted figures in European culture with no symbolic American equivalent. They appear variously as archetypal Amazons, drunkards, and slatterns, but also as maternal figures. Prince chooses to portray them as royalty. The support for the launderer's
coat of arms on the right ("dexter") side is described as a figure of Princess Nau-
sicaa, a character from the Odyssey who encounters Odysseus while she and her
maidens are doing laundry on the beach. Prince remarks that "a modern royal

laundress was Princess Margaret who as a Girl Guide gained her Laundress proficiency badge."[43]

Many of the contradictions in these portrayals have to do with the conflict between the realities of laundresses' lives and the purposes they served as symbols. As noted earlier, British laundrymen often sentimentalized laundresses in complex ways. One strategy was to represent them as representative figures of a preindustrialized past. By doing so, the laundrymen also essentialized history into a female, preindustrial past and a male, progressive, industrial present and future.[44] In the "modern" (one of the laundrymen's favorite terms) world, the only appropriate place for these women was as employees or vestigial remnants of the past.

The Company of Launderers made symbolic use of not only the traditional laundress's persona but also her tools. Besides the flat irons, the coat of arms also features other pieces of technology identified with traditional laundresses. The left-hand supporter of the coat of arms carries a dolly, a long wooden pole with cross pieces at the end traditionally used by British laundresses to stir clothes in a tub or boiling copper. Prince states that the Company's ceremonial mace is also a dolly. One can imagine it being carried in procession at an annual dinner. Prince shows no apparent awareness of the irony that this largely male fraternity adopted the traditional tools of female workers displaced by mechanization. Nor do the launderers seem to have worried about being unsexed by this technological cross-dressing, perhaps because they felt assured in their position or because this was one more means of seizing authority.

As the examples given here suggest, laundrymen's efforts to regender their technology were complex and sometimes contradictory—as complex and contradictory as the meanings given to gender by the broader culture of which they were a part. Like Molière's Bourgeois Gentilhomme who discovers he has been speaking prose all of his life, discourse about gender was a pervasive part of their world even if they never gave it a name.

Certainly, twenty years of feminist scholarship on women (and more recently men) and technology has provided an explanation for *why* men claimed and controlled certain technologies. In brief, this theory argues that patriarchy and capitalism combine to ensure that technologies that are lucrative or culturally significant don't fall into the hands of women.[45] And yet, this explanation is not very explicit about *how* men (or other hegemonic groups) gain and maintain control. If an explanation about process is offered, it is usually materialist: men controlled access to most capital and therefore to capital-intensive production. This explanation also implies that the process is one-sided and coercive, usually through economic and institutional means. Moreover, this model of coercive patriarchy offers no explanation for the complex and extensive discourse about gender evident in the sources discussed here.

A different kind of model borrowed from the Gramscian Marxist idea of cultural hegemony provides a way of explaining why the laundrymen's symbolic discourse was critical to creating their material world. As used by historians and cultural theorists, this theory explains that dominant groups cannot maintain their hegemonic position by outright coercion; they require at least partial consensus from those who are being subordinated. This consensus is created by a shared cultural discourse through which all parties can rationalize unequal power relationships. In order to function properly, the dominant group also has to make some concessions to the subordinate group.[46] This model has most often been used to talk about class or class-like relations,[47] but it works quite well for thinking about relations between the sexes.

Laundrymen's claims about the essential masculinity of what they did were claims to power and authority on cultural grounds. As long as the culture as a whole continued to subscribe to the sex-gender system as a cultural rationalization for unequal power relations between the sexes, the laundrymen's claims needed only to make sense within that context. In return, they were willing to concede a limited amount of authority to women. For instance, women know a clean shirt when they see it. Women are better adapted to the heat, monotony, and attention to detail needed in the ironing room.

That claim to authority was essential. Laundrymen could not have successfully constructed their world, either materially or symbolically, without paying attention to the dynamics of gender. It was essential on the shop floor, it was essential in dealing with other men, not only other laundrymen but the bankers and lawyers and business community that provided support for their enterprises. It was essential in drawing in customers who were paying not for a tangible good but for a service and a partially culturally described quality, cleanliness. It could not be otherwise. Pressed on all sides by an older, more pervasive cultural understanding that laundry was women's work, regendering the technology and asserting their authority over it had to be a dynamic, ongoing process. The broader cultural understanding of the link between masculinity, machinery, and industrial processes provided a useful tool to accomplish that goal.

In the end, the persistent gendering of laundrywork as female outside the bounds of their industry was their undoing. In the late 1920s and 1930s, washing machine manufacturers began to explicitly target women as potential customers. As Ruth Cowan and others have pointed out, their advertisements suggested that women who sent their washing out were abrogating their responsibilities as wives and mothers.[48] They should not trust the laundry (read laundryman) to ensure the cleanliness of their families' personal washing. Laundryowners retaliated with claims about sanitation and science and careful management in vain. Laundry became a rare example of a domestic process, once industrialized, that returned to the home.

In some ways, the laundryman's story is anomalous. It is rare that a technology is so explicitly regendered. In other ways, this case study has broader connotations for thinking about writing the history of technology. Many of the rhetorical strategies that the laundrymen used are awkwardly familiar. We recognize the creation stories, the insistence that technological abilities are innate and gendered, the idea that, as Michael Adas puts it, machines are the measure of men.[49]

Explicit discussions of gender may be new to historians of technology, but implicit notions about gender are pervasive in the sources we use. If this gender talk is functional and not just the cultural detritus of a less enlightened age, then it cannot be ignored. This is particularly true if we accept that technology in both its shaping and uses is socially constructed.

NOTES

Dr. Mohun thanks Nina Lerman, Ruth Oldenziel, Carroll Pursell, Erik Rau, John Staudenmaier, and Angela Woollacott as well as participants in the Johns Hopkins Colloquium in the History of Science, Technology, and Medicine for careful readings and wise advice on previous drafts of this article.

1. Some anecdotal evidence even suggests that men who were forced to do their own washing did it at night or behind closed doors to avoid being observed; see Caroline Davidson, *A Woman's Work is Never Done: A History of Housework in the British Isles, 1650–1950.* (London, 1982), pp. 136–37.

2. For more on domestic laundrywork in the United States in the nineteenth and twentieth centuries, see Susan Strasser, *Never Done: A History of American Housework* (New York, 1982), and Ruth Schwartz Cowan, *More Work for Mother: The Ironies of Household Technology from the Open Hearth to the Microwave* (New York, 1983). For Great Britain, see Davidson, and Pamela Sambrook, *Laundry Bygones* (Ayelesbury, Bucks, 1983). Patricia Malcolmson's *English Laundresses: A Social History, 1850–1930* (Urbana, Ill., 1986) is a unique

study of the lives and experiences of women who worked in both commercial laundries and in taking in washing. There is no equivalent study for American washerwomen, although Tera Hunter, "Household Workers in the Making: Afro-American Women in Atlanta and the New South, 1861–1920" (Ph.D. diss., Yale University, 1990), is one of several recent studies of African-American women who worked as domestic laundresses.

3. For a detailed discussion of the technological development of the industry, see Arwen Palmer Mohun, "Women, Work, and Technology: The Steam Laundry Industry in the United States and Great Britain, 1880–1920" (Ph.D. diss., Case Western Reserve University, 1992).

4. Throughout the rest of this article I will selectively use the word "technology" to refer not just to the machinery but also to the process, the knowledge about the process, and the technological system called collectively a "laundry."

5. For instance, many American middle-class men used tinkering or other kinds of hands-on technological play to assert their masculinity after the turn of the century. Such tinkering seems to have come later to Britain because soiling one's hands on machinery carried a stronger class connotation than in the United States. On middle-class masculinity and tinkering, see Susan J. Douglas, *Inventing American Broadcasting, 1899–1922* (Baltimore, 1987), chap. 6. On multiple masculinities, see Mark C. Carnes and Clyde Griffen, eds., *Meanings for Manhood: Constructions of Masculinity in Victorian America* (Chicago, 1990).

6. For an overview, see Ethel L. Best and Ethel Erickson, *A Survey of Laundries and Their Women Workers in 23 Cities,* Bulletin of the Women's Bureau (Washington, D.C., 1930).

7. See Leigh Eric Schmidt, "The Commercialization of the Calendar: American Holidays and the Culture of Consumption," *Journal of American History* 78 (1991): 891–92; John R. Stilgoe, "Moulding the Industrial Zone Aesthetic: 1880–1929," *Journal of American Studies* 16 (1982): 8; and Claire Badaracco, "Marketing Language Products 1900–5: The Case of Agricultural Advertising," *Essays in Economic and Business History* 8 (1990): 131–46.

8. An extensive search has not revealed surviving examples of the American trade publications predating the 1890s; I have limited my analysis to the period after that.

9. Eugene Ferguson has called for more critical attention to the authorship and editorial context for these journals. See "Technical Journals and the History of Technology," in *In Context: History and the History of Technology: Essays in Honor of Melvin Kranzberg,* ed. Stephen H. Cutcliffe and Robert C. Post (Bethlehem, Pa., 1989), pp. 53–70.

10. Carroll Pursell makes this point in regard to the relationship between technology and masculinity by making the point that masculinity is often defined only by its juxtaposition with an "other," e.g., femininity. See Carroll Pursell, "The Construction of Masculinity and Technology," *Polhem* 11 (1993): 206–19. The growing literature on whiteness as a form of ethnicity also makes this point; one of the most widely cited examples is David R. Roediger's *Wages of Whiteness: Race and the Making of the American Working Class* (New York, 1991).

11. In a rare letter from a female laundry proprietor, Mrs. Lillian G. Barrow wrote to the editor of the *National Laundry Journal* to say, among other things, that she "had several copies of the *Laundry Journal,* and after reading each feel so elevated that I did not recognize myself." "She Will Make a Success of It," *National Laundry Journal,* October 1, 1902, 19.

12. "Don't Get the Wash Too Blue!," Laundry, Warshaw Collection (Stereopticon file), Smithsonian Institution.

13. Warren I. Susman, "'Personality' and the Making of Twentieth-Century Culture," in *Culture as History: The Transformation of American Society in the Twentieth Century* (New York, 1984), p. 274. Susman suggests that this Puritan-republican, producer-capitalist definition of self is characteristic of the nineteenth century and begins to shift in the early twentieth century. If he is correct about the chronology of the shift, the laundrymen (not surprisingly) are way behind the cultural curve. Susman also doesn't discuss the gendered implications of these shifting definitions of self but the implications seem obvious. He's really talking about men.

14. "A Leaf from the Past," *National Laundry Journal*, January 1, 1902, 29.

15. For an extensive sample of American laundrymen's autobiographical stories, see "Leadership through the Years, 1883–1958: American Institute of Laundering 75th Anniversary," William Frew Long Papers, Western Reserve Historical Society, Cleveland, Ohio.

16. "A Leaf from the Past."

17. "She Will Make a Success of It" (n. 11 above).

18. "The New Machine," *National Laundry Journal*, July 1, 1904, 52.

19. On the politics of invisibility with regard to technology see John Staudenmaier, "The Politics of Successful Technologies" in Cutcliffe and Post (n. 9 above), pp. 164–65.

20. Malcolmson (n. 2 above), pp. 5–6 and passim.

21. *The Laundry Journal*, April 13, 1907, 8.

22. "Engineer" in the British context comes closer in meaning to mechanic than to the American meaning of engineer as someone who designs machinery. The role was most fully explicated in the British journal *The Laundry and Institution Engineer*.

23. *The Laundry Journal*, September 15, 1887, 5.

24. *The Laundry Record*, August 2, 1920, 457; *The Laundry Journal*, August 1, 1890, 12.

25. *The Laundry Record*, August 2, 1920, 457.

26. *The Laundry Journal*, March 15, 1899, 11.

27. "Running the 20th Century Limited in a Laundry," *Laundry Age*, April 1, 1924, 24.

28. "Waste Girls," *The Laundry Journal*, September 1, 1887, 5.

29. "Legal and Criminal," *The Laundry Journal*, August 20, 1886, 5; *The Laundry Journal*, August 15, 1887, 11.

30. *The Laundry Journal*, March 1, 1890, 5.

31. "The Laundry Act," *The Laundry Journal*, August 1, 1887, 5.

32. The most significant Supreme Court decision on protective legislation in the progressive period was *Muller v. Oregon*. Muller was the owner of a laundry in Portland, Oregon. The prosecution's arguments were based on the deleterious physical effects of laundry-work on women as presented in what is known as the "Brandeis Brief." That brief and the Court's opinion were published in Louis Brandeis, *Women in Industry: Decision of the U.S. Supreme Court in Curt Muller v. State of Oregon* (New York, 1908).

33. In 1920, U.S. census takers counted an astonishing 383,622 women described as "laundresses (not in laundry)." In comparison, only 78,548 women were employed in com-

mercial laundries. Joseph A. Hill, *Women in Gainful Occupations, 1870–1920*, Census Monograph 9 (Washington, D.C., 1929), p. 184.

34. "The Housewife and the Modern Laundry," *The Power Laundry*, August 20, 1921, 723; "The Customers' Point of View," *Laundry Age*, May 1, 1921, 18; "Housewife Fires Broadside at Laundry," *Laundry Age*, July 1, 1921, 26.

35. *National Laundry Journal*, January 1, 1903, 3.

36. *Laundry Age*, May 1, 1924, 73.

37. *Good Housekeeping*, December 1928, 183.

38. Most of the specifics of de Armond's story date to at least the late 1920s. In composing a history of the industry for a Women's Bureau report, Ethel Best recounted this story of the miner. She also added a gendered comment of her own: "It sounds like a man's idea of a labor saving invention, and it was." See "A History of the American Laundry Industry," typescript, Women's Bureau Papers, Record Group 86, Box 105, National Archives, Washington, D.C. Best was given historical material by the National Laundryowners Association and culled information from the pamphlet "The Brief Story of the Progress of the Power Laundry Industry from Ancient Times": see L.C. Ball to Ethel Best, September 22, 1928, and L.C. Ball to Ethel Best, April 13, 1928, Women's Bureau Papers, Record Group 86, Box 105. Frederic W. Bradshaw, *Power Laundries: The Story of a Five Hundred Million Dollar Industry* (New York, 1926) begins in Egypt and has a variant on the gold fields story as well as a recounting of patents.

39. Fred de Armond, *The Laundry Industry* (New York, 1950), p. 7.

40. It is most likely derivative of textile technology, specifically that used in bleacheries in late-eighteenth-century Britain. See Charles Sylvester, *The Philosophy of Domestic Economy*, (Nottingham, 1819).

41. Malcolmson (n. 2 above), p. 162.

42. Ancliffe Prince, *The Craft of Laundering: A History of Fabric Cleaning* (London, 1970), p. 11.

43. Prince, p. 6.

44. This is a dichotomy that has also been adopted by some feminist historians, making a virtue out of women's differing historical relationship with certain kinds of technology. Notably, see Carolyn Merchant, *The Death of Nature: Women, Ecology, and the Scientific Revolution* (New York, 1980).

45. There is an extensive literature on this subject. I have been most influenced by Judy Wajcman, *Feminism Confronts Technology* (University Park, Pa., 1991); the work of Cynthia Cockburn; Heidi Hartman, "Capitalism, Patriarchy, and Job Segregation by Sex," *Signs* 1 (Spring 1986): 137–170, and "Capitalism and Women's Work in the Home, 1900–1930" (Ph.D. diss., Yale University, 1974); as well as a number of thoughtful review essays by Ruth Cowan, Judy McGaw, and Joan Rothschild (see the introduction to this volume).

46. For useful overviews of Gramscian theory as applied to history and cultural studies, see T. J. Jackson Lears, "The Concept of Cultural Hegemony: Problems and Possibilities," *American Historical Review* 90 (June 1985): 567–93, and John Storey, *An Introductory Guide to Cultural Theory and Popular Culture* (Athens, Ga., 1993). Historians of technology have

also recently discovered the concept. See essays by Rosalind Williams and Philip Scranton in *Does Technology Drive History? The Dilemma of Technological Determinism,* ed. Merritt Roe Smith and Leo Marx (Cambridge, Mass., 1994).

47. One of the most successful uses is in Eugene Genovese, *Roll Jordan, Roll: The World the Slaves Made* (New York, 1974).

48. Ruth Schwartz Cowan, "Two Washes in the Morning and a Bridge Party at Night: The American Housewife between the Wars," *Women's Studies* 3 (1976): 147–72.

49. Michael Adas, *Machines as the Measure of Men: Science, Technology, and Ideologies of Western Dominance* (Ithaca, N.Y., 1989).

SEVEN

Industrial Genders: Soft/Hard

PAUL N. EDWARDS

Constructing, maintaining, or retooling gender systems, as we
have seen, requires a kind of cultural *work,* performed by groups
of people (such as laundrymen, home economists, or car designers) interested in the shape of the gender boundary. Paul
Edwards examines this "boundary work" during the cold war, in
the twin contexts of computer science and the military. From the
early days of electronic computing, both expertise and funding
sources connected militarized computer science tightly to the
increasingly computerized military. In this essay, Edwards explores the categories and boundaries used by the military and
the computer science subcultures, the ways such categories are
gendered, and the ways labels like "hacker" or "strategist" inform
gender identities.

Edwards explores styles of technological activity and the interactions of different kinds of technological knowledge as an approach to understanding the implications of widespread computer
use. He draws on an influential study carried out in the 1970s, in
which sociologist Sherry Turkle identified two styles of computer
programming she called hard mastery—rational, logical, carefully
planned, and highly organized—and soft mastery—more intuitive,
spontaneous, creatively resourceful. Turkle's soft masters were
mostly female, but Edwards finds these same distinctions within
the predominantly male domains he studies. In computer science,
the hacker style is spontaneous, creative, and intuitive; the more
conventional programmer's work is highly organized and ratio-
nally planned. In the military, rational planning is the province of

"The Army and the Microworld: Computers and the Politics of Gender Identity," *Signs* 16 (1990). © 1990 by The University of Chicago Press. Reprinted by
permission.

bureaucrats with desk jobs, and resourcefulness the province of the seasoned and battle-ready soldier. Edwards explicates the ways gender systems are both reinforced and undermined in the process of adopting new technologies, and he examines the degree to which the machines embody the masculinity commonly ascribed to them.

Edwards wrote this piece at the end of the cold war, and he describes a military culture based on several decades of bipolar cold war thinking, in which all communism endangered the free world and military preparedness meant planning to face a superpower enemy. This article, which is both historical and forward-looking, was published just after the fall of the Berlin Wall and before the Gulf War, and also before the heyday of the Internet. Nonetheless, the proportion of women in defense jobs in the United States remains small (not far from the 11 percent figure in this article), and while girls readily use computers, they remain a small minority in programming classes in U.S. schools. What relationships did Edwards find between gender ideologies, military thinking, and computing? What connections does he posit between military and civilian culture, between gender and technology? What are the relationships here between gender ideologies and actual practices? Do you find this piece relevant now or primarily historical?

In contemporary America, computer work—programming, computer engineering, systems analysis—is more than a job. It is a major cultural practice, a large-scale social form that has created and reinforced modes of thinking, systems of interaction, and ideologies of social control. In the 1970s, American women entered the higher levels of computer work in ever-increasing numbers. By 1984, 35 percent of U.S. computer programmers and 30 percent of American systems analysts were women.[1] Across the board, in every field of computer work, these percentages continue to rise at varying rates. Yet in spite of these facts, computer work remains a largely male world, one in which women are perceived as unprepared, alien, or unwilling participants. Women are said to perceive the linear logic of computing and the high-tech radiance of its machinery as a hostile and dominating force. In the United States the reasons behind these perceptions lie in connections between the modes of thought involved in computer science research, the culture of engineering, and the deeply entangled institutions of military service and of masculinity as a political identity in an age of high technology war—connections that provide a bridge between the masculinity of war and the different but related masculinity of the world inside the computer.[2]

Modes of Thinking with Machines

In her book *The Second Self,* Sherry Turkle discusses two approaches to computer programming among children learning to program in the LOGO language at a private school. "Hard masters" employ a linear style that depends on planning, advance conceptualization, and precise technical skills, while "soft masters" rely on a less structured system of gradual evolution, interaction, and intuition. In her words:

> Hard mastery is the imposition of will over the machine through the implementation of a plan. A program is the instrument of premeditated control. Getting the program to work is more like getting "to say one's piece" than allowing ideas to emerge in the give-and-take of conversation. . . . [T]he goal is always getting the program to realize the plan.
>
> Soft mastery is more interactive . . . the mastery of the artist: try this, wait for a

response, try something else, let the overall shape emerge from an interaction with the medium.

It is more like a conversation than a monologue.[3]

"Hard" and "soft" are exceptionally rich words that cover a variety of overlapping conceptual fields. "Hard," according to the dictionary, includes among its fifty-four uses the meanings not soft; difficult; troublesome; requiring effort, energy, and persistence; bad; harsh or severe, unfriendly; sternly realistic, dispassionate. "Soft" may be glossed as not hard, easily penetrated; smooth and agreeable to touch; pleasant; gentle, warm-hearted, compassionate; responsive or sympathetic to the feelings of others; sentimental; not strong, delicate (the example given is "He was too soft for the Marines"); easy; submissive. Hard and soft also have obvious sexual connotations.[4]

The hard master of computers is someone whose major cognitive structures are preconceived plans, specific goals, formalisms, and abstractions, someone who has little use for spontaneity, trial and error, unplanned discovery, vaguely defined ends, or informality. This is also American culture's prevalent image of scientists, often portrayed as unusually disciplined thinkers who deploy long chains of logical and mathematical reasoning to arrive at a subtle, powerful understanding of nature's ways. Men, too, are supposed to be tough-minded, exceptionally rational, unswayed by emotion, good with maps and mathematics, while women outdo them in supposedly more intuitive skills such as nursing and child care. In American ideology, computers, scientists, and men are hard; children, nurses, and women are soft.

In practice, of course, the image is false. Scientists of both sexes, including computer scientists, experience their work as a visceral, creative, social, and unpredictable enterprise, where formal, linear thinking is useful mainly as one part of a larger process involving many kinds of thought and practice. Neither men nor experts in formal thinking hold a monopoly on scientific abilities. Women are fully capable of all the tasks of science and computer work. Equally, men have their own kinds of softness and intuition.

The hard/soft split nonetheless plays a major ideological role. It is reinforced by the popular media and also by professional practices such as the highly impersonal style and the cast of absolute certainty in the language of scientific journals, textbooks, and professional conferences.[5] It can also be found reduplicated within science. There are hard sciences like physics and soft ones like psychology, and within the soft disciplines, at least, there are hard and soft approaches. Often, experimentalism is not the key distinction. Indeed, part of the role of the hard/soft terminology is to distinguish what counts as an experiment from what is merely an observation or an interaction. In fact the split centers primarily around the degree of formalization and mathematization the discipline has achieved. In many sciences contests for legitimacy are staged between those who manage to deploy a hard cognitive approach, using a technical language, mathematical or logical formalisms, and a technical apparatus (including computers), and those who rely more on the resources of nontechnical language, personal observation, and nontechnical methods such as clinical practice.

Today computer scientists enjoy a mystique of hard mastery comparable to the cult of physicists in the postwar years.[6] Computers provide them with unblinking precision, calculative power, and the ability to synthesize massive amounts of data. At the same time, computers symbolize the rigidities of pure logic and the impersonality of corporations and governments. By association with the miracles of its machinery, computer work is taken to require vast mental powers, a kind of genius with formalisms akin to that of the mathematician, and an otherworldliness connected with the ideology and iconography of the scientist. What makes computer scientists archetypal has to do partly with the nature of computers themselves.

Computers are a medium for thought, like the English language or drafting tools. In order to think with a computer, one has to learn its language. All existing computer languages consist of a relatively small vocabulary of admissible symbols (perhaps several hundred to a few thousand in the most sophisticated) and a set of simple but powerful rules for combining those symbols to form lists of sequential instructions to the machine. They are much simpler and more restrictive than any human language, especially insofar as human language tolerates—indeed, relies upon—ambiguity and imprecision.[7] In their forty years of evolution, computer languages have come far closer to approximating human language in vocabulary and grammar; mistakes which would be utterly trivial in an exchange between humans, such as a misspelled word or misplaced punctuation mark, still routinely cause catastrophic failures in computer programs. Attempts to get computers to understand unrestricted spoken or written English have been plagued by precisely this problem.[8] Thus, if a language is a medium for thought,[9] the kind of thinking computer languages facilitate is special, quite different from the reasoning processes of everyday life.

Jonathan Jacky, a computer scientist at the University of Washington, has pointed out that each computer language tends to encourage a particular programming style, as do the subcultures associated with each one. Thus, the Pascal language was deliberately designed to promote a structured, highly specified, self-documenting approach to programming, while artificial intelligence languages like LISP seem to breed a flexible, more interactive approach, with unexpected creative results a part of the goal. Thus certain languages draw their users toward more organic methods of programming, while others pull programmers toward a more structured approach. This corresponds to computer programmers' stereotypes of each other—LISP programmers as "hackers," sloppy but artistic visionaries; Pascal programmers as precise but uncreative formalists, "software engineers."[10] The continual invention and spread of new computer languages is a symptom of the search not only for convenience of interaction, but also for styles of thinking congenial to different kinds of users and their projects.

A second constraint on computers as a medium for thought relates to the hardware itself. Most computers process commands sequentially, so the step-by-step order of a program's instructions must be completely correct and exactly specified. Von Neumann processors, as these are called, thus enforce a pattern and a type of rigor on their users of a sort much like that required of mathematicians in proving theorems. An emerging generation of so-called parallel computers will speed processing by allowing many operations to proceed simultaneously; but programming these systems will be more rigorous, riot less, because of the problem of coordinating the many processors contributing to the solution.[11]

Every program manipulates symbols according to well-defined, sequentially executed rules to achieve some desired transformation of input symbols into output symbols. Rule-oriented, abstract games such as checkers or chess also have this structure. As a result, all computer programming is gamelike. This is one reason programming computers to play chess became a favorite puzzle of computer scientists in the early days of work in artificial intelligence. Rule-based games remain a centerpiece of the culture of computer science—as well as one of its major products—from chess to graphic-display video games.[12] Computer scientist Douglas Hofstadter's 1979 Pulitzer prize–winning book *Gödel, Escher, Bach* exemplified the widespread fascination of computer professionals with art, literature, and music based on recursive and self-referential operations reminiscent of mathematical formulae and certain kinds of computer programs.[13]

Many writers have suggested that these modes of thought—syllogistic, quasi-mathematical logic, and formal gaming—seem more familiar and friendly to most men than to most women.[14] They fit well with a conception of knowledge as an objective, achieved state rather than an ongoing, intersubjective process and with a morality built on abstract principles rather than shifting, contextually specific, emotionally complex relationships.[15] Their deep affinity with mathematics binds them to the western construction of rationality itself, going back to the ancient Greeks.

> In the rationalistic tradition, emphasis is placed upon the formulation of systematic logical rules that can be used to draw conclusions. Situations are characterized in terms of identifiable objects with well-defined properties. General rules that apply to situations in terms of these objects or properties are developed which, when logically applied, generate conclusions about appropriate courses of action. Validity is assessed in terms of internal coherence and consistency, while questions concerning the correspondence of real-world situations with formal representations of objects and properties, and the acquisition of knowledge about general rules, are bracketed.[16]

This standard—rationality as a dispassionate, context-independent, logic-oriented discourse fundamentally separated from its object—has historically provided a kind of master trope for the construction of gender, politics, and science.[17] Computers, which literally embody the split between thought and its objects, act as fundamental paradigms of the rationalistic tradition in the modern West. Thus one reason computers are identified both with hard mastery and with maleness is historical. As a concept and a practice, cognitive rationalism has a very long history of association with masculinity, from Aristotle to Descartes to Locke.

A second reason has to do with the emotional structure of computer work. Programming can produce strong sensations of power and control—part of the reason Turkle's choice of the term "mastery" is so apt. To those who master the required skills of precision, planning, and calculation, the computer becomes an extremely malleable device. Tremendously sophisticated kinds of play are possible, as well as awesome power in the transformation, refinement, and production of many kinds of information. Control of myriad complex systems, such as machines, robots, factories, and traffic flows, becomes possible on a new scale. Many computer subcultures—hackers, networking enthusiasts, video game addicts—are mesmerized by what Turkle calls the computer's "holding power."[18] The phrase aptly describes the computer's ability to fascinate, to command a user's attention for long periods, to involve him or her personally.

What gives the computer this power, and what makes it unique among formal systems, is the simulated world, entirely within the machine itself, that does not depend on instrumental effectiveness. That is, where most tools produce effects on a wider world of which they are only a part, the computer contains its own worlds in miniature. Artificial intelligence researchers have called the simulated worlds within the computer "microworlds."[19]

Problems of simulation are essentially problems of representation — of creating symbolic entities with properties and rules of interaction that correspond to real entities and their interactions. Yet since all representations are formed for some human purpose, the form of any simulation is shaped by that purpose and therefore is, from the point of view of other possible purposes, to some degree arbitrary. A formal representation is never a whole. It can never embody all the human purposes to which it could ever be put. Computer simulations are thus by nature partial, internally consistent but externally incomplete; this is the significance of the term "microworld." Every microworld has a unique ontological and epistemological structure, simpler than those of the world it represents. Computer programs are thus intellectually useful and emotionally appealing for the same reason: they create worlds without irrelevant or unwanted complexity.[20]

In the microworld, as in children's make-believe, the power of the programmer is absolute. Computerized microworlds have a special attraction in their depth, degrees of complexity, and implacable demands for precision. The programmer is omnipotent, but she is not omniscient. Complex programs can lead to totally unanticipated results. Even a simple program may contain logical errors the location and solution of which can require extraordinary expertise and ingenuity. As an imaginative domain, this makes the microworld exceptionally interesting. It is a make-believe world with powers of its own — but limited and unmotivated powers, not the ultimately intractable powers of other people. For men, to whom power is an icon of identity and an index of success, a microworld can become a challenging arena for an adult quest for power and control.[21]

Sally Hacker's interviews with male professors of engineering reveal personal backgrounds of emotional isolation, histories of difficulty in forming intimate relationships, and a relative lack of interest in sensual pleasures, coupled with a preference for rational pleasures. Electrical engineers, a large percentage of whom work with computer technology, were the most marked in this respect. "Engineering faculty," she notes, "also learned to place greater value on hierarchical relations as rational, and on abstract and scientific skills as worthy of greater social rewards than other skills."[22]

Certainly, women too can be loners and outcasts; but the male gender identity is based on emotional isolation, from demands for competitive achievement at others' expense to the organized violence at the center of the male role.[23] It may be that some men choosing engineering careers replace missing human intimacy with what are for them empowering, because fully rational and controlled, relationships with complex machines. Human relationships can be vague, shifting, irrational, emotional, and difficult to control. With a "hard" formalized system of known rules, one can have complexity and security at once: the score can always be calculated, sudden changes of emotional origin do not occur. Things make sense in a way human intersubjectivity cannot.[24]

However, the equation between engineering, hardness, and masculinity, though very real in its exclusionary effects, tends to conceal the fact that the hard/soft split reappears *within* engineering at every level. In computer science the line can be drawn between software engineers, who program according to a disciplined, structured developmental logic, and hackers, undisciplined, often fanatical computer wizards whose nonstandard methods emerge from profound personal involvements with the machine.[25] The anarchistic hacker subculture frowns on orthodoxy. Hackers are often, but not always, extremely intuitive, interactive, and artistic in their approach. In Turkle's terms they are "soft" masters whose motivations derive from the supposedly "hard" quest for power just described. Though many live in a shadow world, unable to cope with the rigidities of institutions, hackers have been among the consummate masters of computers: Steve Wozniak, co-inventor of the Apple computer, was one, and much of the important work of major university computer laboratories such as those at Stanford University and MIT has benefited from hacker contributions.[26] Most hackers, to date, have been men. Thus the soft and the hard intertwine, in complex and sometimes contradictory ways, with gender, cognition, and machines.

Computers and Strategic Thinking

Militarism in the United States of the Reagan-Bush era is, as it has been since 1945, at least as much a matter of socioeconomic organization and political discourse as of the actual exercise of force. In general, the concept of militarism refers to a political order dominated by a nation's armed forces and a social order based on the needs, goals, and interests of the military. Even in the postwar years, the United States has avoided the pitfalls of traditional or Prussian militarism: civilians mostly control the government, the size of the armed forces is relatively small, and military values are not the primary message of major social institutions such as schools. Nonetheless, this way of understanding militarism is inadequate to capture the massive social changes of the post–World War II era, the militarization of society described by Cynthia Enloe: "Militarization can be defined as a process with both a material and an ideological dimension. In the material sense it encompasses the gradual encroachment of the military institution into the civilian arena. . . . The ideological dimension . . . is the degree to which such developments are acceptable to the populace, and become seen as 'common-sense' solutions to civil problems."[27] Now that the vast military presence in American social reality has become "common sense," we must look for a theory of militarism that goes beyond the army as lever.

The conjunction of high technology and military-based bilateralist containment politics is central to the peculiar form of postwar militarism. For us, the military has become much more than a tool of statecraft. It has become a major social system, an integral part of civil society. Technology, and the enormous infrastructure that goes with it, has been the American solution to demands for massive military force without massive commitments of men or cultural resources; but a high tech military needs a high tech society to sustain it. The arms race and technology-based market competition are mutually reinforcing drivers pacing the rate of technological innovation at geometrically increasing speeds. It is this systemic ability to synchronize economic and social goals with military ones that I would name as *our* militarism. It is this ability that links computing to war not just practically but ideologically. This is why maintaining world leadership in computer technology is conceived not only as an economic goal but also as a national security imperative.[28]

Computers are the heart of an extremely widespread process of automation in the U.S. armed forces. In both practical and ideological terms, one major effect of automation has been to erode the boundary between front lines and rear areas in military conflict. This change is the direct result of the extremely rapid development and massive deployment of high technology throughout the American military beginning with World War II. Moreover, automation underwrites the special character of high technology warfare:

1. High technology war takes place beneath the ultimate shadow of nuclear arms. Even with the demise of the cold war and the success of recent disarmament talks, mutual and total annihilation remains a real and ever-present possi-

bility. This must be figured into the strategic calculations of every war involving the superpowers, no matter how apparently local and contained.

2. Three new kinds of weapons systems are key elements of high tech military conflicts: (a) weapons so massively destructive as to reduce the role of the individual soldier in battle nearly to insignificance, except as a support for the weapon itself (both nuclear weapons and so-called near-nuclear conventional weapons such as cluster bombs fall into this category); (b) weapons in which humans are fully integrated components (these systems mediate or augment human sensory or communications processes and/or perform some decision or calculation functions of their own, such as advanced jet fighters, computerized aircraft carriers, and electronic command-and-control networks); (c) self-controlled weapons able to follow complex evasion or attack maneuvers using internal sensory capacities, usually under microprocessor control, such as cruise missiles and "killer satellites."

3. A fourth category of high technology weaponry deserves separate mention: self-controlled, coordinated, total weapons systems able to perform the entire cycle of reconnaissance, threat identification, response planning, execution, and follow-up with minimal human intervention or with none at all. Though not yet commonplace, a number of military planners believe this technology is mature, and pressures to adopt such systems are extreme and steadily mounting. Examples would include any launch-on-warning nuclear defense system and the proposed "Star Wars" space-based ballistic missile defense system.[29]

4. Vastly accelerated speed is one of the major forces shaping the high tech battlefield. Military systems now approach the ultimate horizon of electronic time with the nearly instantaneous processes of electronic command and control.[30]

5. High technology war, though not necessarily total war, involves the total resources of the high tech society. Historically, military research and development has raised the pace of technological change in America and altered its focus. This has been especially important in certain fields of computer technology.[31]

6. Like any highly organized human activity, high technology war must be reproduced. In American society, this involves economic, political, and ideological forces acting in concert to convince citizens of the continuing need for massive military force and to reproduce the political classes of protectors and protecteds. Women are held responsible for the physical reproduction of (male) soldiers. Political, cultural, and military institutions are responsible for the acculturation of men to violence as a primary mode of resolving conflict.[32]

Among other things, this ensemble of defining qualities has vastly magnified war's gamelike structure. With nuclear annihilation its ultimate horizon, the calcu-lated outcome of war, rather than any actual denouement, became the central focus of politics. Technical comparisons of nuclear arsenals and projections of devasta-tion remain possible even when the potential damage surpasses all human scales of estimation or comprehension. The object for each side, then, is to maintain a winning *scenario*—a theatrical or simulated win, a psychological and political effect—rather than actually to fight such a war. It is not so much a true standoff backed by realistic threats as a simulation of a standoff, an entirely abstract war of position. Wars in the Third World and so-called low-intensity conflicts—low intensity, of course, only by the standard of full-scale high technology war—are slotted into the maneuverings for position in a larger struggle. Like a video game, the object of nuclear strategy is not to win but simply to stay in the game as long as possible, at ever-accelerating speeds. The world of nuclear arms becomes by its very grossness and scale a microworld for which everything outside is irrelevant. In a way, war has become a cognitive event.

There is thus a peculiar relation of suitability, a kind of cultural fit, between the cognitive hard mastery of computer science research and the strategic situation of contemporary militarism. The military itself, as an institution, already has the character of a microworld. Its hierarchies, strict laws of conduct, chains of com-mand, orders, uniforms, and precise jargon all contribute to the gamelike structure of a microworld where every goal, entity, and legal move is laid out in advance and clearly defined. The word "regimentation" conveys the sense of a life whose con-texts have been distilled, reduced, and specified, whose every situation has been anticipated, planned, and regulated. As Vietnam veteran William Broyles wrote, "War replaces the difficult gray areas of daily life with an eerie, serene clarity. In war you usually know who is your enemy and who is your friend, and are given means of dealing with both. (That was, incidentally, one of the great problems with Vietnam: it was hard to tell friend from foe—it was too much like ordinary life.)"[33] Like the microworld within the computer, the military microworld holds signifi-cant attractions for men whose emotional isolation makes them want to know where they stand in relation to others, what their purpose is, and what the rules are of the game they perceive their lives to be. As Broyles put it, "War is a brutal, deadly game, but a game, the best there is. And men love games. . . . Nothing I had ever studied was as complex or as creative as the small-unit tactics of Vietnam. No sport I had ever played brought me to such deep awareness of my physical and emotional limits."[34]

The analogy between war and gaming is an old one. High technology has made it possible, even necessary, to take the comparison more seriously than ever before, and historically the computer has played a special role in that process. The rise of the Rand Corporation and, later, of the military managers under Defense Secretary Robert McNamara, eventually culminating in General William Westmoreland's concept of the electronic battlefield, was a direct effect of the confluence of modes of thought derived from computer programming with military strategic thinking.[35] Military planners were drawn by the precision of formal-mechanical models and their great power for simulation and prediction of strategic situations. The assistance of formal modeling techniques such as operations research had played a significant role in World War II, and the novelty and prestige of computers helped make formal models attractive. Nuclear strategy, worked out by Rand's civilian scientists and academics using simulations and formal theories, became a model for strategy on a smaller scale. At the same time, computer science, besides being heavily funded by the military, was conceptually driven by the strategic and technological puzzles the military provided, especially in its early years.[36] Later, scientists whose work at Rand had drawn them into military problems gained political power as the military managers ascended to prominence.[37]

The success of the managerial model during the Vietnam era, and the emergence of a vast high technology strategy and procurement establishment with large stakes in its own reproduction, has led to fierce debates within the contemporary military over the relative merits of two models of the military. One, the managerial model, sees war as a problem of hard mastery. Plans, weapons, and troops must be prepared to anticipate all possible battlefield contingencies. Every situation would meet with an instant, programmed response. Battles would occur like automated drills. Technology, especially computing, is seen as the ultimate solution to strategic problems. The second, the leadership model, sees war as an art—a problem for Turkle's soft masters. Planning is not unimportant, but the real key to military success lies in human factors. Commanders must be true leaders, skilled in the intangible, unformalizable qualities that inspire others to follow them and produce the strokes of intuitive genius and superhuman feats that are the stuff of real— not simulated—battles. Weapons and their technology take second place, in this view, to human factors. Good equipment is necessary but far from sufficient to constitute a battle-worthy force.[38]

In a curious reversal of associations, many, perhaps most, veteran commanders view the managerial approach as "soft" (read "weak"): a civilian bureaucrat's way of war. It is seen as wimpish, disembodied, faceless, effeminate: armchair warfare that pretends to understand complexities and dangers which in reality can only be comprehended through experience. The leadership approach is "hard," bringing back the masculine virtues of heroism and individual toughness. The values of planning, logic, and dominating control are secondary to intuition, interpersonal skills, and flexibility. What Turkle calls hard mastery, a cognitive skill, is coldly logical and dominating; but hard mastery in warfare involves the charisma, intuition, physical presence of the "hardened" veteran—the flexible, adaptable, living skills of the "soft" cognitive master. The cognitive mode of Turkle's hard masters is seen as stifling, bureaucratic, and unrealistic.[39]

Gender, Computers, and Front Lines

Another way high technology has affected war is to render obsolete the image of the front lines of battle as the domain of brave men on foot locked in hand-to-hand combat or shooting at nearby foes with rifles, defending large zones of safety, the rear areas, where women and children could escape danger. The front in a strategic nuclear confrontation would be most of the northern hemisphere, and exactly where the rear would lie during a tactical or local nuclear exchange is unclear. With air warfare and so-called near-nuclear and brilliant weapons, the front/rear distinction has lost its meaning even on the conventional battlefield. In fact, then, the front/rear distinction has largely disappeared. Yet that distinction is the basis for one major political line between the genders, the line drawn between male protectors and their protectees, women and children. Because high technology warfare erodes that political boundary, it has the potential to alter gender roles radically.

During the 1970s women were, for the first time, integrated directly into the regular U.S. armed forces. The Women's Army Corps (WAC) was dissolved, and a large number of military occupational specialties (MOSs) were opened to women. In 1976 the first class of female cadets entered West Point.[40] Lively and often bitter debates about the proper role of women led to a series of policy shifts and reversals. By 1989 the number of women in the armed forces had stabilized—by policy decision, not by availability of recruits—at around 225,000 (11 percent of the total force). Though 88 percent of Army MOSs have been opened to women (up from 35 percent at the end of the Vietnam War), the remaining 12 percent of MOSs comprise 40 percent to 50 percent of all Army jobs. The rationale for women's continued exclusion is that these jobs involve direct participation in combat or "combat support."[41] A combat role for women would require admitting the possibility of reversing or scrambling the protector/protected roles. Thus the debate over this issue is at the heart of challenges to gender ideologies.

These ideologies disguise the historical reality of women's ongoing active involvement in war on many levels. In addition to their activities as nurses, camp followers, workers in war industries, prostitutes, spies, and victims, women have formed small but often significant elements of combat forces throughout recorded history, especially in revolutionary conflicts and guerilla war.[42] Moreover, the definition of "noncombatant" roles is so fluid that at times it has been conveniently readjusted to include, for example, women antiaircraft gunners within the category of "noncombatants."[43] When a military police unit commanded by Captain Linda Bray seized an attack-dog compound defended by Panamanian troops during the U.S. invasion in December 1989, ex-infantryman Brian Mitchell (author of *Weak Link: The Feminization of the American Military*) told *Time* magazine that "the sorts of things they were doing could be done by a twelve-year-old with a rifle."[44]

Contemporary opponents of a female combat presence argue that women would reduce the military effectiveness of combat troops for reasons ranging from physical strength, to the problem of feminine hygiene and privacy under field conditions, to the specter of sex or pregnancy on the front lines. None of these rationales, with the possible exception of inadequate physical strength, have proved valid within the limited experience of the peacetime volunteer forces. But the worry remains that under the stresses of actual combat it would be too late to restructure a force somehow weakened in its effectiveness by women's presence.[45]

The tenacity of these practical objections suggests the massive institutional and popular commitment to thinking of war as an essential test of manhood and a quintessentially masculine activity. It is supposed to be hard, in the multiple senses of difficult, dangerous, bad, harsh, unfriendly—a real man's job. Women's role in the institution of warfare has been constructed as peripheral to a center constituted by violent combat. Women are coded as what Judith Stiehm terms "the protected"—the primary class of persons excluded from direct participation and in whose name wars are fought.[46] They are supposed to be soft—vulnerable, gentle, delicate, and peaceful.

It is from within this framework that some activists have located an intrinsic connection between feminism and pacifism in the image Micaela di Leonardo calls the "Moral Mother": "a belief that women *as mothers,* close to nature and responsible for human reproduction, will reform warring males. . . . The Moral Mother—nurturant, compassionate, and politically correct—the sovereign, instinctive spokeswoman for all that is living and vulnerable . . . will take the toys away from the boys."[47]

The problem with this view, of course, is that it duplicates the innatist mythology it seeks to destroy. As di Leonardo argues, even the attempts of some feminists to use the image as a political strategy, keeping an awareness of its status as a social construction, are problematic. Instead, she proposes that "women may be particularly good at opposing global militarism . . . not because [they] are morally superior nurturers, but because gender is at the center of recurrent contradictions in the militarization process."[48] To grasp these contradictions, it is critical to understand the linkage between masculinity and the military, not just in psychological terms but in terms of the deep politics of gender identity.

Combat and manhood are powerfully interlocked locations of personal, political, and cultural identity, in wars that go beyond tradition to the core of the structure of the modern state.[49] The rise of citizen armies in eighteenth-century revolutionary Europe, contemporaneous with the emergence of universal male citizenship, created a new political equation: to be a man was to be at once a patriot, a soldier, and a citizen.[50] To be a woman was to be at once a political nonentity, mother or wife of soldiers, a member of the protected class, and a symbol of all the brave soldier held dear. The borderlines between public and private, state and family, soldier and innocent, and man and woman were drawn, in the politics of the postrevolutionary world, with the same broad stroke. Thus, whatever the reasons behind the traditional masculine identification of war in prerevolutionary history, in the shift to universal military service the army was converted into a political class whose members were all and only men.[51] Being a warrior, formerly a mark either of aristocracy or economic marginality, became a political requirement of manhood. Male sexuality was bound to the state through violence.

This bit of history reveals the source of a vexing problem for feminists who seek to connect the feminist agenda with antimilitarism or pacifism. On the one hand, eligibility for military service has, since the French revolution, been centrally connected with citizenship in Western democracies. When the question of whether women should be registered for the draft along with men was being debated during the Carter administration, the National Organization for Women (NOW) filed an amicus brief in the Supreme Court urging a connection between women's rights and their responsibilities to defend the country on an equal basis with men.[52] The NOW position was that as long as women cannot be required to serve they can be seen as remaining in a position of debt to their (male) protectors, even if they do not want the protection. On the other hand, if women seek or agree to compulsory military service in order to break the male monopoly on the means of destruction, with its corresponding claim of a deeper right to citizenship based on actual or potential sacrifice, they surrender their ideological status as Moral Mothers—with no guarantee that joining the system will limit the militaristic tendencies of society.

Nonetheless, technology long ago removed many of the barriers to women in combat by changing the kinds of skills required to serve effectively in combat. The infrastructural demands of a high technology military force are so enormous that modern army personnel structures lean quite lopsidedly toward support staff—those behind the lines who operate the administrative and technological infrastructure. Yet with the disappearance of the front, support staff are nearly as likely to give their lives in battle as the gun-toting infantry. Furthermore, machines, centrally including computers, enable anyone, almost regardless of physical ability, to control nearly limitless destructive power.

High technology is not only erasing one of the most potent historical divisions between the genders, it is creating tensions within gender roles as well. Just as computers have eroded barriers to women's militarization and all that this entails for the "moral mother" stance, so, too, the maleness of computers, with their mechanized abstract intellect, collides with the gritty, totally concrete maleness of classical warfare. The massively technology-based character of the contemporary battlefield, American style, has led to a situation in which these tensions must not only coexist but intertwine within the same ideological space. Organized violence becomes an increasingly abstract art—one in which planners, programmers, and engineers are responsible for at least as much damage and destruction as soldiers in the field.

Computers and War Futures

Recently the Pentagon has begun to try to cash in on its enormous past investments in computer research in a wide variety of ways. In 1983 the Defense Advanced Research Projects Agency (DARPA)—historically the foremost U.S. sponsor of research in artificial intelligence since the early 1960s—announced the

Strategic Computing Initiative (SCI), a ten-year plan to develop artificial intelligence (AI) technologies for military systems. Whereas DARPA's past sponsorship of AI mostly had been limited to pure research, with only remote links to weapons systems, the SCI attempts to direct research more vigorously by aiming at specific military applications goals.[53]

The SCI, with a proposed budget of $900 Million over the period 1983–93, is by far the largest single source of funds for artificial intelligence research in the United States today. A funding source that could eventually dwarf even this huge contribution is former President Reagan's Strategic Defense Initiative (SDI). Its uncertain status makes direct assessment difficult, but it has been estimated the SDI might have to invest $10 *billion* in computer science research to achieve its goals.[54] The software required to control the space-based ballistic missile defense system of the SDI would be at least five times larger than the largest existing programs; Strategic Defense Initiative Organization officials have called for advances in AI as aids to writing and testing the programs.[55] Strategic Defense Initiative computer research to date has included efforts in parallel architectures, optical computing, and gallium arsenide microprocessors.[56]

The SCI and the SDI are only the largest and most recent examples in an ongoing process of militarization. In fact, the U.S. computer industry began as a military project of the Army's Ballistics Research Laboratory during World War II. In their early years, computers' value was not immediately obvious to American business and industry (at one time it was estimated that the United States might need at most ten large computers, all for scientific research) with the result that the military and other branches of government were almost the only sources of research funding as well as the major customers for computers until the mid-1950s.[57]

Examples abound of areas where this financial support has allowed the military to shape technology according to its priorities. In 1959 specifications for the programming language COBOL, an acronym for Common Business-Oriented Language, were defined by the Department of Defense (DOD) to insure a standard for computers it purchased. COBOL subsequently became the ubiquitous language for business applications, only recently beginning to lose ground to newer languages.[58] Not all such support results in a commercially profitable product, however. According to Leslie Brueckner and Michael Borrus, a recent DOD program to develop very-high-speed integrated circuits (VHSICs) for military systems, touted as a project with excellent potential for commercial spinoffs, instead resulted in chips that had little commercial utility. They were custom designed rather than standardized and failed to maximize use of the chip surface, and contracts for their development were all awarded to large firms rather than smaller, more innovative companies. All of these factors were likely to retard commercial development.[59] Both the SCI and the SDI have made major investments in gallium arsenide (GaAs) microprocessor designs. The commercial value of GaAs chips is minimal because of their expense and also because silicon technology has not yet reached the limits of its potential efficiency. Yet GaAs is of great military interest because it is radiation-hard (highly resistant to disruption by environmental electromagnetic radiation). This makes it useful in satellite and orbiting weapons applications and also in systems designed to survive the electromagnetic pulse effects of nuclear weapons.[60]

Whether or not key new developments are commercially adaptable and profitable, military priorities continue to shape the goals, directions, and uses of computer technology in the United States. Yet the issue is not whether the computer technologies we get from a military-oriented economy are different from the ones we might otherwise want. Rather, the concern is with the social locations of power to make decisions about technology. For example, the SCI was widely read as the U.S. response to the Japanese government-sponsored "fifth generation" computer project begun in 1981.[61] Many of those scientists who accept money from it and from the SDI do so not because they believe in the widely disputed technical feasibility of the two projects' goals but because they are currently the easiest way to get money for certain areas of research. At a Silicon Valley Research Group conference on the SCI in 1985, David Mizell of the Office of Naval Research told the audience, "People are always asking whether these projects will 'work.' Well, as far as I'm concerned, they already work. We got the money."[62] The process of constructing knowledge about and with computers, in other words, is deeply enmeshed in funding decisions that are not only shaped by military prerogatives but also take place within institutional structures that preclude public debates.

Both computer science and military service are at present culturally coded as both male and "hard," in multiple and ambiguous senses. For computer science, this comes front the long history of ideological association of men with the mathematical mode of reasoning, as well as with technology itself. For military service, the association with men extends to their political role in the modern state, which relates the rights of citizenship to the responsibility of combat. The historical connection between computer science and military technology assists the work of gender construction by coupling one gender-defined domain to another. The SCI and the SDI—which, combined, have placed the military back at the helm of cutting-edge computer science research—thus have this additional ideological import. Strengthening the military's control of computer science, they reinforce one male-dominated arena with another, militarizing computer science while computerizing the military.

This matters to women working in high technology because, especially if the SDI achieves its planned size, a large percentage of computer science research jobs opening in the late 1980s and 1990s will be militarized. The gender-based character of military work may act as a subtle filter excluding women from upper-echelon research work. Even if it does not, women's work with computers will increasingly intertwine them with the future of high technology war.

There are profound gaps between the ideological coding of the work of computer science and military service and the actual social and historical situations of women and men who do that work. In fact, just as with most dichotomies involved in gender construction, what we think of as gender-based cognitive styles are nonexclusive elements of a larger creative process. There is nothing inherently masculine about computer technology. Otherwise women could not have had such quick success in joining the computer work force. Gender values largely float free of the machines themselves and are expressed and enforced by power relationships between men and women. Computers do not simply embody masculinity; they are culturally constructed as masculine mental objects. Similarly, military institutions may be more flexible in gender coding than they seem. They have always depended on and included women. With an 11 percent female All-Volunteer Force (AVF), women's participation is increasingly permanent and official. It is the ideology of combat, rather than genuine gender differences, that dissociates women from organized violence.

The political task of shaping a critical social environment around computer science and militarism thus becomes at once more difficult and more potentially rewarding. In a society built around high technology, women cannot simply reject computers. It is true that this technology—like many others—will require women to learn to think in the styles the culture has constructed as male, as well as to apply other styles to create unprecedented new methods. Still, this may be but a remnant of a dichotomy about to be abandoned. The greater challenge is to reconstruct the gender codes that surround that thinking without reinscribing biological arguments. As Donna Haraway has put it, "we must insist that high technology is for, among other things, the liberation of all women, and therefore usable by women for their self-defined purposes. . . . Feminists must find ways to analyze and design technologies that effect the lives we all want without major dominations of race, sex, and class. Those goals will sometimes lead to insisting on small, decentralized, personally scaled technologies. Such technologies are not synonymous with soft, female, and easy."[63]

If the military were, over the coming years, to acquire the deep connection to women's experiences that it has always had to men's, changes in women's ideological location in the cultural coding would be possible in several directions. Perhaps ending the association of masculinity and the military would create a "defender" mentality,[64] less bellicose precisely because gender identity would be decoupled from organized violence, with responsibilities more evenly shared. It seems to me most sensible to take a skeptical view of arguments that the mere presence of women in the military, no matter what their numbers, will render it less powerful or aggressive. The major source of the present drive to integrate the services is not egalitarian idealism but the developing manpower shortage caused by declines in the number of eligible eighteen-year-old males. Equal opportunity provisions at present are less the harbingers of a more humane military than of a pragmatic search for manpower solutions. The process of militarization has been, if anything, speeded rather than slowed by the AVF.

It is also necessary to avoid a disingenuous stance, like the position Judith Stiehm terms "pseudopacifism": "A truly pacifist position commands respect. It does not flinch from reality; it challenges and risks; it accepts suffering too. In contrast, pseudopacifists choose safety; they refuse to hurt others but they permit others to hurt for them. Women accept, even expect, the police and military to provide protection. They only superficially accept Camus' charge to be neither victims nor executioners. . . . Pacifism is not the issue for women, as most are satisfied to be protectees."[65] Uncomfortable as the balancing of rights with risk and responsibility may be, any serious feminist position on military service must confront it. In a heavily armed world, an unthoughtful, inarticulate opposition to war and violence is not merely useless — it is dangerous.

These are serious political questions. Answers to them must emerge from a full awareness of the background of ongoing historical processes against which gender coding takes place. The ideological collusion of mastery, combat, manhood, and political rights is beginning to come under great pressure, since the actual social structures these concepts grew out of have shifted from underneath them. The significance of women's increasing agency in those shifts depends on how it is read into our cultural codes and on who controls that reading. Openings for change have appeared, but the direction of change now depends on the degree to which alternative gender identities and political strategies are available. Creating these alternatives will be one of the challenging—and exciting—responsibilities of feminist politics in the last years of this century.

NOTES

1. Statistics for 1984 from Jo Schuchat Sanders and Antonia Stone, *The Neuter Computer* (New York, 1986), p. 8. Since 1970 the number of women receiving bachelor's degrees in computer-related fields has increased sharply. In engineering, women accounted for only 0.7 percent of all engineering BAs awarded in 1970, but took 10.1 percent of those granted in 1980. In computer science, women's share of BAs went from 13.6 percent as recently as 1976 to 30.2 percent in 1980. Analysts read into this trend "clear evidence of strong convergence of men's and women's educational choices and decisions in all science fields except mathematics. . . . [A] similarly converging pattern is developing in engineering and computer sciences." National Research Council, *Climbing the Ladder* (Washington, D.C., 1983), 1.13. During the same period women achieved approximate parity with men in terms of the total number of BA degrees awarded by U.S. universities (from 43.2 percent of all BAs in 1970 to 49 percent in 1980). According to the most recent U.S. Census, while women made up 43 percent of the total 1980 labor force, 5 percent of high tech engineers were women—up from 2 percent in 1970. The majority of women involved in the computer industry are in the lower levels of computer work. In 1980, 72 percent of the assemblers of computer equipment and 92 percent of the data entry operators were women, percentages that have changed very little since 1970. As the operation of mainframe computers—at one time a skilled technical craft—became more routinized during the 1970s, the percentage of women computer operators jumped from 29 percent in 1970 to 65 percent in 1984, Sanders and Stone, *The Neuter Computer*, p. 8.

2. My conclusions in this essay are intended to apply only to the United States, where high technology and military power have been profoundly linked since World War II. The lessons of the radically different Japanese experience may point to other possibilities, as yet unexplored, for the articulation of gender identity around computing machines.

3. Sherry Turkle, *The Second Self* (New York, 1984), pp. 104–5.

4. *The Random House Dictionary of the English Language*, ed. Jess Stein (New York, 1973), pp. 645, 1,352.

5. On the professional practices of science, see Bruno Latour and Steve Woolgar, *Laboratory Life: The Social Construction of Scientific Facts* (London, 1979); Evelyn Fox Keller, *Reflections on Gender and Science* (New Haven, Conn., 1985); and Violet B. Haas and Carolyn Perrucci, eds., *Women in Scientific and Engineering Professions* (Ann Arbor, Mich., 1984).

6. See, e.g., Nuel Pharr Davis, *Lawrence and Oppenheimer* (New York, 1986), an account of the rise and fall from political power of physicists lionized in the aftermath of the war.

Paul Boyer's excellent book *By the Bomb's Early Light* (New York, 1985) has a long section on the fickle fortunes of postwar politics for the atomic physicists (pp. 47–106).

7. See, e.g., Geoffrey K. Pullum, "National Language Interfaces and Strategic Computing," Silicon Valley Research Group Working Paper no. 5 (University of California, Santa Cruz, January 1986), also in Paul N. Edwards and Richard Gordon, eds., *Strategic Computing: Defense Research and High Technology* (University of California, Santa Cruz, 1986, typescript).

8. Compare Hubert Dreyfus, *What Computers Can't Do* (New York, 1979), and Terry Winograd, *Language as a Cognitive Process* (Reading, Mass., 1983).

9. On this point, see the works of Ludwig Wittgenstein. The viewpoint that informs this paper corresponds roughly to David Bloor, *Wittgenstein: A Social Theory of Knowledge* (New York, 1983).

10. Jonathan Jacky, "Software Engineers and Hackers: Programming and Military Computing," in Edwards and Gordon, *Strategic Computing* (n. 7 above).

11. Very new methods of "computing," incorporating holograms as information storage and using laser beams as the processing medium, would operate on completely different principles. But these techniques are still in their earliest infancy.

12. Turkle, *Second Self,* esp. 64–92, 196–238 (n. 3 above).

13. Douglas Hofstadter, *Gödel, Escher, Bach: An Eternal Golden Braid* (New York, 1979).

14. For various articulations of the gender coding of modes of rationality, see Turkle, *Second Self* (n. 3 above); Sally Hacker, "The Culture of Engineering: Woman, Workplace, and Machine," *Women's Studies International Quarterly* 4, no. 3 (1981): 341–53; Sandra Harding, *The Science Question in Feminism* (Ithaca, N.Y., 1986); and Brian Easlea, *Fathering the Unthinkable: Masculinity, Scientists, and the Nuclear Arms Race* (London, 1983).

15. On gender and ethical systems, see Carol Gilligan, *In a Different Voice: Psychological Theory and Women's Development* (Cambridge, Mass., 1982).

16. Terry Winograd, "Computers and Rationality: The Myths and Realities," in *Microelectronics in Transition,* ed. Richard Gordon (University of California, Santa Cruz, 1985, typescript), p. 3.

17. Compare Nancy Hartsock, *Money, Sex, and Power: Towards a Feminist Historical Materialism* (New York, 1983); Keller, *Reflections on Gender* (n. 5 above); Harding, *Science Question* (n. 14 above).

18. Turkle, *Second Self,* pp. 13–14 (n. 3 above).

19. Compare Terry Winograd, *Understanding Natural Language* (New York, 1972). For a critique of the microworlds approach, see Dreyfus, *What Computers Can't Do* (n. 8 above).

20. Compare Roger Schank, *The Cognitive Computer* (Reading, Mass., 1984); Joseph Weizenbaum, *Computer Power and Human Reason* (San Francisco, 1976); George R. Kleindorfer and James E. Martin, "The Iron Cage, Single Vision, and Newton's Sleep," in *Technology and Responsibility,* ed. Paul T. Durbin (Boston, 1987); Dreyfus, *What Computers Can't Do* (n. 8 above); Winograd, "Computers and Rationality"; Turkle, *Second Self* (n. 3 above).

21. Compare Turkle, *Second Self,* esp. pp. 196–238.

22. Hacker, "Culture of Engineering," p. 346 (n. 14 above).

23. For a nuanced understanding of the place of isolation in male psychological iden-

tity, see the collective work of the Re-evaluation Counseling Communities, in particular the journal *Men,* nos. 1–4 (1979–84).

24. Compare Turkle, *Second Self.*

25. I owe this distinction to Jonathan Jacky.

26. See Steven Levy, *Hackers* (New York, 1984), pp. 244–61, 203, and passim.

27. Cynthia Enloe, *Does Khaki Become You? The Militarization of Women's Lives* (London, 1983), pp. 9–10.

28. For detailed discussions of this point, see Paul N. Edwards, "Artificial Intelligence and High Technology War: The Perspective of the Formal Machine," Silicon Valley Research Group Working Paper no. 6 (University of California, Santa Cruz, October 1986), and "Border Wars: The Science, Technology, and Politics of Artificial Intelligence," *Radical America* 19, no. 6 (1986): 39–50.

29. Though as a matter of policy the Pentagon will neither confirm nor deny reports about its nuclear strategy, some military watchers believe our nuclear systems have been operating under some form of launch-on-warning scheme for some time. See, e.g., Robert Aldridge, *First Strike! The Pentagon's Strategy for Nuclear War* (Boston, 1983), pp. 245–46.

30. Compare Paul Virilio and Sylvere Lotringer, *Pure War,* trans. Mark Polizzotti (New York, 1984).

31. Compare Kenneth Flamm, *Targeting the Computer* (Washington, D.C., 1987) and *Creating the Computer* (Washington, D.C., 1988).

32. This list is taken from Edwards, "Artificial Intelligence and High Technology War" (n. 28 above).

33. William Broyles Jr., "Why Men Love War," *Esquire* (November 1984): 58.

34. Ibid.

35. On the influence of systems thinking through the Rand Corporation, see Fred Kaplan, *The Wizards of Armageddon* (New York, 1983). A longer discussion of this history may be found in Paul N. Edwards, "The Closed World: Computers and the Politics of Discourse" (Ph.D. diss., University of California, Santa Cruz, 1988). For a critique of management models in the military, see Richard A. Gabriel and Paul L. Savage, *Crisis in Command* (New York, 1978). Westmoreland's "electronic battlefield" speech is reprinted as an appendix in Paul Dickson, *The Electronic Battlefield* (Bloomington, Ind., 1976), pp. 215–23.

36. One example was the vast network of advanced computers and software built for the SAGE (Semi-Automated Ground Environment) continental air defense system in the 1950s. See Paul N. Edwards, "A History of Computers and Weapons Systems," in *Computers in Battle,* eds. David Bellin and Gary Chapman (New York, 1987), pp. 45–60.

37. Kaplan, *Wizards of Armageddon* (n. 35 above), and Gregg Herken, *Counsels of War* (New York, 1987), are rather similar, very thorough accounts of these events.

38. On this debate see, e.g., Edward Luttwak, *The Pentagon and the Art of War* (New York, 1984), and James Fallows, *National Defense* (New York, 1981).

39. On the coding of approaches to war and their relationships to computer science, see Chris Hables Gray, "Artificial Intelligence and Real Postmodern War" (History of Consciousness Program, University of California, Santa Cruz, 1988, typescript).

40. The WAC was a separate-but-equal women's element of the Army, established in

1948. At its high point it accounted for at most 1 or 2 percent of total army personnel. For a brief history, see Martha Marsden, "The Continuing Debate: Women Soldiers in the U.S. Army," in *Life in the Rank and File,* eds. David Segal and Wallace Sinaiko (New York, 1986), pp. 60–65. For an interview-based discussion, see Helen Rogan, *Mixed Company* (Boston, 1981), pp. 129–65 and passim.

41. For useful reviews of the current status and problems of women in the armed services, see Marsden, "Continuing Debate" (n. 40 above), and Rogan, *Mixed Company* (n. 20 above); 1989 figures are from Joannie M. Schrof. "Are Women the Weak Link in the Military?" *U.S. News* 107, no. 20 (1989): 61.

42. See Enloe, *Does Khaki Become You* (n. 27 above); W. Chapkis, ed., *Loaded Questions: Women in the Military* (Washington, D.C., 1981); Judith Stiehm, ed., *Women and Men's Wars* (New York, 1983); and Rogan, *Mixed Company* (n. 20 above).

43. Enloe, *Does Khaki Become You* (n. 27 above), pp. 122–23.

44. Brian Mitchell, quoted in "Fire When Ready, Ma'am," *Time* 135, no. 2 (1990): 29.

45. See Marsden, "The Continuing Debate" (n. 40 above); Rogan, *Mixed Company* (n. 20 above); Brian Mitchell, *Weak Link: The Feminization of the American Military* (Washington, D.C., 1989) presents an extremely negative view of women in the armed forces.

46. Judith Stiehm, "The Protected, the Protector, the Defender," in Stiehm, *Women and Men's Wars,* pp. 367–76.

47. Micaela di Leonardo, "Morals, Mothers, and Militarism: Antimilitarism and Feminist Theory," *Feminist Studies* 11, no. 3 (1985): 599–617, esp. 602.

48. Ibid., 615.

49. Compare Cynthia Enloe, "Women—the Reserve Army of Army Labor," *Review of Radical Political Economics* 12, no. 2 (1980): 42–51; Paul N. Edwards, "Men and Armies: An Essay on the Politics of Manhood" (University of California, Santa Cruz, 1985, typescript).

50. One of the most insightful studies of this phenomenon is Maury D. Feld, "Military Discipline as a Social Force," in his collection *The Structure of Violence: Armed Forces as Social Systems* (Beverly Hills, 1977), pp. 13–30.

51. See Edwards, "Men and Armies" (n. 49 above). Also see Hegel's comments on patriotism in *The Philosophy of Right,* trans. T. M. Knox (New York, 1952), pp. 163–64.

52. The Court sidestepped this issue, claiming that the power to decide that question lay with the Congress, which for obvious reasons allowed the traditional policy of a male-only draft to stand.

53. The three original applications projects proposed by the plan include, first, an automated vehicle able to navigate unfamiliar terrain at high speeds by intelligent interpretation of sensor data. The vehicle might be used for unmanned reconnaissance or, suitably armed, as a kind of robot tank. The second application, an intelligent fighter pilot's assistant, would serve as a sort of automated copilot, keeping track of many functions the pilot must now monitor and able to converse with the pilot in unrestricted spoken English. A third application is a so-called battle management expert system for aircraft carrier battle groups. This device would analyze sensor data from the carrier's huge field of engagement, interpret threats, keep track of friendly forces, and plan response strategies which it would then "suggest" to the group commanders. The speed and complexity of high tech war has brought

these tasks to an asymptote of impossibility for human commanders: the hope is to provide "intelligent" assistance at electronic speeds. Potentially, such a system might control
weapons systems directly. The SCI includes plans to develop prototypes of all three systems
on quite specific timetables as well as to create and refine a base of relevant hardware and
software technologies. Defense Advanced Research Projects Agency, *Strategic Computing—
New Generation Computing Technology: A Strategic Plan for Its Development and Application to Critical Problems in Defense* (Washington, D.C., 1983). Also see the first and second
"Annual Reports," published by the agency in 1986 and 1987, where the program was expanded to cover two additional applications areas. Budget cutbacks have reduced actual
expenditures to about $500 million. The program endured severe criticism from Computer
Professionals for Social Responsibility (CPSR), an organization of high-visibility computer
professionals, beginning with an open letter by Severo Ornstein, B. C. Smith, and Lucy
Suchman in *Bulletin of the Atomic Scientists* 40, no. 12 (1984): 11–15. The organization's out-
spoken opposition to both SCI and SDI as misguided military projects and poor funding
sources for computer science eventually culminated in Bellin and Chapman, *Computers
in Battle* (n. 36 above). On DARPA's history, see Flamm, *Targeting the Computer* and *Creating the Computer* (n. 31 above); Edwards, "A History of Computers and Weapons Systems"
(n. 36 above) and "The Closed World" (n. 35 above); Richard J. Barber Associates, Inc.,
The Advanced Research Projects Agency, 1958–1974 (Washington, D.C., 1975); Don Nielson,
"DARPA's Strategic Computing Initiative: An Overview," in Edwards and Gordon, *Strategic Computing* (n. 7 above); and Jonathan Jacky, "The Strategic Computing Program," in
Bellin and Chapman, *Computers in Battle.*

   54. This back-of-the-envelope estimate emerged in a discussion among computer professionals, in which I participated at the 1987 summer conference of Computer Professionals
for Social Responsibility. The roughly estimated total cost of a full-scale SDI "umbrella defense" (the Reagan administration's original plan, since cut back to the more modest goal
of ICBM defense) is $1 trillion, of which about 10 percent would be required for computers,
programming, and other information-processing needs ($100 billion). Typically, basic re-
search costs for advanced technology projects are approximately 10 percent of the total
budget; thus the $10 billion estimate. The Bush administration has scaled back SDI research
by approximately half, with further cuts likely, so expenditures of this scale now seem less
probable. While hesitant to predict costs with any certainty, the U.S. Office of Technology
Assessment report *Strategic Defenses* (Princeton, N.J., 1986) agrees in rough order of mag-
nitude with this estimate. Commenting on the costs of the High Frontier proposed ballistic
missile defense satellite system, the report predicts "$200 to $300 billion in acquisition costs
alone for the proposed system" (294). Since research and development phases usually add at
least 50 percent to the cost of high technology weapons systems, and since the High Frontier system is among the least costly of those proposed for the SDI, the $1 trillion figure for
total costs seems roughly accurate.

   55. See Eric Roberts and Steve Berlin, "Computers and the Strategic Defense Initiative,"
in Bellin and Chapman, *Computers in Battle* (n. 36 above), pp. 149–69, esp. 158–60; and
Alan Borning, "Computer System Reliability and Nuclear War," in ibid., pp. 101–47.

   56. Flamm, *Targeting the Computer* (n. 31 above), pp. 77, 91.

57. See Herman Goldstine, *The Computer from Pascal to Von Neumann* (Princeton, N.J., 1972); Flamm, *Creating the Computer* (n. 31 above).

58. Flamm, *Targeting the Computer*, 76 (n. 31 above).

59. Leslie Brueckner and Michael Borrus, "The Commercial Impacts of Military Spending: The VHSIC Case," in Edwards and Gordon, *Strategic Computing* (n. 7 above).

60. Edwards, "Border Wars" (n. 28 above), p. 45.

61. See the special issue of the electronics engineers' trade journal, *IEEE Spectrum* 20, no. 11 (1984), for a selection of articles on the parallels between the Japanese and American initiatives.

62. David Mizell during his presentation at the conference on "Strategic Computing: History, Politics, Epistemology," of the Silicon Valley Research Group, University of California, Santa Cruz, March 12–14, 1985. I organized the conference and recorded his words verbatim.

63. Donna J. Haraway, "Class, Race, Sex, Scientific Objects of Knowledge: A Socialist-Feminist Perspective on the Social Construction of Productive Nature and Some Political Consequences," in Haas and Perrucci, *Women in Scientific and Engineering Professions* (n. 5 above), p. 227.

64. Stiehm, "The Protected, the Protector, the Defender" (n. 46 above), p. 369.

65. Judith Stiehm, "Women, Men, and Military Service," in *Women, Power, and Policy*, ed. Ellen Boneparth (New York, 1982), pp. 289–90.

PART III

Industrial Junctions:
Gendering Industrial Technologies

*What makes certain organizations, technologies, or work
"industrial"?*
*How have gender ideologies been embedded in the particular
character of American industrial capitalism?*

Industrial capitalism as a system entwines increased production
and consumer culture. Its development, by definition, brought change to many
realms, as divisions of labor, centralized organizations of work, and the ubiquity
of cash wages replaced a household-based agrarian and artisanal society. The process of industrialization transformed not only mills and factories but also households and the gender relations inside them. By the mid-nineteenth century, urban
Americans could expect some family members to go to work somewhere other
than their own household, farm, or workshop, and after work to return home.
The split between home and work is an industrial opposition — a boundary constructed in a period of profound societal transformation.

Inside the new centralized workplaces, as described in several of the following
essays, division of labor increased output and decreased each worker's involvement in the whole of the production process. Once jobs were broken down into
elemental tasks, it was easier to make machines capable of doing the same things,
and mechanization often increased production even more. Workers contended
with new machines, not infrequently resisting management's efforts to set the pace
of work. Sometimes mechanization provided an excuse to hire an entirely new and
much cheaper workforce, offsetting a capital investment in machines with reduced
labor costs and leaving displaced workers to find other jobs.

Meanwhile, unlike the places called "work," centralization, division of labor,
and wage work were not consistently features of the industrial household. Instead, industrial households each depended on someone to do the varied and un-

paid work of gathering necessities in a cash economy, managing budgets, learning where to buy day-old bread for half price or how to dry the soap to make it last longer—in addition to more traditional tasks like processing foodstuffs into dinner (cooking) or the repair and maintenance of clothing (mending and laundry).

The boundaries between home and work, production and consumption, break down quickly when analyzed with attention to gender and technology. Households engage in both production and consumption; women leave home and go to work; changing technologies and social change were features of both sides of these boundaries in the process of industrialization. The articles in this section focus on paid work outside the home, exploring how changing technologies and changing gender systems intersected in the jobs people did and the technological knowledge they relied on. But in each case production is closely linked to the demands of the consumer market, what historian Ruth Schwartz Cowan has called the "consumption junction."

Cowan's 1987 essay "The Consumption Junction" sought to challenge assumptions among her production-oriented colleagues in the history and sociology of technology. Here we extend and relabel her challenge under the rubric "industrial junctions," to invite interrogation of the constructed and intersecting boundaries of home and work, production and consumption in industrial capitalist society. Is dressmaking production? Is meatpacking? Making soup? Does the military count as a consumer when the product is the calculation of missile trajectories? The categories "production" and "consumption" have been represented as male and female, but as in Oldenziel's case of selling cars, cigars and ballistics were used by men. And dressmakers, among other producers in this story, were women.

Examining these industrial junctions—where production and consumption meet, where home and work overlap—allows us to pursue the questions of Parts I and II, to continue exploring the relationships between gender and technology in this period. How have gender ideologies shaped technological change and industrial goals? How does gender intersect with class in factory settings? Professional ones? Do all actors share expectations about gender categories? About technologies? Do we find invisible technologies and invisible workers in these stories? How do you explain the cultural linkage of masculinity and machines, the use of the word *technology* to portray a predominantly male domain?

EIGHT

Cigarmaking

PATRICIA COOPER

Pat Cooper returns us to the turn of the twentieth century, a time of continued mechanization of craft work and the successful development of "mass production" (as it would soon come to be called) and mass consumption. The commodity in question in this study is the cigar, once made by hand and increasingly made by machine. Cigarmaking relied on male and female workers, who performed different tasks or worked on different types of cigars. Comparing these groups, Cooper explores intersections of skill, pay, unionization, and gender before and after a major reorganization of the industry. How many ways of being manly are discussed here, and what were the characteristics of those masculinities? How are they similar to or different from the masculinities among Oldenziel's model builders or Mohun's laundrymen?

Did women workers value similar traits in each other? What gender traits did managers look for in workers, and how was gender related to the technologies involved in the various jobs Cooper describes? On what gender representations did advertisers rely? Did gender expectations change as the technology changed?

"'What this Country Needs is a Good Five-Cent Cigar,'" *Technology and Culture* 29, no. 4 (1988): 779–807. Reprinted by permission of the Society for the History of Technology.

While the words of the title are a familiar quotation from Thomas Marshall, Woodrow Wilson's vice president, few know their context or significance. Marshall supposedly made the remark to a colleague while presiding over the Senate one day during a rather long and boring speech on the needs of the country. Some think it was his most memorable act in office. In any case, what was a good five-cent cigar and what had happened to make the country, much less Marshall, long for its return?

The cigar began its reign as the country's most popular tobacco product in the 1880s, supplanting smoking tobacco and snuff. Connoisseurs preferred the fine cigars of Cuba, while others with discriminating tastes selected American cigars made from imported tobaccos. By the early twentieth century, however, manufacturers of the mean and cheap nickel cigar, which had always been made from low-grade domestic tobacco, began producing a very good smoke for five cents. Its popularity grew markedly, and more expensive brands soon felt the competition. But during the First World War, with the sharp rise in tobacco prices and taxes, the five-cent cigar disappeared completely. The cheapest smokes rose to seven and eight cents each, although few smokers thought they were worth that price. It was just too costly to produce a good, cheap cigar. Even when the war ended, manufacturers assured unhappy consumers that the nickel cigar was gone forever.[1]

It was here, of course, that Marshall's remark became enshrined in folklore, but the story of the five-cent cigar is more than an obscure anecdote. At its core it is a case study of technology and the sexual division of labor, the fate of a craft union in the machine age, managerial prerogative, and the decline of an entire industry in the twentieth century.

One of the central problems of the wage labor system is discipline and command: securing workers' diligence in doing the work they are hired to do. Beginning with Harry Braverman's *Labor and Monopoly Capital*, many scholars have documented the ways in which capitalists have attempted to augment their control over employees and maximize their profits. This body of work is sometimes referred to as the "labor process" approach. A division of labor fragmented work and permitted employers to hire less skilled, lower-waged workers. Scientific management went one step further, deskilling work and separating mental from manual tasks, thereby depriving workers of any exercise of judgment or authority on the job. Employers have used technology in order to assert (or reassert) authority over the production process and to defeat workers' efforts to control and shape the work they do. Richard Edwards's *Contested Terrain* illustrates the ways in which employers have used different forms of control (simple, technical, and bureaucratic) to achieve these ends. Philip Kraft, Andrew Zimbalist, and others who have studied the labor process, along with David Noble, who has examined the social sources of machine-tool technology, have pointed to the ways in which technology itself is socially constructed and part of employers' arsenals.[2] At the same time, several historians, such as David Montgomery, have examined the ways in which workers attempted to control their work and resist managerial prerogative. Most of these studies have focused on male workers and male occupations.[3]

Another group of historians has documented the lives and work of American women. They have challenged older notions of women workers' passivity and isolation and have examined women's labor militancy, women's distinctive culture, family and community networks, the relationship between women and trade unions, and the ways in which men's and women's experiences have been different.[4] In the continuing effort to recast older models, Carole Turbin and others have recently called for paying more attention to the variability in working-class women's lives: not all women's experiences are the same.[5]

Certainly one of the most enduring aspects of women's work has been occupational segregation, a sort of tracking system which placed men and women—and, I might add, blacks and whites—in different jobs.[6] Yet these jobs are different and unequal—both black and white women's jobs for the most part have lower wages, poorer conditions, and less status than those of white men. As Heidi Hartmann has pointed out, it is no accident that women are consistently assigned to these inferior jobs.[7] Focus on male workers and experience tends to marginalize the importance of the sexual division of labor and likewise women's place in the occupational structure. Scholars have only begun to analyze the process by which "gender labels" are assigned both to new occupations and older ones that have been redesigned.

As "labor-process" theorists have pointed out, the restructuring of occupations often coincides with the introduction of new technologies. Previous studies of technology and labor based on male models have focused on class relations and issues of control, but they have ignored another set of relationships central to these changes: gender. The term *gender* is often misused to mean women, but it really refers to the process (and consequences) whereby biological sex becomes transformed into a socially constructed (as opposed to "natural") definition of what it means to be a man or a woman. Mary Blewett, who has studied shoe workers, and Ava Baron, who is writing about printers, along with others like Judy McGaw, who has examined the Berkshire paper industry, have begun to look at the relationship among technology, work, and gender. They ask questions about the relationship between technology and the definition of skill, about the gender labeling of jobs created or affected by technology; they ask whether changes in the labor processes affect men and women differently.[8] Baron has suggested that studies of work must treat gender "even when the occupation being studied is and has been considered men's work," and she has analyzed the conceptualization of gender and work in the wake of the introduction of the linotype in printing, an occupation generally considered men's work.[9]

Labor-process studies, as Baron has argued, need to be reconceptualized: to understand the transformation and reorganization of work, the nature of "deskilling" and the role of technology in the workplace, it is necessary to treat gender as a central category of analysis. It is not simply a question of adding women to the discussion but of analyzing the ways in which gender — and therefore assumptions about and relations between men and women — has defined job skill and structure and has shaped the nature and use of technology.[10]

This is a broad agenda, and the following analysis of the transformation of work in the cigar industry during the first half of the twentieth century touches on only part of it. The cigar industry case does, however, provide an avenue for combining some of the insights of the labor-process and gender-consciousness approaches. Before 1919, cigarmaking itself was a fairly simple operation, basically a handcraft. Still, there was a degree of differentiation within the industry, so that men and women performed different tasks and were situated in different places in the labor process. The industry itself was segmented into different branches and highly decentralized and competitive. In 1919 the introduction of automatic cigar machinery and its adoption by the industry's largest manufacturers radically transformed work. Examining both class and gender relations permits exploration of the reasons for the introduction of technology, how it affected men and women workers, and the ways in which it shaped the sexual division of labor and the overall structure of the industry. This analysis also underscores the variability in women workers' experience. For, as we shall see, the new technology altered the lives of both men and women, but different groups of women were affected in different ways. At the same time, manufacturers did not all benefit equally or in the same way from automatic machinery. The restructuring of the industry after 1919 was not simply a consequence of technological innovation, but rather a complex interplay of forces in which human design and choice played an important role.

The cigar industry remained decentralized and very competitive throughout the late nineteenth and early twentieth centuries. Thousands of one-man shops operated alongside the growing number of larger factories that employed hundreds, even thousands, of workers. Markets and brand names were regional or local, and the 6.7 billion cigars produced in the United States in 1900 represented about 350,000 different trade names. As late as 1914, there was only one national brand, the Robert Burns. Signs of concentration were appearing, however. The American Tobacco Company's subsidiary, the American Cigar Company, had attempted to gain a monopoly in the market at the turn of the century but never succeeded in gaining more than 16 percent. Still, along with a handful of other firms including United Cigar Manufacturers, Otto Eisenlohr and Company, and the Deisel-Wemmer Company, it gained an increasingly dominant role in the industry during the early twentieth century. In 1912, the ten largest companies (out of a total of about 20,000) accounted for 22.6 percent of production, hardly a monopoly but nevertheless reflective of a trend toward concentration.[11]

Cigar makers and cigarmaking at the turn of the century were highly differentiated: different groups of workers made different types of cigars using different tobaccos. The Clear Havanas were the industry's elite smokes, its most expensive. Employed mostly in Florida, the Cuban, Spanish, and Italian craftsmen (and a growing number of women) who made these cigars worked entirely by hand. The small stogie industry produced a crude cheap cigar-like smoke and was centered in Wheeling, West Virginia, and Pittsburgh, Pennsylvania.[12]

Seed and Havanas and five-cent cigars, the subjects of this article, were the two largest categories of cigars in terms of production totals and employment. Jobs were sexually segregated. The Seed and Havanas were made from a blend of high-grade domestically grown leaf (from Havana seeds) and imported tobaccos (grown in the Vuelta Abajo region of Cuba), and sold for the moderate price of 10 or 15 cents. They were skillfully hand crafted by the members (most of whom were men) of the Cigar Makers' International Union of America (CMIU), Samuel Gompers's union, and one of the founding organizations of the American Federation of Labor (AFL). Seed and Havana factories were usually fairly small, employing up to about 100 workers, although a few employed several hundred.[13]

Five-cent cigars had initially been made from cheap, inferior domestically grown tobacco. By the turn of the century, however, nickel cigars were coming into their own because manufacturers could take advantage of improvements in domestic tobacco leaf and low labor costs that enabled them to buy some imported tobaccos as well. Quality improved markedly enough that five-cent cigars began competing with ten-cent brands. Nickel cigars were made at the turn of the century by men and women. But increasingly over the next two decades, firms manufacturing five-cent cigars — which usually employed several hundred to a thousand or more workers — hired women exclusively. Women, who had been only 33–36 percent of this labor force in 1900, had by 1919 come to outnumber men in the industry by 78,253 to 53,153. Manufacturers sought to cut production costs by hiring young, often immigrant, women at low wages to work under a division of labor sometimes using elementary machinery. The CMIU, a model of male craft Unionism, whose members had long concentrated in the medium-priced Seed and Havanas, increasingly found itself in retreat as the five-cent branch grew and prospered.[14]

The work of cigarmaking was still highly skilled at the turn of the century, but there were important distinctions in men's and women's work. Male unionists were concentrated in the higher-paying jobs making better grade cigars using no machinery or division of labor. The hand process began when the cigar maker first gathered a sufficient amount of the brittle filler leaves, broke them off to the right length, and shaped them in his left hand. Next he rolled them up inside the coarse binder leaf, taking care to give the cigar its special shape and contour. Finally he wound the thin, delicate wrapper leaf around the bunch and finished off the ends, closing one or both of them. It took years to master the technique. Each cigar had to duplicate the others in every respect, including shape and weight. About half the unionists used a wooden shaping mold to press the bunches. Each mold had a top that locked into place. A stack of five or more closed molds was placed in an iron screw press for about twenty minutes until the cigar bunches were all shaped uniformly by the pressure. Molds made bunch making quicker and easier, but they drew moisture from the cigars and created hard spots in the tobacco. The rest of the CMIU's members worked entirely by hand.[15]

Women worked under a division of labor known as the team system and remained outside the union, although the CMIU made periodic and unenthusiastic attempts to organize them. Teamwork meant that one woman made cigar bunches, using the wooden molds, and two others seated on either side of her rolled them in the wrappers. In some factories elementary machinery was used. Bunch machines were manually powered and operated much as a home cigarette roller does today. A cigar maker placed filled tobacco in a canvas trough and laid the binder leaf on the canvas apron above. She pulled a lever toward her that enclosed the binder around the filler. The suction table aided cutting the easily torn wrapper leaf to exact specifications with suction holding the wrapper leaf in place and a sharp cutter moving along a preset track around a raised die. Both hand and teamwork jobs required skill, although somewhat less was needed for teamwork, depending on the use of machinery and whether the teamworker could both bunch and roll.[16]

From a manufacturer's perspective, choices regarding employees related to the need to maximize profits and control the production process. The CMIU was of course to be avoided at all costs. Traditions of craftsmen and male culture clashed with manufacturers' needs and undermined the uneven power relationship between employer and employee embedded in the wage system. Cigar makers' code of behavior at work stressed certain values, shared by other nineteenth- and early-twentieth-century craftsmen, of "dignity, respectability, defiant egalitarianism and patriarchal male supremacy."[17] These they asserted through a maze of shop-floor traditions and customs, some of which were codified in their wage agreements with manufacturers, the Bill of Prices. In it they spelled out the conditions under which they would work. They enforced formal and informal rules within the group: breaking the code of conduct could result in swift retribution. Because of their high piece rates (most earned between $15 and $30 a week), they were able to control their own time and come and go as they pleased throughout the day. Union members also maintained a tradition of geographic mobility whereby new members might go on the road to test their skills and old-timers might travel just to take a break and change the scenery. This helped them compare conditions in different locations and created a national communication network supplementing the news in their monthly journal, the *Cigar Makers' Official Journal (CMOJ)*.[18]

Union cigar makers had clear notions about their class interests vis-à-vis manufacturers, and they supported union brothers in other trades with donations and other assistance. A cigar maker could not cower before or defer to an employer or foreman—anyone who did would be branded a "sucker" and ostracized from the work group. There were numerous conditions under which they would not work at all—bad stock, "obnoxious" shop rules, or any other perceived threat or insult. Resistance could take the form of individual quitting or a spontaneous walkout, for which the cigar makers were notorious. If the cigar makers got angry, noted one observer, "they went home." They also demanded and took three free cigars a day, and any effort to deprive them of this right resulted in serious and unpleasant consequences.[19] While union cigar makers never attempted to restrict output, as many other skilled craft workers of the day did, they did set informal limits on their use of tobacco. Each cigar maker was given allotments in preweighed pads and was expected to make a certain quota of cigars from each pad. Cigar makers sought ways of controlling the weighing including setting their own informal quota—only so many cigars from a pad and no more. In many factories cigar makers fought to increase the allotments of tobacco or to pool all unused tobacco so that an individual accounting could not be made.[20]

By the early twentieth century, the makers of five-cent cigars hired not just nonunion workers but deliberately chose women almost exclusively. They did so primarily because of assumptions about the characteristics of women. They candidly explained that women "don't drink" and were "better adapted and cheaper," "more reliable," "more careful," and "more easily controlled." One manufacturer moved his formerly unionized plant from Kalamazoo, Michigan, to Detroit and hired young Polish women to work in his factory because, he explained, they were "easy to handle" and "orderly." Too, women did not smoke and therefore did not require free cigars. Employers believed that hiring women and restructuring the work process would solve their labor discipline problems and multiply their profits.[21]

Many women worked using the available, although elementary, technology such as the bunching machine or the suction table. These devices did not appreciably increase efficiency, according to a Bureau of Labor investigation, but rather provided "the opportunity of employing girls at low wages." Thus employers used technology to avoid hiring male unionists, whom they perceived to be more troublesome, and to reduce costs and gain more control over the production process. Certainly the division of labor and the use of machinery speeded up and simplified training and permitted the paying of lower wages, but women workers were skilled. It was not skill alone which dictated their lower wages. Rather, it was a basic cornerstone of the sexual division of labor: women merited lower pay than men. Paid by the piece, women averaged anywhere from $5 to $15 a week. The large factories producing these nickel smokes could be found throughout the East and Midwest but were concentrated in New Jersey, Pennsylvania, Ohio, New York, Illinois, and Michigan.[22]

Women workers indeed proved desirable in some respects. For example, they failed to join labor unions (as much a factor of male unionists' exclusionary attitudes as their own reservations). Ultimately, however, they proved to be no real answer to manufacturers' perceived problems. They created patterns of work that often contrasted their interests and those of employers. Women's outlook and actions were based on their notions of what being female meant. Dignity for women looked different than for male unionists, but their strong sense of it fueled opposition to managers' efforts. Low piece rates dictated long hours of steady, diligent work, but many workers resisted this feverish pace and set informal limits on output. A female cigar maker in Lima, Ohio, for example, set her goal at 800 a day. That way she "didn't have to be a nervous wreck." They also used the piecework system to adjust their own hours at work, even when doing so meant thinner pay envelopes. Manufacturers routinely complained about workers' schedules, but, as one noted, "the irregularity in the hours is the workers' own choice. We cannot make them work longer than they want to." Labor shortages in some areas, particularly Detroit, gave women leverage and more freedom to come and go; and the team system tended to promote cooperation and a sense of interdependence. Together these added up to solidarity and a kind of subversive outlook.[23]

Forms of resistance varied. Women cigar makers developed ways to handle the problem of preweighed tobacco. When "the company wasn't looking," they quietly shared tobacco with each other. "You didn't say nothing about that," one cigar maker cautioned. Women also set informal limits on the group as to the number of cigars they would make from a certain weight of tobacco and they also quietly removed cigars to take to fathers, husbands, and boyfriends. In addition to these shop-floor patterns, women also proved able and enthusiastic strikers. By 1910, strikes had begun to multiply in the five-cent cigar branch. A wave of militant confrontations began in 1916 in Detroit and only subsided in 1922 after a strike in Lima, Ohio.

During the war three-quarters of the industry's workers went on strike. Cigar makers, union and nonunion, joined together to demand a 50 percent wage increase, an eight-hour day (which unionists already had), and a shop grievance committee. Women had hardly proved to be the docile, tractable labor force on which manufacturers had been counting.[24]

The wartime strike movement, when thousands of cigar makers left their benches, made earlier confrontations look pale by comparison. Relations between

CMIU men and women making five-cent cigars had never been warm. The union had often tried to organize women, but their efforts had been halfhearted at best. Manufacturers had laughed at the notion of "one big union" in the industry because workers were so hopelessly fragmented. Now, however, cigar makers acted outside the union's formal structure. Leaving behind craft union forms, these workers—union and nonunion, men and women—for the first time created an alliance based on a common set of goals and plan of action. There are hints in the record that gender divisions remained, but overall the strike movement was quite remarkable.[25]

Workers' combined actions negated the industry's traditional labor policy. Most of the larger firms had opened branch factories in several different locations so that workers would find it "practically impossible" to tie up the production of an entire company during a strike. But wartime strikes spread by branches and tied up all operations of Consolidated Cigar Company (formed from the merger of several other firms in 1918), General Cigar Company (formerly United Cigar Manufacturers), Otto Eisenlohr, and others. Following on the heels of the great Detroit strike of 1916, the strike movement raised questions about the docility of women workers, who were striking as much as or more than men, and it explicitly made issues of workers' authority, rather than wages and conditions, the focus of struggle between employers and employees.[26]

In the pages of the industry trade journals, manufacturers debated their course of action and decried the irritating habits of their employees. "Rarely, if ever, do we hear of labor trouble in the cigarette business," one article explained. "Eventually machinery will solve the cigar making problem just as it solved the cigarette-making problem." The prediction became a warning as New York manufacturer John W. Merriam threatened: "If hand labor becomes too unreasonable and costly . . . the manufacturer may see what he can do with machinery."[27]

It was no idle threat. At the peak of the strike movement during the summer of 1919, Waitt and Bond Company, a leading union manufacturer located in Boston, announced that it would close its Boston plant and open two new factories in Newark, New Jersey. There, machines built by the American Machine and Foundry Company (AMF) turned out 480 cigars an hour. Employees were neither union men nor women teamworkers with years of experience; rather, they were young women machine operators with no previous ties to the industry. The strike movement collapsed as favorable wartime conditions, such as an acute shortage of labor, disappeared and morale flagged in the face of the cigar machine. The war had signaled the disappearance of the five-cent cigar, a militant strike wave, and the introduction of new, revolutionary technology. These would transform the industry and alter the nature of labor relations forever.[28]

Yet the machine was not first on manufacturers' minds as the new decade opened. It was still new, untested, and expensive. More immediate were other efforts to reverse labor's wartime gains. An open shop campaign and changed economic conditions put workers "in a more tractable mood." Firms rolled back wage increases won during the war: in New York State wages fell 20 percent in 1920 alone. The U.S. Bureau of Labor Statistics reported that 75 percent of U.S. cigar makers received wage cuts of at least 10 percent during 1921.[29]

Manufacturers were in a position to demand wage cuts not only because of the war's end and the psychological power of their defeat of wartime strikes but also because the industry had entered a period of sharp decline. By late 1920, production levels had fallen visibly as cigar consumption for the first time in its history began dropping. Eight billion cigars were produced in the United States in 1920, but in 1921 production reached only 6.7 billion. Dealers reported to manufacturers that consumers complained of high prices and warned that this was beginning to severely affect cigar sales.[30]

At first industry observers reassured each other that the downturn would be temporary. Over and over the trade journals announced that the industry had "turned the corner," but optimism flagged when the usual Christmas rush in 1921 failed to materialize. The decline continued, and the journals now blamed the "fatalistic attitude" in the trade for the industry's woes. "Confidence will win," they prodded. Soon, however, the cheery homilies disappeared and the industry settled into what became a never-ending decline. By 1933, cigar production had fallen to 4 billion.[31]

While appearances suggested that the decline had been sudden, actually there had been signs for some time that the industry's health was doubtful. Cigarette production had first matched cigar output in 1910. Three years later the ratio was two to one. With the war, smokers' switch to cigarettes became pronounced. Although a woman had been arrested in 1905 on Fifth Avenue for smoking a cigarette in public, now ads directly encouraged women to smoke. The age of the cigar was over, and the flapper was in.[32]

Some manufacturers proposed a simple solution to the problem of plummeting cigar sales — "find the way to make better and cheaper cigars and you will solve the problem of decreasing cigar consumption." A good five-cent cigar. Manufacturers at first protested that the nickel smoke was a thing of the past, "an unperformable trick—it simply can't be done." Attempting to do it would place inferior quality goods on the market and push smokers right into the arms of the cigarette makers. But in 1923 the General Cigar Company introduced the William Penn, the first five-cent cigar since the war. Soon other companies came out with nickel brands, and, by 1929, 55 percent of U.S. cigar output retailed for five cents or less. Within a decade the proportion had reached nearly 90 percent. There was no trick involved. The five-cent cigar reappeared via the cigar machine. In 1926, 18 percent of U.S. production was machine-made; by 1936, the figure reached 75 percent and continued to rise.[33]

The new cigarmaking technology could use high-quality leaf and produce a good cigar for only about one-quarter the cost of hand labor. A team of three hand workers might produce 400–800 cigars a day, while one four-operator cigar machine turned out 4,000 to 4,500 a day. The machine did all the rolling and bunching, while the operators primarily handled the leaf itself. It was a complicated device. One of the four operators fed filler tobacco into a constantly moving tube that carried it to the binder layer's station. There an operator stretched a coarse binder leaf (held in place by suction) over a metal die that cut out the binder. A mechanical arm descended to pick up the leaf and carry it to a station where it could be rolled around the filler leaf. From there the bunch was lifted mechanically and moved to the next station, where a wrapper leaf, cut out in the same way as the binder, could be applied. Finally the cigar moved to the fourth operator, who inspected it.[34]

Machinery offered manufacturers much more control over the production process. With it they could set the exact rate of production, whereas, if "left to the individual operator to set the time," as one machine ad explained, "he [sic] will produce according to his mood." (All ads for machines pictured only women operators.) A roller had only a few seconds to stretch out the wrapper leaf on the metal die "since it is not left to his [sic] judgement as to how much work he should do or how little, all of this having been skillfully worked out by the engineer who designed the system." It also ended cigar makers' annoying habit of working only "as long as they please" since workers had to remain at their machines as long as the power was turned on.[35]

Most important, manufacturers using machinery hired only women workers. They did so in part at the strong suggestion of AMF, but also because they believed a new group of women might be more compliant and they expected to save money by paying women less than they would have to pay men. By itself machinery cut labor costs. It took months or years to learn hand work, while machine operators could be trained in two weeks. One firm estimated in 1930 that machinery had cut the labor cost of making 1,000 cigars from $10 to $3.40. Overall, a 1930 Women's Bureau study of the industry estimated that machinery cut the cost of labor in half while it doubled production. But it was not simply the machine's efficiency alone which produced these savings: the cigar machine also permitted the manufacturer to "utilize labor of a character that he had never been able to draw upon before" — young women, new to the industry. The machine's efficiency and speed coupled with low piece rates for women meant that the unit cost of production was cut dramatically, giving the large firms that used it a tremendous economic advantage.[36]

The cigar machine and the forces it helped to set in motion spelled the end of the competitive, decentralized stage of production and helped to concentrate the industry in the hands of a few giants. In most cases, nickel cigars, now the most popular line, could not be produced at a profit without mechanization, and only the largest firms could afford to lease and operate the required technology. For 100 machines, AMF charged a manufacturer about $380,000 for installation and from $50,000 to $100,000 a year in rental and royalties. Companies had to buy all parts from AMF and pay AMF mechanics full-time to keep the machines running smoothly.[37]

In 1921, 14,578 factories were operating in the United States, but just nine years later there were only 7,452. In 1912, the top 300 companies accounted for 50 percent of production. By 1926, the top twenty-two firms had captured 65 percent of production. The ten largest firms in 1930 — American Cigar, Consolidated, General Cigar, P. Lorillard, Bayuk, Deisel-Wemmer-Gilbert, Webster-Eisenlohr, Mazer-Cressman, Congress Cigar, and Porto Rican American Tobacco Company — together produced just over half of all U.S. cigars. Thousands of medium-sized and small manufacturers, unable to compete, were bought out or forced out of business during the 1920s. A *Barron's* article suggested that the machine and the changes in the industry would, of necessity, "eliminate weak units." The trend toward concentration, the article continued, was nothing more than "a Darwinian process of natural selection," which would ultimately make the industry stronger.[38]

Large firms, using machines and possessing greater financial resources, could make a bad situation work to their advantage. They could actually breathe easier as trade conditions killed off the competition. Prohibition spelled disaster for smaller, independent manufacturers who relied on the saloon for a retail outlet. This was especially true for those using the blue union label. But for large firms that never relied on these local outlets, the Volstead Act could be viewed as a blessing. Indeed, there were rumors (if no concrete evidence) that a number of them had donated liberally to the Anti-Saloon League, hoping thereby to cut out small-time competition. Many of the large firms had preferential agreements with national retail drug chains that put smaller competitors at a disadvantage. Several companies bought out regional jobbers and handled their own distribution, dealing directly with retailers. This disrupted the distribution networks of smaller firms.[39]

Even more detrimental, the big ten used their sales volume, generated by national advertising, to push jobbers into handling their products on an exclusive basis: to get the right to sell the national brand the jobber had to agree to refuse to handle other cigars. A small manufacturer in Warren, Pennsylvania, was one of thousands to feel the blow. Raymond Steber was dumbfounded when the jobber with whom he and his family had dealt for decades broke the news that he would no longer handle Steber's goods. A large cigar company, he explained, had threatened to remove all of its popular brands unless the jobber handled its cigars exclusively. Steber had to find a new jobber, and before long his company closed. Changes in the leaf business added to the troubles of smaller producers. Integrating vertically, several of the big ten set up their own tobacco warehouses and began dealing directly with growers. Independent dealers had a weaker bargaining position and were increasingly unable to offer their clients a reasonable deal on tobacco prices. Everything seemed to fall into place in favor of the industry's new corporate giants, and by the end of the decade their dominance was unquestioned.[40]

The diminishing number of factories resulted partly from business failures and mergers but also from altered management policies. To cut production costs and reduce overhead, major firms now closed branch factories and consolidated operations in one or two huge plants. The Women's Bureau study commented that it was also easier to "lock doors and forsake a town" than to face the embarrassment of publicly firing so many local people in the course of shifting from hand to machine methods. Traditionally the branch factory system had served as one of the industry's chief labor policies — strikes could never shut down a company's entire production. But now manufacturers abandoned this approach because of its failure during wartime strikes, their assumptions about the new workers they hired, their needs given the use of machinery, and the savings they hoped to achieve. Scores of factories closed in Pennsylvania, New Jersey, and Ohio, and thousands of workers lost their jobs. General Cigar went from seventy-two factories in 1920 to forty-six by 1927, while another company which operated in Pennsylvania closed fifty plants between 1922 and 1929, keeping only twenty in operation. Deisel-Wemmer closed all but two of its seventeen plants in western Ohio, and the John Swisher Company, makers of King Edward cigars, closed out its operations in Ohio and moved to Jacksonville, Florida, in 1924.

Relocation of industry was not peculiar to Swisher, however. Many firms began moving away from traditional areas of production in the 1920s. Particularly hard hit were the counties of Pennsylvania's cigar belt running from Allentown south and west to York. In scores of communities, cigarmaking had been the chief industry. By 1930 the factories, still the "most prominent buildings" in town, stood dusty and deserted. The companies never acknowledged any responsibility to former workers or the towns left behind, a fact that prompted the Women's Bureau to decry the companies' "ruthless disregard of human needs." Although Pennsylvania and New Jersey remained the most important centers of production, Florida now joined the list, and, in 1939, 67.2 percent of all workers in the industry worked in these three states.[41]

The new plants hardly resembled their grimy predecessors. They were larger, fully mechanized, and equipped with modern plumbing and, quite often, air conditioning. For maximum efficiency and control, the layout followed the flow of work so that each plant "became a logical unit." Air conditioning not only regulated the temperature and humidity for tobacco and cigars, eliminating much waste, it also affected workers' habits since it reduced labor turnover during the summer, traditionally a difficult time to keep cigar makers at work. Careful cost accounting was introduced into these plants as well as plans for "perpetual inventory" — keeping daily accounts of production, tobacco usage, and sales, rather than the traditional way of settling all accounts in January and July only. For the first time, larger companies opened personnel departments and removed hiring and firing from departmental foremen. Technology had thoroughly reshaped work processes, but it was only one segment of a complete restructuring and reorganization of the industry.[42]

The size and composition of the cigarmaking work force shifted markedly. The squeeze from mechanization and declining production eliminated over 56,000 jobs from the industry between 1921 and 1935. Many more cigar makers were laid off, however, because of branch factory closings and because manufacturers tended to hire a new group of younger women with no previous experience in the industry. AMF, as we have seen, recommended this policy. Thus women workers replaced not only male unionists but also women hand cigar makers. The proportion of women in the labor force climbed from 54 to 58 percent in 1920 to over 80 percent by the late 1930s. "Their temperament," noted one official, "is suited to the work and they are willing to work for lesser wages." Employers tended to prefer young girls, not former teamworkers—some firms would not hire anyone over twenty-five. Average earnings in the industry dropped 27 percent between 1921 and 1935, doubtless doubly affected by the Great Depression on top of the industry's own troubles.[43]

In many other ways the cigar industry wore a new look in the 1920s. For the first time, larger firms began pushing national brands. Companies such as General Cigar soon pared down 150 brands to only five, and by the late 1920s a smoker could walk into a tobacco shop anywhere in the country and buy brands of the leading firms. Smaller companies could never market that way.

In keeping with the burgeoning culture of consumption, advertising became a key element of corporate operations. The largest manufacturers launched massive assaults on consumer consciousness through newspapers, magazines, billboards, and radio. In one case the move proved especially fortuitous. Congress Cigar, owned by Sam Paley, moved to Philadelphia from Chicago in the early 1920s and began advertising its La Palina cigar on local radio. Eventually the family bought the station, which would become CBS and place son Bill as head of a media empire. For most firms the results of advertising were less dramatic, but no less important. Few had felt the need to advertise previously, but with sales down and all the 1920s hype and publicity given to "up-to-date business methods," large companies set up advertising departments and rushed to push their goods. By 1930, both General Cigar and American Cigar each spent $1 million on advertising. Recognizing that a woman was unlikely to smoke a cigar, firms stressed its "masculinity" with such slogans as "Be a Man—Smoke Cigars" and further implied that cigarettes were effeminate. Ads featured women in bathing suits and used other suggestive copy designed to catch a man's eye. None mentioned machinery.[44]

The most daring example of cigar ad copy demonstrated how companies really tried to reshape cultural attitudes. Manufacturers achieved much during the 1920s by adopting the new technology to make five-cent cigars, but they remained wary of letting the public in on the key to their success. The cigar market was shaky, and smokers traditionally held strong prejudice in favor of all hand-made cigars. An unwritten rule in the trade precluded mention of machinery in advertising, and AMF judiciously kept secret the names of its clients. In 1929, however, this would change when George Washington Hill, president of the American Cigar Company, broke the taboo. Searching for a bold new way to advertise the Cremo and convinced that he could sell a machine-made cigar, Hill transferred several ad men from the successful Lucky Strike staff at the American Tobacco Company (American Cigar's former parent) to his own. And the infamous "spit" campaign was born. The first advertising copy appeared in the *New York Times* in late June 1929 and then followed in 100 other city newspapers. One of the ads pictured an old man sitting at a cigarmaking table licking the end of a cigar. "Do you remember the old, filthy shop where the man in the window rolled the leaves . . . with dirty fingers . . . and spit on the ends?" It was true, the ad admitted, that "spit is a horrid word," but it was worse when it was on the end of one's cigar. Cremo cigars were, by contrast, made by machine under modern, sanitary conditions. The slogan for the campaign—"Cremo, the good 5 cent cigar that American needed"— said it all.[45]

The rest of the trade was aghast. The ads would impugn the reputation of every cigar on the market unless all adopted the same approach. A Better Business Bureau investigation termed the ads "unjustifiable" and an "unfair attack," while *Printer's Ink* huffed that machine methods by no means insured sanitation. But no matter. Cremo sales skyrocketed. By 1931, Cremo had bumped William Penn as the number-one-selling five-cent cigar in the nation.[46]

The spit campaign had other results. It hastened the introduction of cellophaning cigars and it precipitated some of the cleanest ad copy in history. Rocky Ford cigars were made in "clean and sanitary factories," and Cinco cigars were "manufactured under the most modern hygienic Factory conditions." York County, Pennsylvania, manufacturers, who had not yet adopted machinery since their labor costs were so low, viewed the ads with great apprehension. To counter any ill effects, they went to great lengths to stress sanitary conditions. Cigar makers there were "good, clean people, high typed and God fearing. From childhood up they are taught the spirit of cleanliness. Thus York County cigars are manufactured under the best sanitary conditions by American people, whose first name is 'clean.'" What could be more sanitary than that?[47] Yet, when an explosive strike of cigar makers erupted there in 1934, most York County manufacturers quickly adopted machinery.

On one level the transformation of the cigar industry appears to have been simply a case of inevitable changes and adjustments. But a closer look suggests a more complex and more orchestrated scenario. By every indicator the cigar industry was in decline during the 1920s. Yet some firms could afford machinery, install air conditioning, and pay thousands of dollars for advertising. The depression in the trade was real, but not all manufacturers suffered equally. Indeed, for the big ten and a few others, the cigar industry's woes during the 1920s turned out to be a boon — the opportunity they had awaited for decades. The cigar depression enabled them to consolidate control of the industry and eliminate competition. In the bitter contest for survival in a shrinking market, large manufacturers using AMF machines held a powerful advantage. Throughout the 1920s the large firms recorded healthy increases in profits. As industry spokesman Charles Barney explained, "from the standpoint of the larger manufacturers . . . a condition of even stationary production is not unfavorable for so long as smokers continue to smoke the same aggregate number of cigars, the larger, well-entrenched manufacturers . . . are going to continue to expand their business." It was true that the industry had been declining, but "the large cigar manufacturers have been enjoying substantial prosperity as a result of increasing individual production, the utilization of cigar machines, and related improvements in manufacturing processes." The smaller companies, struggling to survive amid relentless decline in consumption, thus bore disproportionately the burden of the industry's crisis.[48]

Certainly the large companies were adversely affected by the Great Depression, as profits of even the largest firms fell sharply from 1930 to 1933. Most responded with cost cutting, which only eliminated more manufacturers and further consolidated the industry. Ten-cent cigars suddenly sold for five cents or nickel smokes went for two for five cents, jumps that smaller firms just could not make. Big firms cut or even eliminated dividends to shareholders, but all continued to operate in the black, and Bayuk hit an all-time record profit of $1.5 million in 1939. The leading firms drafted the codes of fair competition under the National Recovery Administration, where again they came out on top. Small firms, joined by the CMIU, petitioned the code authority for revisions and for some kind of regulation on the use of cigar machinery, but no action was ever taken. In 1935, 5,100 small companies produced 5 percent of U.S. cigars and the top fifteen firms accounted for 66 percent of production. By the late 1930s, over 80 percent of production was five-cent cigars, made by women workers in large mechanized plants.[49]

If the big winners in the industry's mechanization were the large cigar companies which now dominated it, the greatest losers were the men and women who had made cigars by hand only a few years before. Mechanization and the new structure of the industry resulted from conscious decisions on the part of corporate managers. For workers these changes meant staggering personal disaster. Yearly wages fell from an average of $800 in 1920 to $500 in 1933. The Women's Bureau study concluded that it was no longer a man's industry because "it is impossible to earn a family wage in it now." In 1936, 20 percent of cigar workers earned less than $16 a week. As more companies moved to the South and others saw the economic advantages of breaking the color line, more black workers, usually women, found jobs in the industry, particularly in stemming departments. The wage differentials by race could be startling. For example, in 1938, white cigar workers earned an average of $16.65 a week, while black workers earned $7.10.[50]

Perhaps the CMIU was the most visible casualty of the machine age. "The cigar makers' union, once one of the most powerful of organized labor groups," crowed a *Barron's* article in 1931, "is today almost history." The observation was accurate. Union members, fearful of the threats to their jobs, had closed ranks behind CMIU president George W. Perkins, who had taken office in 1892 and did not retire until 1927. The following year, members agreed to organize machine workers and permit the union label on these goods, but by then it was too late and their label had little remaining value anyway. Membership dropped from 40,000 in 1919 to 15,000 in 1933. Scores of union factories simply went out of business, while others declared an open shop.[51]

In Rochester, New York, Manchester, New Hampshire, and elsewhere, union members hung on by accepting repeated wage cuts and other concessions after companies threatened to move away or adopt machines. Cigar makers at the huge union factory of R. G. Sullivan in Manchester, New Hampshire, took wage cut after wage cut in return for company promises to keep machines out. In 1930, however, they gave up. The first "iron cigar makers" began to arrive, and by early 1931 only ten hand workers remained, out of what had been a force of 700. "A lot of men died after that," explained one former stemmer whose father and husband had both been fired. "A lot of the old-timers. They were just heartbroken. They couldn't work." Many of those laid off tried opening their own small cigarmaking shops, but 64 percent of the fired men were still unemployed in 1936.

In Boston, the famous union factory of H. Traiser had "not an out-and-out operator in the place. . . . This is scarcely believable," remarked one union member. Boston, long the stronghold of the CMIU, was "shot and shot plenty." The CMIU was of no use to unemployed members and nearly as useless to those who held onto their jobs but found their earnings, now lower than that of "semiskilled" machine operators, inadequate at best. Members voted to drop the traveling loan system in 1928, and soon after they had no choice but to eliminate the death benefit too. It was a dying industry and a dying union, a point graphically driven home by one statistic: the average age of a union member in 1928 was sixty-four.[52]

But what of the women who had always made five-cent cigars? With some important exceptions, they fared little better. Scores of branch factories in Pennsylvania, Ohio, New York, and New Jersey shut down as companies consolidated operations and moved to new locations. In many places manufacturers dared not announce the closing of a factory in advance for fear of sabotage or violence, and often they shut down operations without notice just after a Fourth of July or Thanksgiving holiday. Each factory closing, one woman explained to a Labor Department investigator, was "like a funeral," while another explained, "it broke our hearts to leave that cigar factory." Given the depressed economy and limited opportunities, women had few other options. A Polish cigar maker in Baltimore spoke about the disappearance of the cigar factories there. "Where am I to go? There's no cigars anywhere." Noting her alternatives, she added, "I am not going to scrub."

In some areas displaced women workers could find jobs in textile and sewing factories or other traditionally female fields. A Lancaster, Pennsylvania, woman lost her job as a roller in 1926 when she was twenty-five. After that she lost three other jobs as companies closed. After sporadic unemployment, she found a job in a "5-and-10-cent store" where she made $10 a week, down from the $15 to $16 she had earned making cigars. An Ohio cigar maker secured a cleaning job in a hotel, which she hated. The work, she told an investigator, "almost kills me." But even when firms continued operations and adopted machinery, women teamworkers had problems because manufacturers preferred young, native-born women. "We rarely hire those beyond the twenties," one manufacturer explained, "and most are in their teens." Some fortunate teamworkers got a chance at the new machine jobs, though there were never enough positions to take care of everyone. The Women's Bureau found that they often disliked the new work. One Lancaster cigar maker lost her position as a roller after plant closings, but she quit the job she found working at the cigar machine after only three days. "It is a continual rush," she explained, and the new girls who had never worked by hand simply "did not know any better." Another preferred employment in a hand factory because there one could "work in peace."[53]

The return of the five-cent cigar via machinery meant something quite different for the industry's newest workers. For women who had formerly worked as tobacco stemmers or for those new to the trade altogether, the machines offered an opportunity. As the cigar maker from Lancaster had observed, they had little with which to compare their present work experiences and everything to gain from taking these new jobs. In Richmond, Virginia, shortly after the birth of her daughter in 1935, Maude Lenz decided to answer an ad that P. Lorillard had placed in the paper for machine operators — no experience necessary. She could use the extra income to buy a Frigidaire. For her and others, these were jobs, and jobs were hard to find during a depression.[54]

Yet with opportunity came costs, and the contours of shop culture shifted. All machine workers encountered a carefully managed workplace. Unlike their predecessors, they had no control over time. Leased machinery eliminated flexible hours, and now cigar makers had to be at work the entire time a factory was open. A cigar maker could leave the machine to take a break or use the restroom only when someone could fill in for her — officially a foreman or "forelady." As the Women's Bureau investigation pointed out, a machine operator had "hardly a moment to lift one's eyes or to speak to a neighbor; sometimes there is not even convenient opportunity to go to the dressing room." Bad relations with a supervisor could mean a long wait. Machines broke down frequently, and workers lost money while they were repaired since they were still paid by the piece. One needed to be on good terms with the mechanic in order to get prompt service. Many women simply learned to repair the machines themselves, but they had to do so out of the sight of a mechanic or supervisor. Informal pacing of work was difficult because the timing on a machine was pre-set. One had to hurry to keep up. A former hand worker named Anna Bartasius, who was born in Lithuania but worked in Philadelphia, explained that "you have to follow machine. Don't stop. . . . That's more hard."

The first models of AMF machinery had no safety features, so fingers could get caught in the mechanism if the operator did not move them away quickly enough. Talking to pass the time was, in most cases, no longer possible. "You maybe could talk with your partner once in a while," noted Laura Carter, a machine buncher for the Swisher company in Jacksonville. "But you could not do this very much and produce quality cigars." Besides, there was "too much noise." Monitoring the amount of tobacco used could now be much more precise since the machine regulated tobacco usage, although bunchers and rollers still had to try to get as many good cuts out of a leaf as possible, avoiding the holes and tears in each piece. To preclude workers' sharing tobacco, employers offered bonuses to those who hit above the quota. Workers could be fired at any time for work which was viewed as substandard.[55]

But like their predecessors, machine workers also searched for and discovered ways around the factory regimen and its structured environment. They learned to repair their own machines, adjusted the pace themselves when possible, and tried to find ways to help each other. One New Hampshire operator explained that she and her partner on the machine talked it over "and finally I said, 'Well, why don't you come and learn wrappers. I'll sit for you and you sit for me.' And she'd say, 'We'll get caught.' And that will be it. . . . So when we knew our boss was gone somewhere else and the floorlady was busy on the other side of the room, we'd swap for a few minutes and . . . we developed speed for both jobs that way." Then they could relieve each other. As one inspector noted, "I could go off to the ladies room or I could go visit them at the next machine and come back and if the wrapper girl is fast, she'd pick up for me and look at all them in the rack." Even the tradition of labor militancy reemerged. During the 1930s, strikes in Jacksonville, Florida, and Richmond, Virginia, along with scores of others demonstrated that technology could not do all that employers hoped—put an end to workers' efforts to control and adjust their work lives.[56]

The cigar industry continued to decline despite a brief resurgence during World War II. Most of the companies diversified or merged with still larger conglomerates, and many moved operations to Central and South America during the 1950s and after.[57] The five-cent cigar disappeared once again, and in 1974 so did the CMIU. With only 2,000 members left, officers signed a merger agreement with the Retail, Wholesale and Department Store Union: Sam Gompers's union quietly ended its 110-year history.

By itself, study of the design and operation of cigar machines tells us little. But examined in the context of the history of the industry and its workers, both men and women, the story of the five-cent cigar and the machinery used to make it raises serious questions about work, gender, and technology. American cigar manufacturers wanted to assert control and discipline over their employees. Both men and women resisted employers' prerogatives, however, and tried to find ways to control their own work. Yet the struggle taking place in the cigar industry before 1919 was not simply between workers and employers, it also related to the sexual division of labor. Five-cent cigar manufacturers hired women based on assumptions that they would make docile, cheap workers. Women, who were skilled, although they were usually termed "semiskilled," worked under poorer conditions and for lower pay than their male union counterparts. Until the wartime strikes, male and female workers rarely cooperated and economic and social circumstances often pitted them against each other. They joined forces, especially during 1919, to forge a new kind of movement, apart from union structure, which sent chills through the industry.

Manufacturers wanted a tractable labor force and full control over the production process, especially in view of the threat posed by the united wartime strike movement. To gain both, these larger-sized firms adopted machinery and hired a new group of workers, all women. When it suited their needs, they shut down operations and moved to more attractive sites, leaving behind severely depressed towns and communities.

The "labor-process" discourse has focused primarily on what happened to men displaced by technology, but it is clear in the cigarmaking case that such an approach is inadequate. When technology was introduced, it was not simply union men who were affected. While some women teamworkers managed to move into machine jobs, most did not, and thousands of women were displaced along with men. They both watched as their jobs disappeared or were given to new workers altogether. At the same time, not all women in the industry shared the same experiences of technology. For women who now found jobs in the industry, technology was not a bane but an opportunity that they eagerly embraced. They were thankful to have work, but the wages were low and the conditions so regulated that they had to maneuver carefully even to use the restroom. The restructuring of the labor process and the nature of work in the new machine factories were defined by class and gender relations.

Technology had not been neutral. Large corporate manufacturers had introduced it, controlled it, and used it to serve their own interests, particularly increased control over the production process. Another key concern of theirs, of course, was to stay in business in the midst of a precipitous drop in cigar consumption. But these manufacturers never experienced hard times. The burden of decline was borne by smaller firms that simply could not compete under new conditions. The end result was a mechanized industry dominated by a few large corporate enterprises.

The nation easily recovered from and soon forgot the mechanization and economic decline of the cigar industry, but the individuals touched by these changes could not dismiss them so easily. Thomas Marshall had called for a good five-cent cigar and he got it, but the cost for cigar makers proved to be quite a bit more than a nickel.

NOTES

Patricia Cooper thanks Ava Baron for her constructive criticism.

1. See Patricia A. Cooper, *Once a Cigar Maker: Men, Women, and Work Culture in American Cigar Factories, 1900–1919* (Urbana, Ill., 1987), chaps. 1 and 10, for a discussion of these changes.

2. Harry Braverman, *Labor and Monopoly Capital: The Degradation of Work in the Twentieth Century* (New York, 1974); Richard Edwards, *Contested Terrain: The Transformation of the Workplace in the Twentieth Century* (New York, 1979); David M. Gordon, Richard Edwards, and Michael Reich, *Segmented Work, Divided Workers: The Historical Transformation of Labor in the United States* (New York, 1982); Andrew Zimbalist, ed., *Case Studies on the Labor Process* (New York, 1979). Making similar arguments are Harley Shaiken, *Work Transformed: Automation and Labor in the Computer Age* (New York, 1984), and Mike Cooley, *Architect or Bee? The Human/Technology Relationship* (Boston, 1980). A very different view of technology from the Left is Paul Adler, "Technology and Us," *Socialist Review* 85 (January–February 1986): 67–96.

3. David Montgomery, *Workers' Control in America: Studies in the History of Work, Technology and Labor Struggles* (New York, 1979). Other research includes David Bensman, *The Practice of Solidarity: American Hat Finishers in the Nineteenth Century* (Urbana, Ill., 1984); Alan Dawley, *Class and Community: The Industrial Revolution in Lynn* (Cambridge, Mass., 1976); and Sean Wilentz, *Chants Democratic: New York City and the Rise of the American Working Class, 1788–1850* (New York 1984). Philip Kraft's "The Industrialization of Computer Programming: From Programming to 'Software Production,'" in Zimbalist, *Case Studies in the Labor Process* (n. 2 above), explicitly deals with gender definition of jobs.

4. This body of material is immense, and only passing reference can be made to it here. See, e.g., Susan Porter Benson, *Counter Cultures: Saleswomen, Managers, and Customers in American Department Stores, 1890–1940* (Urbana, Ill., 1986); Sarah Eisenstein, *Give Us Bread but Give Us Roses: Working Women's Consciousness in the United States, 1890 to the First World War* (London, 1983); Thomas Dublin, *Women at Work: The Transformation of Work and Community in Lowell, Massachusetts, 1826–1860* (New York, 1979); Mary Blewett, *Men, Women and Work: Class, Gender, and Protest in the New England Shoe Industry, 1780–1910* (Urbana, Ill., 1988); Ruth Milkman, ed., *Women, Work and Protest: A Century of U.S. Women's Labor History* (Boston, 1985); Alice Kessler Harris, *Out to Work: A History of Wage-earning Women in the United States* (New York, 1982); and Carol Groneman and Mary Beth Norton, eds., *"To Toil the Livelong Day": America's Women at Work, 1780–1980* (Ithaca, N.Y., 1987).

5. Carole Turbin, "Beyond Conventional Wisdom: Women's Wage Work, Household

Economic Contribution, and Labor Activism in a Mid-Nineteenth Century Working Class Community," in Groneman and Norton, *To Toil the Livelong Day,* pp. 47–67.

6. There is an enormous literature on the sexual division of labor, but few studies have actually analyzed how this takes place. See, e.g., Valerie Oppenheimer, *The Female Labor Force in the United States* (Berkeley, Calif., 1970), and the special issue of *Signs* on occupational segregation by sex, vol. 1 (Spring 1976).

7. Heidi Hartmann, "Capitalism, Patriarchy, and Job Segregation by Sex," *Signs* 1 (Spring 1976), and "The Marriage of Marxism and Feminism: Towards a More Progressive Union," *Capital and Class* 8 (Summer 1979): 1–33.

8. Ava Baron, "Women and the Making of the American Working Class: A Study of the Proletarianization of Printers," *Review of Radical Political Economics* 14 (Fall 1982): 23–42. See also Mary Blewett, "Work, Gender and the Artisan Tradition in New England Shoemaking, 1780–1860," *Journal of Social History* 17 (Winter 1983): 221–48, and Ruth Milkman, *Gender at Work: The Dynamics of Job Segregation by Sex during World War II* (Urbana, Ill., 1987).

9. Blewett, *Men, Women and Work* (n. 4 above); Ava Baron, "Contested Terrain Revisited: The Social Construction of Gender and Skill in the Printing Industry, 1850–1920," in *Transformations: Women, Work and Technology,* ed. Barbara Wright et al. (Ann Arbor, Mich., 1987); Elizabeth Baker, *Technology and Women's Work* (New York, 1964); Judith McGaw, *Most Wonderful Machine: Mechanization and Social Change in Berkshire Paper Making, 1801–1885* (Princeton, N.J., 1987); McGaw, "Women and the History of American Technology," *Signs* 7 (Summer 1982): 798–828; Margery W. Davies, *Woman's Place Is at the Typewriter: Office Work and Office Workers, 1870–1930* (Philadelphia, 1982); Cynthia Cockburn, *Brothers: Male Dominance and Technological Change* (London, 1983). See also articles by Roslyn Feldberg and Evelyn Glenn, Nina Shapiro-Perl, and Louise Lamphere in Zimbalist, *Case Studies in the Labor Process* (n. 2 above).

10. Baron, "Contested Terrain Revisited" (n. 9 above); Baron, "Gender and Labor History: Notes towards a New Historical look at Women, Men and Work," unpublished ms., December 1985.

11. Willis N. Baer, *The Economic Development of the Cigar Manufacturing Industry in the United States* (Lancaster, Pa., 1933), p. 257; Meyer Jacobstein, *The Tobacco Industry in the United States* (New York, 1907), pp. 87–88; U.S. Bureau of Corporations, *Report of the Commissioner of Corporations in the Tobacco Industry,* Part 1, *Position of the Tobacco Combination in the Industry* (Washington, D.C., 1909), pp. 423–27.

12. Carl Avery Werner, *A Textbook on Tobacco* (New York, 1909), pp. 38–43; Baer, *Economic Development* (n. 11 above), pp. 85–87; U.S. Commissioner of Labor, *Eleventh Special Report: Regulation and Restriction of Output* (Washington, D.C., 1904), pp. 565, 573, 577, 583; John P. Troxell, "Labor in the Tobacco Industry" (Ph.D. diss., University of Wisconsin, 1931), p. 46. Nancy Hewitt is currently writing a book on women cigar makers in Tampa.

13. See note 12 above.

14. U.S. Department of Commerce, Bureau of the Census, *Census of Manufactures 1900,* Vol. 9, *Special Reports on Selected Industries* (Washington, D.C., 1902), p. 653; Bureau of the Census, *Census of Manufactures 1909,* Vol. 8, *General Report and Analysis* (Washing-

ton, D.C., 1913), p. 254; Bureau of the Census, *Census of Manufactures 1919,* Vol. 8, *General Report and Analytical Tables* (Washington, D.C., 1923), p. 490; Bureau of the Census, *Census of Occupations 1900: "Special Reports"* (Washington, D.C., 1904), p. 12; Bureau of the Census, *Census of Population 1920,* Vol. 4: *Occcupations* (Washington, D.C., 1923), p. 38; *Cigar Makers' Official Journal (CMOJ),* September 1901, p. 8; September 1912, pp. 14–15; April 1920, p. 3; U.S. Commissioner of Labor, *Eleventh Special Report* (n. 12 above), p. 568; Baer, *Economic Development* (n. 11 above), p. 85; *Lima News,* February 9, 1922; U.S. Commissioner of Labor, *Thirteenth Annual Report 1898: Hand and Machine Labor* (Washington, D.C., 1899), pp. 391–95; Cooper, *Once a Cigar Maker* (n. 1 above), chaps. 6 and 7.

15. See Cooper, *Once a Cigar Maker,* chap. 2.

16. Ibid., chap. 6.

17. Montgomery, *Workers' Control* (n. 3 above), p. 13.

18. Interview with John Ograin, May 17, 1980, Chicago; "By-Laws," Local 192, Manchester, N.H.; "Bill of Prices," Local 192, Manchester, 1915, Manchester City Library; "List of Shops and Bill of Prices under the Jurisdiction of Union 97, CMIU, Boston, Massachusetts," 1904, Tobacco Investigation, file 3073, U.S. Bureau of Corporations, Record Group (RG) 122, National Archives (NA), Washington, D.C.; "Bill of Prices," Local 25, Milwaukee, Wis., 1903, State Historical Society of Wisconsin (SHSW); "Bill of Prices," Local 1, Baltimore, Md., ca. 1900, U.S. Department of Labor Library (USDL). Ograin is one of several dozen people I interviewed in the course of my study of cigar workers between 1900 and 1919. Unless otherwise noted, they are tape-recorded and in my possession. See also Cooper, *Once a Cigar Maker,* chap. 3.

19. Cooper, *Once a Cigar Maker;* chap. 5; interview with Ograin, May 17, 1980; *CMOJ,* September 1904, p. 8; May 1906, p. 4; interview with Herman Baust, March 24, 1977, North Haven, Conn.; interview with T. Frank Shea, June 26, 1979, Manchester, N.H.; *Tobacco,* June 1, 1900, p. 4; U.S. Congress, House, Committee on Ways and Means, *Cigars Supplied Employees by Manufacturers,* Hearings before the Committee on Ways and Means on H.R. 17253, H.R. 121357, and H.R. 219513, 62d Cong., 2d sess., 1912, pp. 5, 40; interview with Ograin, May 17, 1980; Elizabeth M. Hennessey, "Report on Cigar and Cigarette Making Industry in Boston," p. 4, typescript, report from the Massachusetts Bureau of Labor and Industries, ca. 1918, U.S. Women's Bureau, Box 23, Unpublished Surveys, General Correspondence, RG 86, NA; A. M. Simons, "A Label and Lives the Story of the Cigar Makers," *Pearson's Magazine,* January 1917, p. 73.

20. Cooper, *Once a Cigar Maker,* chap. 5; interview with Shea, June 26, 1979; U.S. Bureau of Corporations, Tobacco Investigation, file 3073, RG 122, NA; *Tobacco Leaf,* September 24, 1902, p. 38.

21. *Tobacco Leaf,* October 1, 1902, p. 24; *CMOJ,* January 1908, p. 3; Edith Abbott, *Women in Industry: A Study in American Economic History* (New York, 1913), p. 196; Lucy Winsor Killough, *The Tobacco Products Industry in New York and Its Environs, Present Trends and Probable Future Developments,* Regional Plan of New York and Its Environs, Monograph no. 5 (New York, 1924), p. 26; U.S. Bureau of Corporations, Tobacco Investigation, file 3073, "Labor Conditions in the Cigar Industry," RG 122, NA; *Tobacco,* November 8, 1906, p. 22; *Tobacco Leaf,* July 23, 1908, p. 18.

22. See Cooper, *Once a Cigar Maker,* pp. 170–71; *Rates of Wages in the Cigar and Clothing Industries, 1911 and 1912,* U.S. Bureau of Labor Statistics, Bulletin 135 (Washington, D.C., 1913), p. 7; *U.S. Tobacco Journal,* March 24, 1917, p. 8; Baer, *Economic Development* (n. 11 above), p. 84; U.S. Congress, Senate, *Report on Condition of Woman and Child Wage-Earners* vol. 18, pp. 85–88; U.S. Bureau of Corporations, Tobacco Investigation, file 3073, "Machinery in Cigar Manufacture," RG 122, NA.

23. Cooper, *Once a Cigar Maker,* chap. 7; interview with Agnes White, February 4, 1982, Evansville, Ind., conducted by Glenda Morrison for the Indiana State University Evansville, Indiana Labor History Project (ILHP); interview with Stella Sutton, August 19, 1982, Lima, Ohio; interview with Rose Purzon, September 21, 1978, Detroit, Mich.; interview with Frances Salantak, September 20, 1978, Detroit; interview with Norman Wieand, June 21, 1984, Quakertown, Pa.; interview with Pearl Hume, August 23, 1982, Lima, Ohio; *Tobacco Leaf,* August 29, 1906, p. 46; March 27, 1907, p. 28; November 3, 1910, p. 34; T. Drier Miller, "A Sociologic and Medical Study of Four Hundred Cigar Workers in Philadelphia," *American Journal of Life Medical Sciences* 155 (February 1918): 165; New Orleans Factory Inspection Department, *Report, 1923,* p. 2; *Tobacco Leaf,* July 20, 1900, p. 7; June 10, 1903, p. 26; *Tobacco,* July 6, 1900, p. 8; July 13, 1900, p. 6.

24. Interview with Lucille Speaker, August 20, 1982, Lima, Ohio; interview with Sutton; interview with Pauline Stauffer, June 21, 1977, Hanover, Pa.; interview with Purzon; *Report of the Michigan State Commission of Inquiry into Wages and the Conditions of Labor for Women and the Advisability of Establishing a Minimum Wage* (Lansing, Mich., 1915), p. 422; Cooper, *Once a Cigar Maker,* chap. 7; interview with Lima foreman (no name owing to nature of the information), August 24, 1982, Lima, Ohio; interview with Speaker; interview with Jane Hollenbach (pseudonym), March 16, 1983, Sellersville, Pa.; interview with Hume; interview with Neva Fake, Dec. 17, 1982, Windsor, Pa.; *Tobacco,* November 8, 1906, p. 22; *Tobacco Leaf,* July 23, 1908, p. 18; interview with Salantak; interview with Mary Diehl (pseudonym), January 12, 1984, Quakertown, Pa.; *Tobacco,* January 7, 1903, 12; Cooper, *Once a Cigar Maker,* chap. 7; *CMOJ,* January 1917, p. 19; U.S. Commissioner of Labor, *Twenty-First Annual Report, 1906: Strikes and Lockouts* (Washington, D.C., 1907); Florence Peterson, *Strikes in the United States, 1880–1936,* U.S. Department of Labor, Bulletin 651 (Washington, D.C., 1938); New Jersey Bureau of Statistics of Labor and Industries, *Annual Report* (Camden, 1900–17). On the strike wave of 1917–19, see Cooper, *Once a Cigar Maker,* chap. 10; *New York Call,* June–September 1919; and *U.S. Tobacco Journal,* July 26, 1919, p. 10; August 2, 1919, p. 7.

25. In addition to the above, see also *Tobacco Leaf,* March 28, 1900, p. 6.

26. *Tobacco,* July 17, 1919, p. 20; August 7, 1919, pp. 6, 22; August 14, 1919, pp. 5, 6; September 4, 1919, p. 5; *U.S. Tobacco Journal,* July 5, 1919, p. 4; July 12, 1919, p. 3; August 2, 1919, p. 3; *Tobacco Leaf,* June 26, 1919, p. 4; September 4, 1919, p. 11.

27. See note 26.

28. U.S. Bureau of Labor Statistics, "Technological Changes in the Cigar Industry and Their Effects on Labor," *Monthly Labor Review* 33 (December 1931): 12; "Rufus Lenoir Patterson's Cigar Machine," *Fortune,* June 1930, p. 58; Circular, August 5, 1919, box 1, folder 3, and Circular to Members of Local Unions, August 16, 1919, box 2, folder 3, both in papers of

Local 162, Green Bay, Wis., SHSW; *New York Call,* September 9, 23, 30, 31, 1919; *Tobacco,* October 30, 1919, p. 6; *CMOJ,* November 1919, p. 2.

29. "When Labor Goes Too Far," *Printer's Ink,* May 20, 1920, pp. 3–4, 166, 169; *U.S. Tobacco Journal,* July 5, 1919, p. 4; July 12, 1919, p. 3; August 2, 1919, p. 3; January 1, 1921, p. 46; January 22, 1921, p. 14; February 12, 1921, p. 14; March 5, 1921, p. 4; April 16, 1921, p. 38; April 30, 1921, p. 4; May 14, 1921, p. 17; June 4, 1921, p. 42; July 30, 1921, p. 4; January 28, 1922, p. 4; February 18, 1922, p. 14; June 3, 1922, p. 7; May 6, 1922, p. 43; January 1, 1926, p. 2; *Tobacco,* July 17, 1919, p. 20; August 7, 1919, pp. 6, 22; August 14, 1919, pp. 5, 6; September 4, 1919, p. 5; *Tobacco Leaf,* June 26, 1919, p. 4; September 4, 1919, p. 11.

30. Jack J. Gottsegen, *Tobacco: A Study of Its Consumption in the United States* (New York, 1940), p. 14; *U.S. Tobacco Journal,* March 5, 1921, p. 26; April 9, 1921, p. 32; May 14, 1921, p. 4; Oct. 15, 1921, p. 3; March 15, 1922, p. 5.

31. *U.S. Tobacco Journal,* March 18, 1922, p. 5; August 2, 1924, p. 4; March 14, 1925, p. 4; March 27, 1926, p. 4; Gottsegen, *Tobacco,* p. 14.

32. *Tobacco Leaf,* January 2, 1913, p. 6; April 30, 1914, p. 6; August 13, 1914, p. 6; August 20, 1914, p. 5; November 12, 1914, p. 5; December 24, 1914, p. 5; *U.S. Tobacco Journal* April 27, 1912, p. 4.

33. *U.S. Tobacco Journal,* March 5, 1921, pp. 4, 6; April 30, 1921, p. 4; June 25, 1921, p. 18; July 16, 1921, p. 4; January 21, 1922, p. 4; March 18, 1922, p. 4; August 9, 1924, p. 4; January 17, 1925, p. 8; February 17, 1926, p. 4; June 12, 1926, p. 5; "Smokers Set Advertising Theme of the 5-Cent Cigar," *Printer's Ink,* March 9, 1933, p. 6; "Competition in Cigar Machinery," *Barron's,* May 12, 1931, p. 20; U.S. Department of Labor, Wage and Hour Division, *The Cigar Industry* (Washington, D.C., 1941), pp. 29–31.

34. "Technological Changes" (n. 28 above), pp. 11–17.

35. William A. McGarry, "A Good Five-Cent Cigar by Mass Production," *Forbes,* May 1, 1926, pp. 16–18; J. H. McMullen, "Machinery Brings Back the Five-Cent Cigar," *Commerce and Finance,* January 26, 1927, pp. 217–18; "Net of American Machine and Foundry Up 85%," *Barron's,* May 5, 1930, p. 10; "When Gears and Levers Replace the Cigar Makers' Adept Fingers," *Scientific American,* November 1922, pp. 312, 336; *U.S. Tobacco Journal,* May 13, 1922, p. 32; *Tobacco Leaf,* August 3, 1929, p. 13; "Technological Changes" (n. 28 above), pp. 11–17; Caroline Manning and Harriet Byrne, *The Effects on Women of Changing Conditions in the Cigar and Cigarette Industries,* U.S. Department of Labor, Women's Bureau, Bulletin 100 (Washington, D.C., 1932), pp. 18, 27.

36. *U.S. Tobacco Journal,* December 3, 1921, p. 39; "Technological Changes" (n. 28 above), pp. 12–17; *U.S. Tobacco Journal,* May 13, 1922, p. 32; *Tobacco Leaf* (August 3, 1929): 13; McMullen, "Machinery Brings Back the Five-Cent Cigar" (n. 35 above); "Tobacco Prospers," *Business Week,* February 2, 1935, p. 12.

37. See note 36.

38. *U.S. Tobacco Journal,* December 3, 1921, 39; January 3, 1925, p. 8; "Technological Changes" (n. 28 above), p. 12; Killough, *The Tobacco Products Industry* (n. 21 above), p. 18; Reavis Cox, *Competition in the American Tobacco Industry, 1900–1932* (New York, 1933), pp. 90–91, 135; "Tobacco Prospers," *Business Week,* February 2, 1935, p. 12; *U. S. Tobacco Journal,* January 10, 1925, p. 5; January 22, 1921, p. 12; Joseph C. Robert, *The Story of Tobacco*

(New York, 1952), p. 227; "Competition in Cigar Machinery," *Barron's,* May 12, 1931, 20; "Readjustments in the Cigar Industry," *Barron's,* July 20, 1931, 18–19.

39. Cox, *Competition,* pp. 110–11, 135; "Bucking the Cigar Industry's Trend," *Barron's,* August 14, 1939, p. 13; "Concentrating on One Brand Builds Volume for This Jobber," *Sales Management,* April 3, 1926, p. 515; interview with Raymond Steber, July 24, 1978, Warren, Pa.

40. See note 39.

41. Manning and Byrne, *The Effects on Women* (n. 35 above), pp. 11–20, 24, 37; Wilmoth D. Evans, *Mechanization and Productivity of Labor in the Cigar Manufacturing Industry,* U.S. Department of Labor, Bureau of Labor Statistics, Bulletin 660 (Washington, D.C., 1939), pp. 1–4, 43–64; *U.S. Tobacco Journal,* August 2, 1924, p. 4; Killough, *The Tobacco Products Industry* (n. 21 above), pp. 30–33; McGarry, "A Good Five-Cent Cigar" (n. 35 above); *Congressional Record,* 22d Cong., 1st sess., February 29, 1932, p. 5070; "La Corona Native," *Business Week,* March 8, 1933, p. 12; "Costs Dropped and Sales Increased When We Cut 152 Brands to 5," *Sales Management,* August 6, 1927, pp. 201–2; Cox, *Competition* (n. 38 above), pp. 50–57, 90–91, 135; Charles D. Barney, *The Tobacco Industry: Annual Review* (New York, 1927), pp. 5, 18–25, 34–35.

42. *Tobacco Leaf,* August 3, 1929, p. 13; "Does Refrigeration Pay?," *Management and Maintenance,* May 1932, pp. 196–97; Rush D. Touton, "Air Conditioning, a Production Too," *Heating, Piping and Air Conditioning,* January 1936, pp. 14–18; M. L. Wurman, "Management Insists on Safety," *National Safety News,* May 1940, pp. 10–11; *U.S. Tobacco Journal,* January 3, 1925, p. 36; W.J.B. Kress, "Accounts of a Cigar Manufacturer," *Journal of Accountancy,* April 1935, pp. 266–82.

43. Manning and Byrne, *The Effects on Women* (n. 35 above), pp. 11–20, 27–129; Baer, *Economic Development* (n. 11 above), p. 258; Evans, *Mechanization and Productivity of Labor* (n. 41 above); McMullen, "Machinery Brings Back the Five-Cent Cigar" (n. 35 above).

44. *U.S. Tobacco Journal,* December 18/25, 1975, p. 28; "Readjustment in the Cigar Industry" (n. 38 above); David Halberstam, *The Powers That Be* (New York, 1979), pp. 21–23; *U.S. Tobacco Journal,* March 28, 1925, p. 4; "Costs Dropped and Sales Increased," *Wall Street Journal,* September 7, 1979, pp. 201–2; "Scrapping One Hundred and Fifty Sectional Brands and Building Six National Brands Instead," *Printer's Ink,* April 13, 1922, p. 34; "Craftsmanship and the Return of the Five-Cent Cigar," *Printer's Ink,* August 17, 1922, pp. 17–18; "How to Awaken a Slipping Industry," *Printer's Ink,* July 19, 1923, pp. 25–28; "Is There Something Wrong with Cigar Copy?," *Printer's Ink* October 13, 1927, pp. 233–34; *U.S. Tobacco Journal,* April 18, 1925, p. 4; July 5, 1924, p. 4; July 12, 1924, p. 4.

45. "'Cremo' Picked for Sustained Campaign in Cigar Field," *Printer's Ink,* June 20, 1929, p. 102; *Tobacco Leaf,* June 29, 1929, p. 7; September 28, 1929, pp. 30–31; T. P. Headen, "Cremo Sales Soar to Record Heights during Newspaper Drive," *Sales Management,* September 21, 1929, p. 523; Cox, *Competition* (n. 38 above), p. 237; *U.S. Tobacco Journal* July 27, 1929, p. 8.

46. *Tobacco Leaf,* August 10, 1929, p. 4; "Is Advertising by Attack to Be Made a Recognized Policy?," *Printer's Ink,* April 11, 1929, pp. 25–26; C. J. DuBrul, "Hand-Tipped, Spit, Machine-Made-Well, What of It?," *Printer's Ink,* March 20, 1930, pp. 132–34; "National Better Business Bureau vs. American Tobacco, et al.," *Printer's Ink,* February 27, 1930, pp. 65; "And Now Another Advertising Battle Begins," *Printer's Ink,* January 2, 1930, p. 10.

47. *Tobacco Leaf,* September 21, 1929, p. 17; August 24, 1929, p. 36; August 10, 1929, p. 40; August 3, 1929, p. 36.

48. *U.S. Tobacco Journal,* January 3, 1925, p. 8; "Costs Dropped and Sales Increased When We Cut 152 Brands" (see n. 41 above), pp. 201–2; Barney, *The Tobacco Industry* (n. 41 above), pp. 15, 18–25; "General Cigar Company Maintains Its Strength Despite Decline in Industry," *Magazine of Wall Street,* February 18, 1933, p. 491. See also *Moody's Industrials* for specific firms throughout the 1920s.

49. "Some Cigar Companies May Cut Dividends," *Barron's,* June 16, 1930, p. 27; "Cigars, a Classic Ten-Center Becomes a Nickel Cigar," *Business Week,* January 25, 1933; "1930 Burned Up Fewer Cigars, Burned Out More Cigar Makers," *Business Week,* February 11, 1932, p. 10; Evans, *Mechanization and Productivity of Labor* (n. 41 above); *Barron's,* August 14, 1939, p. 13; July 20, 1931, p. 18; "The Cigar Attempts a Comeback, Advertising Budgets Increase," *Sales Management,* February 22, 1930, p. 350; "Five-Cent Cigar Increases Lead," *Barron's* May 5, 1930, p. 26; *Moody's Industrials,* 1930–39; "Transcript of Hearings," Cigar Industry Code, 7225, RG 9, National Recovery Administration, NA; "Hand and Machine Production of Cigars in 1940," *Monthly Labor Review,* July 1941, pp. 95–98.

50. Manning and Byrne, *The Effects on Women* (n. 35 above), pp. 1–20, 27–129; Baer, *Economic Development* (n. 11 above), pp. 216, 258; Russell Mack, *The Cigar Manufacturing Industry* (Philadelphia, 1933), pp. 77–79; Evans, *Mechanization and Productivity of Labor* (n. 41 above); "Earnings and Hours in Cigar Industry, March 1936," *Monthly Labor Review,* April 1937, pp. 953–67; "Machinery Brings Back the Five-Cent Cigar," *Commerce and Finance* January 26, 1927, pp. 217–18; McGarry, "A Good Five-Cent Cigar" (n. 35 above); U.S. Women's Bureau, *Women in Florida Industries,* Bulletin 80 (Washington, D.C., 1930), pp. 2–74.

51. In addition to the above, see "Competition in Cigar Machinery," *Barron's,* May 12, 1931, p. 20; J. H. Korson, "The Technological Development of the Cigar Manufacturing Industry: A Study in Social Change" (Ph.D. diss., Yale University, 1947); Mack, *The Cigar Manufacturing Industry* (n. 50 above), p. 77; George W. Perkins, "Women in the Cigar Industry," *American Federationist* 32 (September 1925): 808–10; John P. Troxell, "Machinery and the Cigarmakers," *Quarterly Journal of Economics* 48 (February 1934): 338–47.

52. Gladys V. Swackhamer and Daniel Creamer, *Cigar Makers—After the Lay-off: A Case Study of the Effects of Mechanization on Employment of Hand Cigar Makers,* Works Progress Administration (Philadelphia, 1937), pp. 34–59; interview with Theresa J. Shea and Jennie Kramsz, June 28, 1979, Manchester, N.H.; interview with Alice Lavoie, June 26, 1979, Manchester, N.H.; Letter from "J.W.M." to CMIU President Ira M. Ornburn, ca. August 1927, Series 1, Box 6, Cigar Makers International Union of America Collection, McKeldin Library, University of Maryland, College Park; David J. Saposs, *Left Wing Unionism: A Study of Radical Policies and Tactics* (New York, 1926), pp. 104–9, 155; Marion Savage, *Industrial Unionism in America* (New York, 1971[1922]), pp. 289–93; Sumner H. Slichter, *Union Policies and Industrial Management* (New York, 1968 [1941]), pp. 218–22.

53. Manning and Byrne, *The Effects on Women* (n. 35 above), pp. 37, 58–61.

54. Untaped interview with Maude Lenz, March 25, 1976, Richmond, Va.; Manning and Byrne, *The Effects on Women* (n. 35 above), pp. 31, 51, 60.

55. Interview with Anna Bartasius, March 31, 1982, Philadelphia; interview with Laura Garter Messer, June 20, 1978, Jacksonville, Fla.; "Rufus Lenoir Patterson's Cigar Machine" (n. 28 above); "Cigar-Making Machine That Is More than Human," *Scientific American,* December 1925, p. 386–87; *Jacksonville Journal,* March 28, 1933; interview with Hume; interview with Sutton (both n. 23 above); interview with Valetta Leiphart, June 17, 1982, Red Lion, Pa.; interview with Bessie Ray Henry, August 21, 1982, Lima, Ohio; interview with Purzon (n. 23 above); interview with Katharine Russell (pseudonym), April 1, 1982, Inglis Home, Philadelphia; interview with Lavoie; interview with Shea and Kramsz (all n. 52 above). Safety features were eventually installed on machines; particularly important was an electric eye that shut the machine off if movement was detected.

56. Interview with Lavoie; interview with Shea and Kramsz (all n. 52 above). Regarding strikes, see, e.g., files 176-1246, 182-1813, 199-51, 199-442, 176-117, 182-2233, 199-917, RG 280, Federal Mediation and Conciliation Service, NA; *CMOJ,* June 1934, p. 11; *York Gazette and Daily,* July 27–31, 1934; *Detroit News,* February 18, 1937, and clipping files of Maurice Sugar Papers, Walter Reuther Labor Archives, Wayne State University, Detroit, Mich.

57. *U.S. Tobacco Journal,* December 18–25, 1975, p. 20.

NINE

Dressmaking

WENDY GAMBER

In dressmaking, unlike cigarmaking, the most highly paid artisans were women. Nonetheless, in the turn-of-the-century era of
increasing mass production, women's clothing, like other commodities, was less and less often produced in small shops. The
case of dressmaking, explains Wendy Gamber, was complicated
by the apparent ubiquity and naturalness of the skill involved:
all females in the nineteenth century were taught to sew. Not all
females could sew equally well, but the dressmakers' competition came both from factory-made and homemade garments. For
much of the century, nonetheless, high fashion demanded the
knowledge of highly trained artisans; skirts and flounces had to
fall correctly, and smooth bodices had to be fitted precisely over
corseted torsos.

In the last decades of the nineteenth century, however, new
drafting and pattern systems—often patented by men—undermined the appreciation of dressmakers' knowledge. These new
technologies claimed to make any woman a competent dressmaker by "scientific" methods. What was scientific about these
methods, and how did they differ from traditional approaches?
Were they labor-saving devices comparable to vibrators and cigar
machines? What, for Gamber, made a "feminine" or a "masculine" technology in this period? How many ways of being female
are discussed here, and what were the characteristics of these
femininities? Would you say that the flapper of the 1920s was an
industrial style?

"'Reduced to Science': Gender, Technology, and Power in the American Dressmaking Trade, 1860–1910," *Technology and Culture* 36, no. 3 (1995): 455–82.
Reprinted by permission of the Society for the History of Technology.

"In Germany," writer-reformer Virginia Penny mused in 1863, "many dress makers are men, and there is one on Broadway, New York." Penny's observation—one of many such comments sprinkled throughout her massive *Cyclopedia of Woman's Work*—reveals much about cultural distinctions between men's and women's work in nineteenth-century America. The very incongruity of a "man dressmaker" affirmed a rigidly decreed sexual division of labor that assigned the making of feminine apparel to women; indeed, "man milliner" was an epithet leveled not only at men who made women's dresses and hats but at all who deviated from established norms of "masculine" behavior.[1] On the surface, at least, there was little conflict between ideology and reality. Women constituted 98 percent of the nation's "milliners, dress and mantua makers" in 1870, a proportion that remained largely unchanged as the century drew to a close.[2]

More often than not, sex segregation has been the bane of wage-earning women's existence, confining them to poorly paid jobs that promised few prospects for advancements.[3] But in this respect dressmaking contradicted conventional wisdom. The custom manufacture of feminine apparel (not until the late 1890s did ready-made clothing pose a serious challenge to "bespoke" work) held out rarely equaled opportunities to women of the working classes: highly skilled work, creative labor, relatively high wages, the very real possibility of someday opening an establishment of one's own. For dressmakers labored in a largely female world, one in which the lowly apprentice, the experienced veteran, even the shop's proprietor (who very likely had never married) were women.[4] Ironically, the most "feminine" of trades presented a compelling alternative to the middle-class standard of domesticity and the working-class ideal of the family wage.[5]

While they often ignored its more radical implications, contemporaries attributed this state of affairs to the "natural" consequence of biological factors—after all, was not sewing woman's "natural" vocation? But as several scholars have demonstrated, sexual divisions of labor are neither fixed nor natural but are continually redefined.[6] Indeed, the very flexibility of gender ideology provided male innovators with a means of challenging craftswomen's monopoly over dressmaking skills during the last quarter of the nineteenth century. Shrewdly manipulating the popular notion that garment making was an ability that *all* women "naturally" possessed, the proponents of new dress-cutting techniques sought to replace female craft traditions with "scientific" methods made and marketed by men.

Inventions such as pattern drafting systems and proportional patterns have been viewed as part and parcel of the "democratization of fashion," a process, culminating in the appearance of mass-produced, ready-made clothing that placed stylish apparel in the hands of ever-increasing numbers of eager consumers.[7] While this interpretation can be faulted for its uncritical acceptance of consumer culture, it also fails to acknowledge that "progress" did not benefit all parties equally. For if patterns and drafting systems served the interests of middle-class home sewers, they undermined the foundations of the dressmaking trade.

Equally significant, the "democratization of fashion" approach obscures the role that gender played in the manufacturing and merchandising of dressmaking innovations.[8] By marketing their creations to "ladies" and dressmakers alike, the proponents of new techniques employed a particular definition of "women's work," one that ignored the very real distinctions between the dressmaker's workshop and the middle-class home. This blurring of boundaries had unhappy consequences for professional clothiers. Tradeswomen not only lost work to would-be customers who fashioned their own clothes, but became increasingly vulnerable to amateurs who believed that their domestic skills qualified them for the marketplace. "Scientific" methods also represented a potent challenge to female authority. The designers and promoters of patterns and drafting systems firmly believed that dressmaking was women's work. But most of them assumed that men — uniquely endowed with "scientific knowledge" and tailoring skills — were the best-qualified teachers of the dressmakers' art. What contemporaries saw as a feminine skill was increasingly defined by men.

A "Feminine" Skill

As late as 1911, an investigator for the Women's Educational and Industrial Union pronounced dressmaking one of "the two most highly skilled trades for women" (the other was millinery). Recent scholarship has cast doubt on such assertions; all too often, feminist historians have noted, work done by men has been labeled as "skilled," work done by women as "unskilled." But if the anonymous social worker inadvertently maligned a large proportion of the female workforce, her assessment probably was correct. For dressmaking required mental as well as manual labor, "designing" as well as making. The vagaries of feminine fashion and the endless variety of bespoke work precluded the manufacture of a standardized product; each dress was an original creation, "a work of art." Even the "basics" were not easily grasped, a fact reflected in the time it took—five years at the very least—to ascend from apprentice to proprietor. Dressmaking was indeed a "highly skilled trade," and women had much to lose from its decline.[9]

The most difficult part of making a dress was cutting, that is, fashioning a "shape" that would fit the wearer—a skill that distinguished dressmakers from seamstresses, tailoresses, and other needlewomen, who typically stitched together garments that had been cut from the cloth by male tailors. In the late nineteenth century, an era when women's clothing tended toward the elaborate, cutting was no easy matter. The tightly fitting bodice (also called a "basque" or "waist"), a staple of Gilded Age fashion, posed a particularly vexing problem. According to one expert, it should fit "like wall paper."[10] (See fig. 9.1.)

Cutting a garment that fit "like wall paper" or "like the skin"—already made difficult by the variability of female forms—was further compounded by a host of seemingly mundane problems. A dress that lay smoothly across the shoulders when the wearer was standing was apt to wrinkle when she sat down. Similarly, the customer who stood ramrod straight during fittings was surprised that her dress hung differently when she assumed her "natural" posture.[11] Perhaps no item caused greater aggravation than the corset. More often than not, this frequently maligned undergarment created a shape quite different from the wearer's natural form. A dress that fit perfectly over one corset might collapse into unsightly folds over another; a garment cut for a tightly laced client no longer fit when the corset strings were loosened.[12]

Corsets or not, the dressmaker's first priority was to cut a garment that fit her patron. By the early nineteenth century, tailors had developed theories of bodily proportions, enabling them to create general patterns that could be altered to suit individual clients. But they were loath to share their knowledge. "We do not see why the plan used by tailors, of fitting by measure, is not more generally applied to dress fitting," Penny, ordinarily a perceptive observer, complained. The answer was simple: tailors kept their secrets to themselves. Denied access to "male" skills, dressmakers made do with a different method, one that costume historian Claudia B. Kidwell has called the "pin-to-the-form" technique. The modiste draped and pinned paper or inexpensive fabric (such as cambric or muslin) about the "form" of her client.[13] From the resulting "pattern," she cut the inside lining of the dress; if she used fabric instead of paper, the cloth itself became the lining. As

waistband, made of a wide sash of the material, tied in a large bow at the back. Percales, sateens, ginghams, linen lawns, may all be made after these simple styles.

No. 3.

No. 1—Is a model for a walking-costume, made of plain and pin-striped woolen goods, summer camel's-hair cloth, cashmere or beige. The short round skirt has three deep kilted flounces of the plain material, each with a narrow knife-plaited ruffle of the pin-stripe showing from under the edge, about two inches to show. The long cuirass bodice buttons close all the way down the front; the side seams at the back are left open half-way from the waist-line, and a kilted flounce is added on the back. Over this is tied a wide sash, with

bow and ends. This may be of gros-grain, satin, or watered ribbon, as the taste may decide. The Mother Hubbard cape is adjustable, and is made of the pin-stripe, using it lengthwise of the material. It is simply a straight length, gathered in to fit the neck and shoulders. It should be lined with soft Florence silk. Cuffs of the pin-stripe. Eight yards of plain material, in double-fold goods, and three yards of striped goods, same width, will be required.

No. 2—Is a toilette of figured and plain sateen.

No. 4.

It is very easy to make up, and can be worn for morning promenade. The skirt is of the plain sateen; and has, first, a flounce six inches deep; over this, six puffs, separated by narrow bands of

Figure 9.1. "Every-day dresses" from *Peterson's Magazine*, 1882. Cutting tightly fitting garments such as these required considerable expertise. (Courtesy Indiana University, Bloomington, Library.)

Figure 9.2. Illustration from page 187 of *The Book of English Trades* (1827) depicting the pin-to-the-form technique. As the text explains: "The plate represents the Dress-maker taking the pattern off from a lady by means of a piece of paper or cloth." (Courtesy Old Sturbridge Village, Sturbridge, Mass.; photo by Amanda Richardson.)

a precautionary measure, she basted the lining together and asked her customer to try it on—often several times—before touching her shears to the more costly materials that made up the outer garment. In the wake of numerous innovations, many tradeswomen clung to the pin-to-the-form method because it remained the most accurate—albeit the most time-consuming—means of obtaining a fit.[14] (See fig. 9.2.)

Compared to the mysterious art of cutting, the process of stitching together the pieces of a garment represented a matter of minor importance. But sewing, too, was a task that involved considerable talent, for ineptitude or inexperience easily ruined the "lines" of a dress. A skillful craftswoman knew that bodices required tiny, interlocking stitches; skirts, longer and looser ones, for seams that were pulled too tightly caused dresses — intended to fall gracefully to the floor — to "hang" awkwardly. "Skirt seams," one observer explained, "do not bear the strain of bodice seams, and sit better if the long stitch is employed." And in an era when bodices fit "like wall paper," it was in the dressmaker's best interest to make her "sewing last while the garment does."[15]

The distinction between cutting and sewing furnished the organizing principle of the dressmaking shop. Workers who cut garments from the cloth stood at the top of the trade hierarchy; in most establishments this was "Madame's" prerogative. Those who cut often bore responsibility for the final fitting as well; hence, "cutters" and "fitters" usually were synonymous terms. Workers who sewed the dress together called themselves finishers; as this nomenclature implied, they also added the final decorative touches. When not running errands or making deliveries, apprentices performed the less critical functions of basting, overcasting, "making" (that is, stitching together) linings, attaching hooks and eyes, and sewing skirt seams — "straight-away work."[16]

Because of the difficulties involved in cutting, observers placed a great deal of emphasis on the "mechanical" aspects of dressmaking. Still, many believed that dressmaking was an "art." Catherine Broughton's *Suggestions for Dressmakers* drew a clear distinction between the "artist" who possessed "a clear comprehension of the laws of beauty in dress" and the "artisan" who "knows only the technical part of her trade."[17] Broughton implied that "artistic" dressmaking was something that had to be learned. But others attributed this skill only to the naturally gifted. "You must have a natural taste for decorating the female form artistically; and thousands of girls have it," William Drysdale's *Helps for Ambitious Girls* explained. "Some girls are so expert at this that they produce better effects without any special training, than most of the dressmakers can produce."[18]

Thus, some argued that dressmaking was a natural skill, others that it was an acquired talent. This seemingly trivial debate had significant consequences. As Drysdale's remarks suggested, women with "natural taste" — but no training — might be tempted to enter the professional arena, lowering trade standards and exposing craftswomen to competition from amateurs. Natural taste apparently was not restricted to the few; "thousands of girls have it." Enterprising inventors — most of them men — carried this idea to its logical extreme by arguing that dressmaking was the natural employment of all women, an idea that found ready acceptance in a culture that equated women's labor with domesticity. And they sought to turn ideology into reality by making and marketing products that were intended to endow amateurs with the same skills as professionals.

Home and Workshop: The Double Meaning of "Women's Work"

Most dressmakers learned their trades through a more or less traditional system of apprenticeship; long after new ways of learning had appeared, the workshop remained the best teacher. But this fact did not deter an untold number of amateurs. "There is probably no occupation in which there are so many incompetent persons as that of dressmaking," Penny complained. "Many persons take it up without having learned that trade at all, and many who become reduced in circumstances immediately resort to it without any preparation, and are destitute not only of experience, but of skill, ingenuity and taste."[19]

Why did "incompetent persons" assume that they could easily transfer their domestic skills to the marketplace? They too believed that women were "naturally" endowed with dressmaking skills. This belief was not without foundation: in an era when ready-to-wear garments were unavailable, many women—perhaps the majority fashioned their own clothing. "I worked on Lucie's Basque and my own dress very steadily all day," Sarah Preston Everett Hale recorded in her diary on March 22, 1859. Hale's accomplishments should not be minimized. Home sewing, like all housework, was no more natural to women than blacksmithing was to men; on the contrary, it required considerable skill.[20] And well before the introduction of techniques that promised to simplify their endeavors, housewives might transgress the always artificial boundaries between home and marketplace by making garments for friends and neighbors, sometimes for cash, sometimes in exchange for payment in kind.[21]

Still, a considerable gulf separated amateur from professional. The typical dressmaker, a single woman of working-class origins, had spent years perfecting her craft—in the workshop, not in the home. The typical home sewer, engaged in performing domestic labor in the middle-class household, was a generalist, not a specialist; as Nancy Page Fernandez has ably demonstrated, she "made" her dress, not by designing a pattern anew, but by laboriously picking apart an older garment and using its pieces as her guide.[22] Historical and contemporary impressions notwithstanding, different life courses, different priorities, and, above all, different levels of expertise distinguished the professional dressmaker from her domestic counterpart.

Aspiring amateurs found support in the pages of advice manuals such as *The Ladies' Hand-Book of Millinery and Dressmaking* and *Beadle's Dime Guide to Dress-Making and Millinery,* and fashion periodicals. "Hints to Dressmakers and Those Who Make Their Own Dresses" was a regular feature of *Godey's Lady's Book; Demorest's Monthly Magazine,* established three decades later, offered "Hints in Regard to Dress-Cutting and Fitting."[23] "Hints" of this sort provided valuable aids to women who could not afford the services of professionals; no doubt they contributed to what Margaret Walsh has called the "democratization of fashion." But at the same time they deprived tradeswomen of actual and prospective customers. "I take it as a matter of economy," a dedicated subscriber wrote to *Demorest's Monthly Magazine,* "for its suggestions, hints, and good advice are worth more than its price, and save more than that in a dressmaker's bill."[24]

The next step was predictable: advice writers and fashion editors increasingly argued that by following their "hints," any woman could turn her natural talents to profit. To be sure, early manuals addressed themselves primarily to home sewers; antebellum "how-to" writers seldom implied that the knowledge they imparted was sufficient preparation for professionals.[25] But a change was evident as early as 1860. In that year Marion Pullan recommended her "dime guide" both to dressmakers "who would be glad of an opportunity to learn the principles of their art" and to "private individuals who . . . can not afford to employ a dress-maker." By the turn of the twentieth century, how-to manuals presented themselves as effective substitutes for apprenticeship. Simply by mastering the "twenty complete lessons" found in her *Dressmaking Self Taught,* Madam[e] Edith Marie Carens explained, "ambitious girls and women" could become professional dressmakers: "The author has aimed to make this work a school in itself—taking the place of . . . actually coming in contact with a sewing establishment.[26]

Those who argued that "anyone" could learn dressmaking had a point: technological innovations *were* making it easier for the untrained to fashion stylish apparel.

"Masculine" Technologies: The Transformation of Dressmaking

"Scarcely any dress-maker is now without a Wheeler & Wilson, however small her establishment," an observer noted in 1860. This statement, an exaggeration in its day, was an undisputed fact ten years later.[27] Much has been written of the devastation sewing machines wreaked on seamstresses, outworkers who sewed men's shirts, vests, and pants for a mass market. "If the sewing machine accomplished the work quicker," a perceptive analyst noted, "the amount, for a given amount of pay, was increased in precisely the same ratio." But the vagaries of feminine fashion prevented a similar upheaval in the world of dressmaking. As Fernandez has noted, sewing machines neither reduced the time and skill involved in cutting nor eliminated the need for hand sewing. While admirably suited to stitching the long, straight lines of the skirt, they were ill-adapted to the curved seams of the bodice. Long after the sewing machine had become a fixture of the dressmaking shop, "finishing" — overcasting seams, attaching trimmings — continued in the traditional manner. And in some instances knowledgeable modistes eschewed mechanization entirely, for machines easily ruined fragile fabrics.[28]

But if the impact of the sewing machine was less than revolutionary, a second development had more ominous implications. In the last quarter of the nineteenth century, hundreds of "systems" for drafting patterns challenged the dressmaker's most precious skill: her monopoly over cutting. Armed with a belief in progress and a desire for profits, the authors of treatises such as *The Scientific Lady Tailor System, The Science and Art of Cutting and Making Ladies' Garments,* and *Scientific Dress Cutting* sought to replace trade secrets with "scientific principles," feminine skills with masculine technologies.[29]

Much of this furious invention took place outside of the dressmaking shop. The majority of innovators were either tailors or men with little practical experience in the clothing trades. Charles Hecklinger, a member of the first group, denounced the second as charlatans who combined already discredited methods "with a few superficial ideas learned from their wives." As Hecklinger's criticism (itself an interesting indication of male attitudes) implied, women were not entirely absent from the ranks of systematizers. The "Buddington dress cutting machine," invented by Mr. and Mrs. F. E. Buddington, was a joint venture (although Mr. Buddington held the copyright); so was the "Excelsior Square," brainchild of Mr. and Mrs. B. T. Phelps of Bellows Falls, Vermont. And women, some of whom were dressmakers, accounted for more than one-fourth of the creators of drafting systems.[30]

But men, who also manufactured and marketed the most financially successful inventions, dominated system making. Condescending at best, deprecating at worst, they sought to "reduce" the feminine art of dressmaking to a science. The instructions to the 1896 edition of *S. T. Taylor's System of Dress Cutting* included a heartfelt dedication to "the inventor of our system: Mr. S. T. Taylor." The venerable founder was aptly named: "His profession was that of a men's tailor, with a thorough knowledge of cutting as it is done for men by the best artists." Taylor, his proponents insisted, was motivated by benevolence alone: "He saw the difficulties under which the dressmakers labored, and applied a life's study to elaborate a way of dress-cutting which would be on a par with cutting done by tailors."[31]

These humble pretensions aside, many inventors presided over vast commercial enterprises. Samuel T. Taylor and his corporate heirs not only marketed at least sixteen versions of his dress-cutting system between 1850 and 1915, they also sold patterns and published several fashion magazines. The McDowell Garment Drafting Machine Company created the *French Dressmaker;* in the early twentieth century the J. J. Mitchell Company—manufacturer of garment-cutting systems and "correct and perfect fitting" patterns—countered with the *American Dressmaker.*[32]

Equally significant, even the lowliest inventors did not purport to teach women the secrets of tailoring. Rather, they used their skills to devise various methods of cutting women's garments that could be learned without training, a process that displayed little respect for the abilities of dressmakers *or* home sewers. The McDowell Garment Drafting Machine Company claimed that its invention was "so simple . . . that a child can use it." The creators of Taylor's "system of dress cutting" believed that they had found no better demonstration of the ease with which their system could be mastered: "Pupils have learned the system well, and use it successfully who are both deaf and dumb, and foreigners who did not know a word of English, but had to be taught by sign."[33]

Bolstering their authority with repeated references to the at least implicitly masculine arena of science, systematizers rarely missed an opportunity to assert the superiority of masculine (tailoring) over feminine (dressmaking) skills.[34] "We frequently meet cutters," Hecklinger wrote, "who have practiced the greater part of their lives, and yet, who . . . work by such a crude system, compared to that sanctioned by experts in the art, that they can scarcely be mentioned beside them." "Among the many thousands of professional dressmakers in this country the number who excel in their profession are comparatively few," Caleb H. Griffin and David Knox, inventors of "the Great American Draughting Machine," regretfully acknowledged.[35]

What did "dress cutting . . . reduced to science" entail? Most "systems"—characterized by odd assortments of "tools," "charts," "scales," and "machines"—were

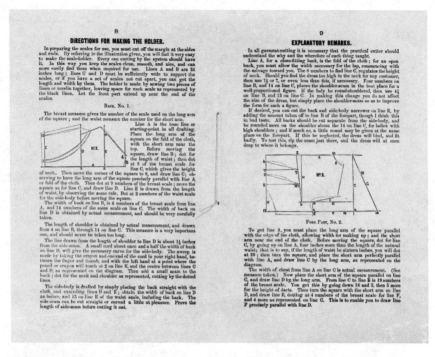

Figure 9.3. S. T. Taylor's illustrated instructions for drafting the front and back of a bodice provides an example of a "hybrid" system that combined proportional techniques with a series of direct measurements. Taylor was one of the most prolific and successful inventors of drafting systems; his emphasis on "mechanical" step-by-step instructions sharply contrasted with the spirit and rationale of the pin-to-the-form method. (S. T. Taylor, *A System for Cutting Ladies' Garments* [New York, 1871]; courtesy Library of Congress.)

designed with the difficult task of cutting bodices in mind; they attained their greatest popularity in the 1880s and 1890s when basques were at their tightest (see figs. 9.3 and 9.4). There was no shortage of ideas; between 1841 and 1920 the U.S. Patent Office bestowed its blessing on more than 360 systems. A detailed account of their workings cannot be given here, but, as Kidwell, the author of the most extensive study of the subject, has shown; each was a variation on one of three basic themes: proportional, hybrid, and direct measure. As Kidwell's nomenclature implies, proportional systems reflected inventors' conceptions of the relative dimensions of the female form. A single measurement — usually the breast or "bust," occasionally the waist — determined the remaining proportions. Hybrid

Figure 9.4. Mr. and Mrs. F. E. Buddington's "Improved Dress Cutting Machine" offered a relatively simple but less than accurate method of cutting dresses. The machine at left has been "set" in accordance with a series of measurements; the user would have traced around its outlines to draft the pattern. (Mr. and Mrs. F. E. Buddington, *Instruction Book for Using the Buddington Improved Dress Cutting Machine* [Chicago, 1896]; courtesy Library of Congress.)

systems determined some drafting points by proportional means, others by direct measurement. Direct measure systems depended on measurements alone.[36] Whatever the particular scheme they endorsed, drafting systems differed from the pin-to-the-form technique in a crucial respect, for they relied, not on acquired skill and the ability to mentally envision the "shape" of a garment, but on numeric measurements and "scientific" theories of bodily proportions.

Systematizers' claims to the contrary, science rarely triumphed over art; systems rarely produced professional results. Proportional techniques ("generally *invented* by a so-called *'professor'* and warranted to fit anything from the Venus of Milo to a baboon monkey," a proponent of a rival procedure complained) served the few rather than the many: "fit" depended on how well the wearer conformed to the inventor's notions of bodily proportions. More often than not, even hybrid methods proved no match for the tightly fitting bodices of the 1880s and 1890s.[37] Direct-measure systems offered greater accuracy but were difficult and time-consuming to use. Consider, for example, Elizabeth Gartland's *Original American Lady Tailor System*. After taking thirteen measurements, the user was ready to draft a "plain basque." She then confronted a bewildering list of instructions:

1. Draw line 1, ten inches above the bottom of paper, the entire length of square, for *waist line*.
2. Draw line 2 from centre of line 1, according to *length of back*.
3. On line 2 make a dot above the waist line, for the *under-arm measure*, and draw a line parallel to line 1, for line 3.
4. Place centre of circle on line 2, resting on line 3, and *draw a circle* according to *arm's-eye measure*.
5. Draw line 4 though the centre of circle, the same length as line 3. Also line 5 parallel to line 4, so it will touch the top of circle on line 2.
6. Draw line 6 touching right of circle from line 5 to line 1.

When she had reached step number 40, the pattern was complete (see fig. 9.5). Whether systems such as Gartland's represented an improvement over older methods of garment cutting is doubtful. And their success depended on how skillfully the user wielded her inch tape; inexact measurements resulted in ill-fitting garments.[38]

These drawbacks failed to discourage numerous purchasers lured by the promise of dressmaking made easy. No accurate sales statistics exist, but the sheer num-

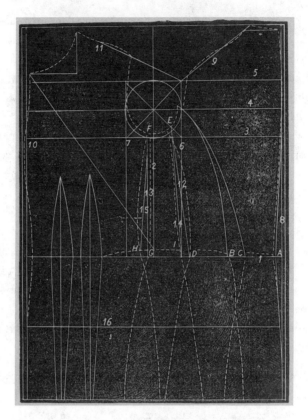

Figure 9.5. Elizabeth Gartland's diagram for drafting a
"plain basque" illustrates the complexity of direct-measure
systems for drafting patterns. (Elizabeth Gartland, *The
Original American Lady Tailor System* [Philadelphia, 1884];
courtesy Library of Congress.)

ber of systems available provide some indication of their popularity. Who used
them? Some inventors catered to amateurs, others to professionals, still others to
both. *Every Lady Her Own Dressmaker* and *A Trade Worth Fifty Dollars to Any
Family* clearly intended themselves for thrifty consumers. Taylor, in contrast, offered his invention to "the practical dress maker that wishes to please her customers." Mrs. H. A. Ross was more ecumenical; she dedicated her "Tailor System"
"to the practical dress-makers of the United States and to all who wish to master
the science of garment cutting." Those who testified to the merits of the "Standard
Garment-Cutter" included four dressmakers and seven home sewers; among the

latter was twelve-year-old Hermene Crome, who announced that "since learning the System [I] can cut and fit my mamma's and my own clothes. I like to use it because it is so easy."[39]

By the 1890s, inventors who marketed their creations to amateurs and professionals alike probably were in the majority. This was a shrewd business tactic; aiming for a dual market vastly increased the number of potential buyers. But by simultaneously recommending their systems "to dressmakers" and "to ladies in private life," the makers of systems increasingly blurred the boundaries between home and workshop. Consciously or not, male inventors redefined not only the nature but the meaning of "women's work." By identifying dress cutting with middle-class domesticity, by classifying it as a variant of the housewifely labor that "all" women presumably performed, systematizers obscured the artisanal origins of the dressmaking trade.[40]

What is more, they threatened established traditions. For two centuries, novices had honed their skills at the workplace; "Madame" was the supreme arbiter of dressmaking knowledge. Advocates of scientific methods challenged her authority, for they argued not only that every lady could become her own dressmaker but that with the aid of their systems every lady could become a professional dressmaker.

The recipient of "hundreds of letters from young girls and widows who have their living to make, and oftentimes from two to six children to support," Will C. Rood implored his readers to contribute to a fund that would supply poor women with his "Magic Scale." However charitable his motives, Rood preached a subversive doctrine. The lucky possessors of Magic Scales, he implied, were ready for business; they could set up shop without the benefit of apprenticeship. Certainly, this was the intention of Alice M. Avis of Blue Rapids, Kansas, a graduate of one of the "schools" sponsored by the Standard Garment Cutting Company. "In my future business as a dressmaker, I would not do without the System at any price."[41]

To be sure, systems in part represented a response to the apprenticeship crisis that plagued the dressmaking trade. "Mistress" dressmakers—usually operating under considerable economic stress—were infamous for failing to impart promised knowledge. While complaints that apprentices learned nothing but "plain sewing" surface as early as the 1840s, the situation clearly had worsened by the closing decades of the nineteenth century. By every indication, "madames" evinced particular reluctance to divulge the secrets of cutting. As one critic observed in 1885, "after thirty days" the novice "may know how to cut the inner side of a sleeve, instead of the whole dress which it was promised she should know how to fit"; ten years later another complained that "many a woman spends month after month in stitching fells till she has acquired a purely mechanical accuracy, and who could by no possibility either cut, fit, or make an entire garment."[42]

But scientific methods compounded the very problems they purported to solve. Much of dressmakers' ill will toward their apprentices could be attributed to the competitive nature of their trade; refusing to train novices was one way of limiting the number of future contenders. By providing a substitute for apprenticeship, systems only made matters worse, increasing the competition in an already over-crowded field. And given their technical limitations, they hardly ensured commercial success for the inexperienced, a consequence that weighted particularly heavily on women of slender resources who were unexpectedly obliged to support themselves. All too often, one suspects, systematizers' promised evaporated into cruel hoaxes.

Even when incorporated into long-standing trade practices, drafting systems had an injurious effect. The dressmaker who informed the S. T. Taylor Company that "I have used your system of dress-cutting for about six years and . . . teach it to my girls" preserved the form if not the substance of apprenticeship. But she had exchanged a centuries-old female tradition for a scientific system made and marketed by men. In a sense, she had abdicated her own authority. And nothing prevented her "girls" from acquiring Taylor's methods on their own.[43]

Because they lacked a forum in which to air their opinions, dressmakers' response to cutting innovations is difficult to gauge.[44] If testimonials can be believed (Taylor insisted that "every letter is, of course, genuine"), many wholeheartedly embraced scientific methods. "For fourteen years I have used your system, always giving perfect satisfaction to customers and myself," a New York tradeswoman wrote to Taylor in 1896. Hattie E. Gillespie of Cincinnati was equally enthusiastic: "I have used your system for the past eight years and don't think there is anything like it."[45]

In one respect Gillespie's decision made a great deal of sense, for (most) scientific methods had a crucial advantage over older techniques: speed. Systems and sewing machines were a powerful combination; together they enabled dressmakers to fashion garments more quickly than ever before, simultaneously promising to satisfy the demands of impatient customers and to increase potential profits.[46] Perhaps no prospect was more tantalizing than that systems did away with fittings.

Never one to minimize the significance of his achievements, Taylor declared: *"By my system dresses can be cut to fit so perfectly, as to avoid any necessity of fitting!"*[47]

At least one observer suggested that such claims should be taken with a grain of salt. Broughton, the author of *Suggestions for Dressmakers* and a staunch defender of traditional methods, maintained that no system could do away with the need for fittings; "'trying on' is bound to continue to be thought one of the bug-a-boos of dressmaking. . . . Dress fitting is not a mathematical problem that can be solved by learning a few rules," she warned. "Parisian dressmakers" — the model to which Broughton believed all American modistes should aspire — "are famous for their pooh-poohing of systems." These apostles of perfection apparently showed little interest in time-saving techniques: "If the Parisian dressmaker requires your presence for ten fittings, she tells you so and gets them."[48]

Broughton's observations suggest that a quiet battle raged in the hearts and minds of dressmakers between quantity and quality, industry and art. Quite likely the dilemma resolved itself along class lines. Taylor argued that the pin-to-the-form technique, while appropriate for French modistes who could "afford to charge small fortunes for their creations," ill suited the mass of American dressmakers, most of whom depended on less affluent customers.[49] Although his objectivity can be questioned, Taylor had a point. The "best" shops in both France and America were likely to disdain systems. Wealthy patrons could afford to pay "small fortunes," and the leisure to endure innumerable fittings was perhaps as much a symbol of conspicuous consumption as the dress itself.[50] More often than not, the customer with less time on her hands and less money to spend settled for "slovenly" work and — systematizers' claims to the contrary — a less than perfect fit.

Drafting systems appear to have been distributed by geography as well as class. A surprising number of inventors hailed from the small towns of the Midwest: Quincy, Illinois; Baraboo, Wisconsin; Steubenville, Ohio; and Webster City, Iowa, were places where apprenticeship traditions were likely to be weak. Perhaps not coincidentally, most of the North Dakota dressmakers interviewed for one study relied on "charts."[51] Enthusiastic proponents of systems could be found in New York and Chicago as well as Beattie, Kansas, and Maryville, Missouri. But the pin-to-the-form technique probably survived longest in urban areas. Virtually every Cleveland dressmaking shop visited by social investigator Edna Bryner in 1916 used a variant of this traditional method: "The system of dressmaking used in all establishments consists in making a cambric lining, called a foundation or guide, to fit the customer; putting this foundation on a form somewhat smaller and stuffing it out so as to duplicate the customer's figure; fitting the lining proper upon this foundation; 'building' the dress upon it in the proper lines; and fitting the draped dress upon the customer to make finer adjustments before completing the finishing operations."[52]

But if systems were not universally adopted, they gave rise to another, perhaps even more troubling, invention. "Graded" — that is, sized — patterns were the logical outgrowth of proportional methods. These fragile pieces of tissue paper offered considerable benefits to home sewers — women who faced great pressure to dress fashionably *and* cheaply — but they had tragic consequences for the art of dressmaking. For patterns eliminated the need for drafting altogether; the modiste who relied on them was little more than a seamstress.[53]

Patterns were not new to Gilded Age Americans; unsized versions appeared in fashion periodicals as early as the 1850s. But these at best provided only limited assistance; *Godey's* patterns, for example, were intended to fit "a lady of middle height and youthful proportion."[54] Graded patterns were not invented until 1867, a distinction that Ebenezer Butterick, a former tailor, rightfully claimed. (Indeed, he parlayed his creative efforts into one of the longest-lived enterprises in corporate history; the Butterick Company makes patterns today.) Butterick's creations followed the same principles as proportional drafting systems, that is, they were sized according to bust measurements. "Designed for the use of persons not very familiar with making garments," they clearly intended to eliminate the need for skill. Butterick's arch rival, Madame Demorest, made much the same point; her patterns, she claimed, were "so accurately cut and notched that any novice can put them together."[55]

Like drafting systems, patterns primarily were a male invention. Herman and Schamu Moschcowitz, makers of "Bazar Cut Paper Patterns," presided over a New York establishment that employed ninety "man dressmakers." James McCall, a former tailor and the creator of the "Royal Chart," produced another extraordinarily long-lived project, the McCall pattern. Ellen Curtis Demorest, who with her husband created a fashion empire that included *Demorest's Monthly Magazine,* the Demorest Sewing Machine Company, Demorest's patterns, and Madame Demorest's Excelsior System, constituted the sole exception; she evidently played some role (how significant is unclear) in designing the innovations that bore her name.[56]

Gender mattered little to consumers; patterns were a resounding commercial success. "Last year this firm sold over four millions of patterns," E. Butterick and Company declared in 1871; while exaggerated, this was probably not far from the truth. And no wonder. Once possessed of graded patterns, home sewers and amateur craftswomen did not need to cut or to draft, but only to sew.[57]

Patterns offered a substitute not only for the dressmaker's skill but also for her judgment. For unlike drafting systems, which required the user to exercise her imagination, they provided a finished template, a paper version of the actual garment. The pattern, in other words, *was* the fashion. (See fig. 9.6.) Inexpensive and easy to use, patterns provided stylish apparel to women who could not afford the services of professionals. But the same development that heralded "democracy" for the consumer threatened penury for the dressmaker. Like drafting systems, patterns encouraged erstwhile patrons to make their own clothes. "I went a few days ago since to get a dress made, but the dressmaker could not make it at once, and I don't like to wait, so I came home, looked over the many patterns sent with the Demorest number, and with aid of those . . . have cut and fitted my dress," a satisfied subscriber informed *Demorest's Monthly Magazine* in 1879. "Country" was more blunt: "Until last year I employed a dressmaker, but I find I can, with a good pattern, suit myself better."[58]

Tradeswomen themselves might use patterns, but only the most inexperienced relied on them entirely. Like drafting systems, they met the needs of women unexpectedly thrown on their own resources; like systems, they were popular among hinterland dwellers who had few opportunities to "learn" their trades.[59] But the "fashionable" dressmaker who enthusiastically embraced patterns was likely to lose customers, for standardized styles failed to satisfy discriminating patrons. Manufacturers' protestations to the contrary, their "newest" designs aped last year's Paris fashions; despite the illusion of variety, their catalogs often betrayed a dull uniformity. "No woman wishes a gown made in a fashion already in common use when the pattern maker selects it," Broughton explained, noting that the typical designer "gets his inspiration . . . at second or fifth hand." (Significantly, Broughton's references to designers invoked the masculine pronoun.) And patterns by their very nature limited creativity; modistes who feared "'making two dresses alike' . . . a fear for which . . . their customers are responsible," wisely avoided them.[60]

Skilled practitioners had even greater reason to eschew them, for patterns, because they were based on the same theories, exhibited all the weaknesses of pro-

Figure 9.6. Illustration from *E. Butterick & Co's. Catalogue, Summer, 1875.* Perhaps because they recognized the deficiencies of the proportional techniques on which their products were based, pattern manufacturers' selections emphasized looser garments such as house dresses and wrappers. (Courtesy American Antiquarian Society, Worcester, Mass.)

portional drafting systems. However "correct" the "principles" on which they were "graduated," however "mathematically wrought," they seldom rendered a perfect fit. More often than not, the wearer faced two alternatives: an ill-fitting dress or a spate of alterations.[61]

Conclusion

The persistence of age-old practices and the not-inconsiderable deficiencies of modern devices suggest that, by the turn of the twentieth century, scientific methods had won an incomplete victory at best. We have no means of learning how much business dressmakers lost to newly emboldened home sewers, but despite the availability of paper patterns and drafting systems, the number of professionals continued to increase.[62] Some dressmakers clung to traditional methods; others embraced "scientific principles." Some continued to learn their trades from "Madame"; others—with the help of patterns and systems—taught themselves. The ratio of one to the other no doubt will remain a mystery. But one thing is certain: craftswomen who could not cut without the aid of patterns and systems were far more prevalent in 1900 than in 1860. Indeed, one suspects that they accounted for many of the "incompetent" dressmakers of whom nineteenth-century Americans were so fond of complaining.

Perhaps more significant, new technologies—eagerly proffered to "ladies" and dressmakers alike—challenged the tradeswoman's special expertise and eroded her unique authority. Female clothiers increasingly competed with male innovators to control the dissemination of craft knowledge, to determine who could become a dressmaker. They also competed with amateur craftswomen and home sewers. Theoretically, aspiring modistes no longer needed to apprentice themselves to mistress dressmakers; theoretically, the purchasers of patterns and systems could do without the services of professionals, even if they yielded less than perfect results. But innovations were not an unequivocal benefit—even for consumers. Less work for the dressmaker, Ruth Schwartz Cowan reminds us, meant "more work for mother."[63]

Viewed from the perspective of custom clothiers, new technologies had far more ominous implications, for they paved the way for ready-made clothing. Constrained by fashion, they remained within the confines of custom production through the nineteenth century. But "sizes" based on proportional theories, wholly inadequate for Gilded Age vogues, were entirely appropriate for the more loosely fitting shirtwaists and tea gowns of the 1910s and, especially, the "flapper" dresses of the 1920s.[64] To be sure, mass production replaced a cumbersome and expensive system with one that offered immediately available ready-to-wear clothing at hitherto unheard-of prices. But if the democratization of fashion represented an unmitigated blessing to consumers, it provided few advantages to the newly appointed producers of feminine apparel. Ready-made clothing did not simply relieve middle-class women from domestic drudgery, it also "relieved" female artisans from relatively remunerative and satisfying employment. In place of highly skilled, creative labor, garment manufacturing offered monotonous, repetitive work at low wages. Nor were women able to maintain control over the sexual division of labor; inside the male-operated factory was a "mixed" workroom that barred women from all but the humblest positions. "Scientific principles" heralded a future without dressmakers, a future that largely would be controlled by men.[65]

NOTES

Dr. Gamber thanks Patricia Cooper, Nancy Page Fernandez, Morton Keller, Glenna Matthews, Carole Srole, and especially Carole Turbin for their comments on earlier drafts of this article, and Patricia Trautman for allowing access to her impressive collection of instructional pamphlets. The research for this article was supported by fellowships from the Women's Studies Program, Brandeis University; the Smithsonian Institution; the Arthur and Elizabeth Schlesinger Library on the History of Women in America; and the Massachusetts Historical Society.

1. Virginia Penny, *The Employments of Women: A Cyclopedia of Woman's Work* (Boston, 1863), p. 324. See Roscoe Conkling's comments on reformers, quoted in Morton Keller, *Affairs of State: Public Life in Late Nineteenth Century America* (Cambridge, Mass., 1977), p. 248.

2. According to the census of 1870, there were 92,084 "milliners, dress and mantua makers" in the United States, of whom 90,480 were female and 1,604 male. See *A Compendium of the Ninth Census* (June 1, 1870) (Washington, D.C., 1872), p. 612. For 1900 figures, see *Statistics of Women at Work* (Washington, D.C., 1907), pp. 70, 75.

3. As historians are beginning to recognize, sex segregation did have advantages, for it gave women access to skilled jobs within "their" trades. See esp. Mary H. Blewett, *Men, Women, and Work: Class, Gender, and Protest in the New England Shoe Industry, 1780–1910* (Urbana, Ill., 1988), pp. xvi, 104–10, 138–39, 324–25.

4. This is not to say that all dressmakers lived lives of prosperity and ease; rather, they enjoyed significant advantages *in comparison to* other wage-earning women. See Wendy Gamber, "The Female Economy: The Millinery and Dressmaking Trades, 1860–1930" (Ph.D. diss., Brandeis University, 1991), esp. chaps 1 and 2, and "A Precarious Independence: Mil-

liners and Dressmakers in Boston, 1860–1920," *Journal of Women's History* 4 (Spring 1992): 60–88.

5. The literature of domesticity and "woman's sphere" is vast. See esp. Barbara Welter, "The Cult of True Womanhood, 1820–1860," *American Quarterly* 18 (1966): 151–74; Gerda Lerner, "The Lady and the Mill Girl: Changes in the Status of Women in the Age of Jackson, 1800–1840" in *A Heritage of Her Own: Toward a New Social History of American Women,* ed. Nancy F. Cott and Elizabeth H. Pleck (New York, 1979), pp. 182–96; Nancy F. Cott, *The Bonds of Womanhood: "Woman's Sphere" in New England, 1780–1835* (New Haven, Conn., 1977); and Mary P. Ryan, *Cradle of the Middle Class: The Family in Oneida County, New York, 1790–1865* (Cambridge, 1981). For interpretations of the family wage, see Martha May, "Bread before Roses: American Workingmen, Labor Unions, and the Family Wage," in *Women, Work, and Protest: A Century of U.S. Women's Labor History,* ed. Ruth Milkman (Boston, 1985), pp. 1–21; and Alice Kessler-Harris, *A Woman's Wage: Historical Meanings and Social Consequences* (Lexington, Ky., 1990), esp. pp. 8–12, 19–20.

6. On this point, see Julie A. Matthaei, *An Economic History of Women in America: Women's Work, the Sexual Division of Labor, and the Development of Capitalism* (New York, 1982); Ava Baron, "Contested Terrain Revisited: Technology and Gender Definitions of Work in the Printing Industry, 1850–1920," in *Women, Work, and Technology: Transformations,* ed. Barbara Drygulski Wright et al. (Ann Arbor, Mich., 1987), pp. 58–83; and the essays in Ava Baron, ed., *Work Engendered: Toward a New History of American Labor* (Ithaca, N.Y., 1991).

7. Margaret Walsh, "The Democratization of Fashion: The Emergence of the Women's Dress Pattern Industry," *Journal of American History* 66 (1979): 299–313; for general discussions of the "democratization of fashion," see Jessica Daves, *Ready-Made Miracle: The American Story of Fashion for the Millions* (New York, 1967); Daniel Boorstin, *The Americans: The Democratic Experience* (New York, 1974), pp. 91–100 (Boorstin limits his analysis to men's clothing); and Claudia B. Kidwell and Margaret C. Christman, *Suiting Everyone: The Democratization of Clothing in America* (Washington, D.C., 1974). For criticisms of this approach, see Stuart Ewen and Elizabeth Ewen, *Channels of Desire: Mass Images and the Shaping of American Consciousness* (New York, 1982); Susan Porter Benson, *Counter Cultures: Saleswomen, Managers, and Customers in American Department Stores, 1890–1940* (Urbana, Ill., 1986), p. 78; Susan Strasser, *Satisfaction Guaranteed: The Making of the American Mass Market* (New York, 1989); and William R. Leach, *Land of Desire: Merchants, Power, and the Rise of a New American Culture* (New York, 1993).

8. By "gender," I mean the social construction of "masculinity" and "femininity" as opposed to biological sex; see Joan Scott's now-classic discussion, "Gender: A Useful Category of Historical Analysis," *American Historical Review* 91 (1986): 1053–75.

9. Women's Educational and Industrial Union, Research Department, "Industrial Opportunities for Women in Somerville" (typescript), 1910–11, box 7, folder 16, p. 46, Records of the Women's Educational and Industrial Union (hereafter cited as WEIU), collection B-8, Schlesinger Library, Radcliffe College, Cambridge, Mass. For analyses that document the sexism inherent in skill labels, see Benson (n. 7 above), p. 229; Judith A. McGaw, "No Passive Victims, No Separate Spheres: A Feminist Perspective on Technology's History," in

In Context: History and the History of Technology, ed. Stephen H. Cutcliffe and Robert C. Post (Bethlehem, Pa., 1989), pp. 178–84; Anne Phillips and Barbara Taylor, "Sex and Skill: Notes towards a Feminist Economics," *Feminist Review* 6 (1981): 79–88; and Joan Wallach Scott, "Work Identities for Men and Women: The Politics of Work and Family in the Parisian Garment Trades in 1848," in her *Gender and the Politics of History* (New York, 1988), pp. 93–112.

10. [Catherine Broughton], *Suggestions for Dressmakers* (New York, ca. 1896), p. 4; Claudia B. Kidwell, *Cutting a Fashionable Fit: Dressmakers' Drafting Systems in the United States* (Washington, D.C., 1979), pp. 16–18; Martha Louise Rayne, *What Can a Woman Do?* (1893; reprint, New York, 1974), p. 210.

11. [Broughton], pp. 13, 15, 17; Kidwell, p. 18. See also *Demorest's Monthly Magazine,* September 1881, p. 729.

12. [Broughton], p. 13; Kidwell, p. 18; *Demorest's Monthly Magazine,* July 1885, pp. 600–601; Amelia Des Moulins, "The Dressmaker's Life Story," *Independent* 56 (April 28, 1904): 945.

13. Penny (n. 1 above), p. 325; Jo Anne Preston, "'To Learn Me the Whole of the Trade': Conflict between a Female Apprentice and a Merchant Tailor in Ante-Bellum New England," *Labor History* 24 (1983): 259–73; [Broughton], p. 21; Kidwell, pp. 11–13. Of course tailors might employ women to sew. But they rarely taught them how to cut.

14. Kidwell, p. 20; Penny, p. 326.

15. [Broughton], p. 29; Rayne (n. 10 above), p. 213.

16. Penny, p. 324; [Broughton], p. 29; Edna Bryner, *Dressmaking and Millinery* (Cleveland, 1916), p. 79; May Allinson, *Dressmaking as a Trade for Women in Massachusetts* (Boston, 1916), p. 33. To be sure, dressmakers labored under a wide variety of circumstances. At one extreme was the modiste who "went out by the day," obliged to "know how to do all parts, from the fitting to the finishing off." At the other, evident by the late nineteenth century, was the large shop, characterized by minute subdivisions of labor. Employees in such an establishment specialized on a particular part of the garment: waist, skirt, sleeves. But as late as the 1910s, the small shop predominated. See Penny, p. 325; [Broughton](n. 10 above), p. 29; *Girls Trade Education League, Vocations for Boston Girls, No. 5: Dressmaking* (Boston, 1911), p. 6; Allinson, pp. 31–42; and Bryner, *Dressmaking and Millinery,* pp. 28–35.

17. [Broughton], pp. 1–3. See also Gertrude G. de Aguirre, *Women in the Business World: Or Hints and Helps to Prosperity by One of Them* (Boston, 1894), p. 37.

18. William Drysdale, *Helps for Ambitious Girls* (New York, 1900), pp. 428–29. See also Mary Cadwalader Jones, "Opportunities for Women," *Delineator* 98 (July 1896): 106.

19. *Demorest's Monthly Magazine,* August 1880, p. 473; Penny (n. 1 above), p. 325.

20. Horace Greeley et al., *The Great Industries of the United States* (Hartford, Conn., 1872), p. 587; Sarah Preston Everett Hale, Diary, March 22, 1859, Sophia Smith Collection, Smith College, Northampton, Mass.; Jeanne Boydston, *Home and Work: Housework, Wages, and the Ideology of Labor in the Early Republic* (New York, 1990); Glenna Matthews, *"Just a Housewife": The Rise and Fall of Domesticity in America* (New York, 1987); Susan Strasser, *Never Done: A History of American Housework* (New York, 1982).

21. For example, Mary Ann Laughton occasionally made dresses for other women (at

least sometimes for cash payments) before she decided in August 1837 to "learn [her] trade." See Mary Ann Laughton, Diary, February 1837–August 1837, Essex Institute, Salem, Mass.; see also Persis Sibley Andrews, Diary, January 14, 1842, Dabney Papers, Massachusetts Historical Society, Boston. See Boydston for a critique of the home/work dichotomy.

22. See Gamber, "Female Economy," chap. 1, and "A Precarious Independence," pp. 66–70 (both n. 4 above), for a more detailed discussion of dressmakers' careers. See also Nancy Page Fernandez, "'If a Woman Had Taste': Home Sewing and the Making of Fashion, 1850–1910" (Ph.D. diss., University of California, Irvine, 1987), pp. 103–5, and "Women, Work and Wages in the Industrialization of American Dressmaking, 1860–1910" (paper presented at the ninth Berkshire Conference on the History of Women, June 1993, Vassar College, Poughkeepsie, N.Y.), pp. 8–10.

23. *The Ladies' Hand-Book of Millinery and Dressmaking, With Plain Instructions for Making the Most Useful Articles of Dress and Attire* (New York, 1844); Marion M. Pullan, *Beadle's Dime Guide to Dress-Making and Millinery: With a Complete French and English Dictionary of Terms Employed in Those Arts* (New York, 1860). For do-it-yourself advice in fashion magazines, see, e.g., *Godey's Lady's Book* 53 (October 1856): 307–8; 53 (November 1856): 433–35; 53 (December 1856): 527–29; 56 (May 1858): 413–14; and 56 (June 1858): 523–24; and Demorest's Monthly Magazine, January 1870, p. 18; April 1880, p. 231; and July 1885, pp. 600–601. See also *Delineator* 37 (August 1891): vii; and Frank Luther Mott, *A History of American Magazines*, vol. 1 (Cambridge, Mass., 1939), pp. 580–94, and vol. 3 (Cambridge, Mass., 1957), pp. 325–28.

24. Walsh (n. 7 above), pp. 299–313; *Demorest's Monthly Magazine*, February 1877, p. 107.

25. When moralist T. S. Arthur suggested that all "young ladies" learn dressmaking and millinery just in case they were someday forced to support themselves, he did not provide instructions; he recommended a term as an apprentice in a dressmaker's or milliner's shop. See T. S. Arthur, *Advice to Young Ladies on Their Duties and Conduct in Life* (Boston, 1849), pp. 28–35.

26. Pullan, p. 8; Madam[e] Edith Marie Carens, *Dressmaking Self Taught in Twenty Complete Lessons* (Toledo, Ohio, 1911), pp. 3, 120–27. See also Sara May Allington, *Practical Sewing and Dressmaking* (Boston, 1913).

27. Pullan, p. 10.

28. Fernandez, "If a Woman Had Taste" (n. 22 above), pp. 233–35. Kidwell has suggested that the time-saving features of the sewing machine may have provided an impetus for creating faster methods of cutting garments from the cloth (Kidwell [n. 10 above], p. 20). See also Elizabeth Gartland, *The Original American Lady Tailor System* (Philadelphia, 1884), p. 36; Bryner, *Dressmaking and Millinery* (n. 16 above), p. 105; Boston Public Schools Trade School for Girls, *Bulletin*, no. 20 (November 1927).

29. This sampling of titles was taken from Patricia A. Trautman's *Clothing America: A Bibliography and Location Index of Nineteenth-Century American Pattern Drafting Systems* (n.p., 1987), pp. 32, 41, 54.

30. Charles Hecklinger, preface to his *The Dress and Cloak Cutter: In Two Parts* (New York, 1883); Mr. and Mrs. F. E. Buddington, *Instruction Book for Using the Buddington Improved Dress Cutting Machine* (Chicago, 1896); Trautman, pp. 64–65. Eighty-four (27.4 per-

cent) of the 307 systems for drafting women's garments listed in Trautman's bibliography were invented or co-invented by women.

31. *S. T. Taylor's System of Dress Cutting* (New York, 1896), pp. 21, 23.

32. Trautman, pp. 82–84; Samuel T. Taylor, *A System for Cutting Ladies' Garments* (New York, 1871), pp. 1–3, 18; *S. T. Taylor's System of Dress Cutting*, pp. 42–43; *Instruction Book with Diagrams for S. T. Taylor's System of Cutting Ladies' Garments* (New York, 1915). See, e.g., *French Dressmaker* 10 (January 5, 1894); and *American Dressmaker* 1 (June 1910) and 1 (December 1910): 5.

33. *The McDowell Standard System of Garment Cutting* (New York, [1886?]), p. 2; *S. T. Taylor's System of Dress Cutting*, p. 27. All systems cited (i.e., instruction booklets for drafting systems) are in the collection of Patricia Trautman, School of Family Studies, University of Connecticut, Storrs.

34. See Evelyn Fox Keller, *Reflections on Gender and Science* (New Haven, Conn., 1985), esp. pp. 7–13, 33–65, 75–94, for a discussion of historical and contemporary linkages between masculinity and science. Drafting systems appear to be less than "scientific" from the vantage point of the late twentieth century. But as several scholars have shown, the line that separated science from pseudoscience in the nineteenth century often was a thin one. See esp. Arthur Wrobel, Introduction, pp. 1–20, and John L. Greenway, "'Nervous Disease' and Electric Medicine," pp. 46–73, in *Pseudo-Science and Society in Nineteenth-Century America*, ed. Arthur Wrobel (Lexington, Ky., 1987); and Sally Shuttleworth, "Female Circulation: Medical Discourse and Popular Advertising in the Mid-Victorian Era;" in *Body/Politics: Women and the Discourses of Science*, ed. Mary Jacobus, Evelyn Fox Keller, and Sally Shuttleworth (New York, 1990), pp. 47–68. Women, too, might appropriate the language of science, as did Madame Mallison, the author of *Dressmaking Reduced to a Science: The Eclectic Lady-Tailor System of Dress Cutting* (Washington, D.C., 1886); see Kidwell (n. 10 above), p. 50.

35. Hecklinger, preface (n. 30 above); Caleb H. Griffin and David Knox, *The Science and Art of Cutting and Making Ladies' Garments, as Demonstrated by Griffin & Knox's Great American Draughting Machine* (Lynn, Mass., 1873), p. 6.

36. Samuel T. Taylor, *Dress Cutting Simplified and Reduced to Science* (Baltimore, 1850). In the early 1890s, when fashion called for tightly fitting skirts, inventors obliged by creating methods for drafting the latter as well (Kidwell, pp. 2, 25–74, 99–100, 127–50). During this period 363 systems for drafting women's garments received patents; of course not all of them actually reached the marketplace. Much of the following discussion relies on Kidwell's pioneering study. For additional information on the workings of systems, see Fernandez, "If a Woman Had Taste" (n. 22 above), pp. 122–61; and Gamber, "Female Economy" (n. 4 above), pp. 293–322.

37. *S. T. Taylor's System of Dress Cutting* (n. 31 above), p. 16; Kidwell, pp. 28, 45.

38. Gartland (n. 28 above), pp. 11–17, 33; Buddington (n. 30 above), p. 3; [Broughton] (n. 10 above), p. 8.

39. Trautman (n. 29 above), pp. 53–54, 85–86; Taylor, *Dress Cutting Simplified and Reduced to a Science*, p. 2; Mrs. H. A. Ross, *Instructions for Using Mrs. Ross' Tailor System for Cutting Ladies' and Children's Garments of All Kinds* (Battle Creek, Mich., n.d.); Standard

Garment Cutting Company, *Standard Square Inch Tailoring System* (Chicago, 1896), pp. 38–39; Louis Molpoer, *Every Lady Her Own Dressmaker: The Scientific Lady Tailor System for Cutting Ladies' Dresses and Coats,* 2d ed. (Washington, D.C., 1891); Mrs. Bertha Thompson, *A Trade Worth Fifty Dollars to Any Family* (n.p., 1885).

40. Kidwell has suggested that amateurs and home sewers were more likely to use proportional and hybrid methods, and professional dressmakers, direct-measure systems. See Kidwell (n. 10 above), pp. 80–81, 90–93. But the inventors of systems were far more ecumenical in their marketing (Gartland, p. 8).

41. Will C. Rood, *Supplement No. 2: Advanced Studies for Those Using Dressmakers' Magic Scale* (Quincy, Ill., 1889), p. 16; Standard Garment Cutting Company, p. 39.

42. Helen L. Summer, *History of Women in Industry in the United States, Vol. 9: Report on the Condition of Women and Child Wage-Earners in the United States,* 61st Cong., 2d Sess., 1910, S. Doc. 645, p. 117; *Report of the Women's Educational and Industrial Union for the Year Ending May 1, 1885* (Boston, 1885), p. 36, carton 1, vol. 1, WEIU (unprocessed); Helen Campbell, "Darkness and Daylight in New York: Life in the Great Metropolis by Day and Night as Seen by a Woman," in *Darkness and Daylight: Or, Light and Shadows of New York Life,* by Helen Campbell, Thomas W. Knox, and Thomas Byrnes (1895; reprint, Detroit, 1969) p. 260. "Stitching Fells" meant to finish seams by turning under the raw edges.

43. *S. T. Taylor's System of Dress Cutting* (n. 31 above), p. 57.

44. Because dressmakers by and large remained aloof from the nineteenth-century labor movement, they left behind no newspapers that might have reported their attitudes toward drafting systems. Trade journals such as the *American Dressmaker* and the *French Dressmaker,* published by system manufacturers, are equally useless on this score. See ibid., p. 18.

45. Ibid., pp. 44, 47, 49.

46. Buddington (n. 30 above). The pin-to-the-form method was not necessarily a time-consuming process in the hands of skilled practitioners. "Some of the leading cutters," Broughton noted, "use no guide but their heads. They cut the lining for each dress bodice upon the figure. They throw a piece of slightly stiffened muslin over the shoulders, pin it here, and snip it there, and in a few minutes they have a lining guide of surprising accuracy." See [Broughton] (n. 10 above), p. 7.

47. Samuel T. Taylor, *A System for Cutting Ladies' Dresses* (New York, 1875), p. 4; emphasis in original.

48. [Broughton], pp. 8, 11–13. While Broughton defended the pin-to-the-form technique, she should not necessarily be considered an advocate of traditional methods of training, for her *Suggestions for Dressmakers* represented part of a larger trend that substituted "book learning" for craft traditions.

49. *S. T. Taylor's System of Dress Cutting* (n. 31 above), pp. 7–11.

50. Thorstein Veblen, *The Theory of the Leisure Class* (Boston, 1973), esp. pp. 68–72, 118–31.

51. Trautman (n. 29 above), pp. 17, 63, 64, 85–87. See also pp. 12, 16, 21, 44, 71, 91, 93, 96–98; Linda Novak Jonason, "Dressmaking in North Dakota between 1890 and 1920: Equipment, Supplies, and Methods" (M.S. thesis, North Dakota State University, 1977), pp. 19, 23.

52. *S. T. Taylor's System of Dress Cutting,* pp. 44–45; Standard Garment Cutting Company (n. 39 above), pp. 38–39; Bryner, *Dressmaking and Millinery* (n. 16 above), p. 29.

53. The best discussion of the evolution of patterns is Kidwell (n. 10 above), pp. 81–90. See also Walsh (n. 7 above), and Fernandez, "If a Woman Had Taste" (n. 22 above), pp. 161–200, 238–57.

54. *Godey's Lady's Book (July* 1855), quoted in Fernandez, "'If a Woman Had Taste,'" p. 174.

55. Kidwell, pp. 83–85; *Ladies' Report of New York Fashions* (1867), quoted in Kidwell, p. 83; *Mirror of the Beautiful or Catalogue of Mme. Demorest's New and Reliable Patterns of Ladies' & Children's Dress, Fall and Winter Fashions, 1872–1873,* Trade Catalog Collection, American Antiquarian Society, Worcester, Mass.

56. Kidwell, pp. 85–86; Isabel Ross, *Crusades and Crinolines: The Life and Times of Ellen Curtis Demorest and William Jennings Demorest* (New York, 1963), pp. 20–25.

57. *Catalogue of E. Butterick & Co's Patterns for Winter, 1870–71,* p. 1, Trade Catalog Collection, American Antiquarian Society, Worcester, Mass.

58. *Demorest's Monthly Magazine,* February 1879, p. 112, and July 1883, p. 591.

59. *Demorest's Monthly Magazine,* October 1885, p. 818. Perhaps not surprisingly, some North Dakota dressmakers relied entirely on patterns. See Jonason (n. 51 above), p. 20.

60. [Broughton] (n. 10 above), p. 9. Broughton implied that some pattern makers catered exclusively to dressmakers; these apparently offered more up-to-date styles. See also *American Dressmaker 1* (December 1910): 5; *Demorest's Monthly Magazine* (July 1885), p. 606.

61. Kidwell, pp. 86–90; *Catalogue of E. Butterick & Co's Patterns for Summer, 1870,* Trade Catalog Collection, American Antiquarian Society, Worcester, Mass. Walsh (n. 7 above) overstates the extent to which patterns solved the problem of "fit."

62. Boston's dressmaking history is a case in point. Three hundred tradeswomen graced the pages of the city directory of 1870; 454 in 1880; 325 in 1890 (the reason for the decline is unclear); 625 in 1900; and 845 in 1910. These figures include proprietors of establishments and dressmakers who went "out by the day"; they do not include employees. Hence, the total number of dressmakers in the city was actually quite a bit higher. See *The Boston Directory* (Boston, 1870, 1880, 1890, 1900, 1910).

63. Ruth Schwartz Cowan, *More Work for Mother: The Ironies of Household Technology from the Open Hearth to the Microwave* (New York, 1983), esp. pp. 64–65.

64. Kidwell (n. 10 above), pp. 1, 94–98.

65. See Elizabeth Beardsley Butler, *Women and the Trades: Pittsburgh, 1907–1908* (1909; reprint, Pittsburgh, 1984), pp. 101–40, 184–250; Edna Bryner, *The Garment Trades* (Cleveland, 1916), esp. pp. 17–69; and Allinson (n. 16 above), pp. 131, 133–35, for discussions of wages and conditions in garment factories.

TEN

Meatpacking

ROGER HOROWITZ

Like cigarmaking and clothing manufacture, meatpacking
underwent further reorganization and centralization in the late
nineteenth century, in this case because effective refrigeration
techniques enabled larger-scale operations and long-distance
shipping. Roger Horowitz presents the result as a "master machine" meatpacking plant, into which raw materials walked on their
own four legs and out of which came products from butcher-sized
cuts of meat to ancillary packages of sausage and bacon. The
division of labor in these plants was based on race, ethnicity, and
gender, but the distribution of work as men's work and women's
work was based on the value of the meat being processed more
than on the ascribed gender characteristics of the tasks. Men
worked on the expensive cuts, women worked with the scraps. As
such, gender and ethnic divisions were embedded in the spatial
organization and structure of the plant, which helped to preserve
them until new factory buildings, with new technologies and a different layout, were built nearly a century later. How did gender
shape the technologies of work in this example? Did technologies
shape ideas about gender? Is there a role for the household or
the consumer in this story? What does Horowitz mean by "experience" as opposed to language? What comparisons can you draw
with cigarmaking?

"'Where Men Will Not Work': Gender, Power, Space, and the Sexual Division of Labor in America's Meatpacking Industry," *Technology and Culture* 38,
no. 1 (1997): 187–213. Reprinted by permission of the Society for the History of
Technology.

Tillie Olsen lived in Omaha's stockyards district in the 1930s. In her novel, *Yonnondio of the 1930s,* she vividly portrayed work in meatpacking's predominantly female offal and casings rooms: "Year-round breathing with open mouth, learning to pant shallow to endure the excrement reek of offal, the smothering stench from the bloodhouse below. Windowless: bleared dank dark . . . Heat of hell year round . . . feet always doubly in water—inside boots, outside boots . . . And over and over, the one constant motion . . ." Olsen concluded her account insightfully describing these departments as places "where men will not work."

Her powerful and accurate description forcefully poses the question, what made these unpleasant jobs women's work? Greatly complicating an explanation is the seemingly unrelated array of tasks performed by female packinghouse workers for much of the twentieth century. Their jobs ranged from the brutal toil in offal and casings, to the skilled knife jobs in the frigid "ice hell" of the pork trim, to the simple, repetitive operations in the immaculate sliced bacon department. How did such different settings all become places dominated by women workers?[1]

The answer developed in this article operates at two levels of analysis: how the sexual division of labor in meatpacking originally took shape and why it remained so stable for almost fifty years. I argue that the pattern of women's work in meatpacking emerged from a complex interaction between the objectives of male and female packinghouse workers, production-floor supervisors, and the personnel departments of packing companies. These struggles took place within a larger context of changes in technology and patterns of meat production, driven by new habits of food consumption. The competing agendas and needs of different groups of workers, and different management factions, produced an outcome which no one party contemplated at the beginning of the process. Conflict, rather than consensus, about the appropriate role for women shaped where the women worked.

The contours of women's work in meatpacking emerged at the beginning of the twentieth century. Expanding demand for meat and processed meat products led to innovations in production methods and technology that in turn created an entirely new range of jobs. To fill these positions, meatpacking firms had to develop mechanisms to recruit workers who were available. For women interested in finding work, their family needs and labor market opportunities drastically shaped whether, and where, they obtained employment in meatpacking. In addition to their sex, women's opportunities were profoundly influenced by race and ethnicity. Male workers influenced this process by clinging to jobs they traditionally had held, although the defeat of labor unions in the early twentieth century greatly weakened their capacity to interfere with management's hiring decisions. Between 1890 and 1919, the proportion of women grew from 2.7 percent to 10.5 percent of meatpacking's work force; moreover, this period established the sexual division of labor that would last, with few changes, until the 1960s. An extraordinary study of 6,000 female packinghouse workers by the Women's Bureau of the Department of Labor, published in 1932, allows us to identify with great precision where women worked.[2]

Identifying origins, however, does not explain persistence, especially as unionization of packing firms in the 1940s drastically altered the balance of power between labor and management. The stability of the boundaries between men and women's roles in meatpacking resulted from the way the packinghouse "master machine" (to adopt Lindy Biggs's formulation), grounded those notions in the everyday physical and social experience of workers and management. As a result, the social relations of gender, race, and ethnicity among workers became embodied in the actual structure of the buildings, departments, and production processes that comprised the meatpacking factory. The indelible, ordinary experience of where men and women worked entrenched notions of gender among those who labored in America's slaughterhouses.[3]

The accepted procedures for making meat, and the traditional association of particular types of workers with specific varieties of work within the industry, largely explain why neither workers nor management questioned the peculiar placement of women workers until the disruption of those arrangements. By the early twentieth century, women in meatpacking's "master machine" played a subordinate role in the production process and filled ancillary jobs located along "side roads" off the main avenue for the flow of meat — and wealth — through the plant. When men and women worked in the same department, men generally operated any machinery. Women only had opportunity for advancement within the limited confines of female departments. Even though many could, in theory, have performed jobs elsewhere in the plant, their auxiliary roles in production established firm limits on women's job options.

In meatpacking it was the production of specific types of meat in distinct precincts of the plant which became gendered, rather than particular artifacts, tools, or work procedures. The gendered character of individual operations was the result of their location in the process of making meat, rather than intrinsic features of the job. Consequently, gender was encoded into the very physical structure of a packinghouse, especially the complicated flow of the production process and the location of female departments. The tight bonds between gender, task, production process, and physical location are an important reason why the sexual division of labor seemed "axiomatic" to both workers and management. Gender was simply part of the physical landscape, a seemingly intrinsic element of the production process reflected in the places where men and women worked and the way meat flowed through the plant.[4]

Investigating the particular features of women's work in meatpacking necessitates reconstructing the dynamics of this industry as well as the struggles for power among different actors. My argument builds on the work of scholars such as Ruth Milkman who have shown that the nominal "idiom of sex-typing" in particular industries generally did not correspond to the actual pattern of men's and women's work. While Milkman accurately emphasized the role of industrial structure and managerial strategies on the sexual division of labor, I also consider the impact of workers' initiatives, especially their efforts to secure employment and to form labor organizations. Carole Turbin's work shows the power of women to aspire to, and obtain, a "good job" in industry and to use labor organizations to advance their interests. Studies such as those in Ava Baron's anthology *Work Engendered* indicate how male workers also shape this process through their efforts to reserve a distinct array of occupations for themselves.[5]

By showing how gender operated as an integral feature of factory design and the production process, my argument also draws on the work of scholars who have examined the "gender-ing" of consumer markets, work, technology, and domestic space, and others who have identified the spatial dimensions of racial identity and race relations in American life. This approach thus links older concerns with industrial structure and production processes to recent emphasis on the ways that racial and gender attitudes influence the organization of our society.[6]

While contemporary explanations for the sexual division of labor in this industry are available, the rhetoric of gender in meatpacking does not in itself provide an adequate guide to the development of distinct domains for women and men in America's packinghouses. On the contrary, assigning too prominent a role to the notions of gender conveyed in this language can obscure conflict among different agents in the constitution of a system of gender relations in meatpacking. The architects of the formal rhetoric, officials in packing firms' employment and public relations offices, generally had little power over the actual allocation of work to men and women by their immediate supervisors, or the utilization of new technology and production methods by plant superintendents. Thus, analytic use of the language of gender in meatpacking requires linking the rhetoric with the capacity of distinct social groups to realize their agendas and to implement their preferences and prejudices within the walls of meatpacking's "master machine."

Contemporary explanations for the sexual division of labor, to the extent offered at all, generally asserted that women were innately clean workers who were more dexterous than, and not as strong as, men. As Tillie Olsen's account indicates, this rhetoric did not reflect the actual range of women's work. It was an attempt to justify, inaccurately and after the fact, a pattern that stemmed from a complex confluence of pressures rather than coherent strategies or a priori notions of appropriate roles for men and women. It was the experience of the gendering of work which created the powerful ideology of gender in meatpacking, not the other way around.

The development of gendered patterns of work in meatpacking thus also contains insights pertinent to the debate among scholars concerning the relationship between language and experience. By identifying gender as a deeply embedded phenomenon intrinsic to the experiences of men and women in meatpacking, the paper accepts that social constructs such as gender can heavily influence how experiences are understood. However, the paper also criticizes how a focus on language, rather than social processes, reflects the unfortunate tendency of discourse analysis to elide issues of agency and causation. By locating the development of gendered notions of work in the industry's development and the conflict among employers and employees, and attributing the perpetuation of these ideas to the daily working experience, my argument ultimately places primary weight on the impact of experience on ideology.[7]

The language of gender in meatpacking does, however, provide important clues which can inform subsequent investigation. The explanations for where women worked were imbricated with assumptions about race and ethnicity, as well as gender. This suggests that the gendering of work in meatpacking was intimately linked with the way race and ethnicity shaped the allocation of jobs and the production of meat. And ironically, the very poverty of the language indicates that the peculiar pattern of men and women's work in meatpacking needed little defense because it seemed so natural. Instead of words, it was the seemingly immutable structure of the packinghouse, and shape and flow of meat production, which conveyed and reproduced gendered notions of where men and women should labor.

This article first discusses the consolidation of separate "spheres" of male and female employment in meatpacking in the first half of the twentieth century. It then shows how dramatic changes in technology and production methods between 1955 and 1965 undermined the sexual division of labor and generated tremendous conflict over the redefinition of "gender at work."

Gender and Women's Work in Meatpacking, 1890–1955

The contours of women's work in meatpacking took shape in a predominantly midwestern industry dominated by a handful of companies. The Big Four firms of Armour, Cudahy, Swift, and Wilson controlled the heartland of meat production, a swath reaching from St. Paul in the north to Ft. Worth in the south, stretching as far west as Denver and as far east as Chicago. Rail spurs extending into the countryside transported live animals from farms to central stockyards in large urban hubs, where buyers from adjacent packinghouses made their purchases for the day. Direct rail connections to the east sped the meat to consumers in major cities. In 1916 these firms (along with Morris, acquired by Armour in 1923) killed 94.4 percent of the cattle processed in the twelve cities (all in the midwest except for New York) which produced 81 percent of the nation's beef. They also controlled 81 percent of the hog slaughter in those centers.[8]

Meat production grew significantly in the early twentieth century, although the rate of increase for processed products such as sausage outpaced traditional fresh cuts. Between 1904 and 1925, production of fresh meat grew by 66 percent, to 9.5 billion pounds. In the same period, sausage production by packinghouses almost tripled to 900 million pounds. Despite the growth of processed products, however, fresh meat and related cured pork products like ham remained the dominant products of the meat industry.[9]

The expansion of the Big Four's midwest plants rested on their ability to significantly increase the productivity of labor and to develop an extensive distribution and sales network. Prior to 1860, local firms provided most meat in the United States because there was neither the refrigeration nor transport capacity to allow for the perishable product to be shipped significant distances. Refrigeration, both of the packinghouses and railroad cars, allowed firms employing economies of scale to displace local concerns. The growing popularity of processed meat allowed a second tier of midwestern firms, such as George A. Hormel & Co. and Oscar Mayer, to slowly expand their pork operations in the 1920s and 1930s.[10]

Tracing the historical construction of women's work in this industry requires attending to both where women could not work and which women came to dominate certain departments. To begin with, deeply entrenched notions of what constituted men's work profoundly influenced where women could work. Men dominated the killing, dismembering, and butchering of livestock and controlled the flow of fresh meat though the plant. As no work can begin in a packinghouse until the animal is slaughtered (thus providing raw materials for subsequent factory operations), male workers exercised far more power than women. While firms had eliminated the all-around butcher by fragmenting the butchering process, technological change in fresh meat departments was limited to automatic mechanisms which moved the carcass from one work station to the next. The irregular size of animals, and the physical subtlety of many butchering tasks, made introduction of machinery exceedingly difficult. Skill in meatpacking, as John R. Commons observed in 1904, was "specialized to fit the anatomy." A study of five packinghouses in the early 1930s determined that machine operators or helpers comprised only 20 percent of the work force. The remainder were stratified between workers who wielded knives and cleavers to make precise cuts in the animal carcass and workers who performed various rote operations with their hands.[11]

Male domination of the production of fresh meat provided important benefits at work and in the communities of packinghouse workers, as it gave men who acquired skill with a knife the opportunity to climb a firm's internal labor market. The hoped-for pattern of advancement for a male worker was to rise from a laborer relying on his brawn to a butcher performing critical tasks in the slaughtering of livestock and the cutting of carcasses. Becoming a butcher not only meant obtaining a solid working-class job; the skill provided an income and status that carried over to the community. Even black men, who entered the industry in large numbers only during World War I, moved into knife jobs in the 1920s and 1930s. Black packinghouse worker William Raspberry aspired to become a butcher because he was "impressed" by a calf skinner who lived in his neighborhood. "He was clean, he was dressed nice," recalled Raspberry. "He had a neatly manicured lawn, and his wife kept flowers around it." Neither workers nor management questioned whether men should continue to dominate those jobs and have access to the community status and family role which they made possible.[12]

Employment options dramatically expanded for women in the 1890s and early

1900s as firms used new technology to broaden their portfolio of processed meat products. Moreover, these opportunities for women occurred at the same time as millions of immigrants arrived in midwestern cities. While immigrant men were able to move into jobs that had always been male, ethnic (and in some cases black) women were restricted to departments where the character of work fundamentally changed or in newly created operations where they were the first employees. Even though management could and did pay women less than men, female substitution only occurred in areas undergoing substantial changes in production methods, such as sausage. Management either resisted or perhaps never considered introducing women into what had been male departments devoted to the production of fresh meat.[13]

Women's jobs carried not only less pay than men's work but also far less social status. Unlike the male butchers, female packinghouse workers "were kind of ashamed working in the yards," recalled Chicagoan Estelle Zabritsky. The physical toll of the jobs, such as cracked and broken fingernails and rough hands from the hard toil, made it impossible to conceal the embarrassing truth from other women or a prospective date. While male jobs provided upward mobility and community respect, women (especially the young) felt humiliated for having no other option than work in a packinghouse.[14]

Accounting for what was men's work provides only a starting point for understanding the distinction between male and female jobs. While the 1932 Women's Bureau study found that 75 percent of the women in meatpacking were concentrated in only six departments, it also noted significant differences among the women in accordance with their marital status, race, and ethnicity. The significant variations among women workers illuminate how the family needs and labor market opportunities of different racial and ethnic groups interacted with gender dynamics to shape the parameters of women's work in meatpacking.

East European ethnic women dominated the women's jobs that entailed the most training and skill. The close proximity of ethnic neighborhoods to the packinghouse district, and the family needs of the women, made them a reliable and flexible source of labor for frontline supervisors. For example, when foreman Tommy Megan at the Chicago Armour plant needed more workers, he would recruit neighbors and relatives from the ethnic Back of the Yards community adjacent to the stockyards. Typically, ethnic women worked prior to marriage, left when they had children, and reentered the work force when their sons and daughters entered school or if their husband could no longer support the family. One woman who also worked at the Chicago Armour plant, Sophie Kosciowlowski, first found work in packing in 1918, when she was only thirteen, and left work to get married. She was able to get her old job back in 1931, after obtaining a divorce, because she knew the Polish foreman and many of the women in her old department. The intermittent, yet persistent participation in the labor market of women like Kosciowlowski resemble the work histories of the Irish women studied by Carole Turbin in Troy, New York. While company publicists might impugn East European ethnics as "lesser races," in fact ethnic women with family responsibilities dominated skilled female jobs because they provided the most return to the foremen for the months spent in training.[15]

Black women predominated in the least desirable women's jobs, performed under poor working conditions and for low pay. Like female ethnics, black women were reliable long-term employees because their families needed the income; however, they had far fewer job opportunities than white women. While female ethnics could find work in a variety of factory settings, meatpacking was one of the few industrial jobs open to black women before World War II. Even low-paid positions in a packinghouse offered three to four times the remuneration of domestic service, the area of employment in which most black women worked. Management routinely took advantage of black women's poor labor market position to assign them to the worst women's jobs. Moreover, employers in meatpacking typically harbored a racist assumption that black men and women could handle jobs where, as one company publicist wrote in 1913, "the heat is intense and the smell uncongenial" for workers of more "sensitive" disposition.[16]

The women with the highest turnover rate, young native-born women, predominated in cleaner departments that required little training. Native-born women worked in jobs that actually paid as poorly as those for black women, but they labored under conditions that were among the best in meatpacking. These women generally were unmarried, in their teens or early twenties, living at home and contributing their wages to their family's income. Unlike ethnic or black women, they had a short-term perspective on wage work, correctly anticipating that they would not return to the packinghouses after marriage. They also had far more employment options and could obtain clerical work at higher pay than available in a packinghouse. As meatpacking was only a temporary job for most young native-born women, ethnic and black women actually were more reliable and valuable employees. Hence the central employment office of packing firms generally hired these women to perform simple operations (such as packing sliced bacon) and placed them in clean departments that were "showplaces" for visitors to aid their public presentation of meatpacking as a modern business suitable for the employment of young, "American" girls.[17]

The considerable variation among the experiences of female packinghouse workers must be balanced by a common feature of their jobs: women overwhelmingly labored in areas linked to making processed meat products. Women worked in departments which had been either transformed or created in response to the growth of the mass consumer market for preserved products. Their ancillary relationship to the flow of fresh meat production reflected how women's work had been literally spliced onto preexisting patterns of meat production, and their departments physically added onto meat producing facilities. Gendered notions of what constituted women's work took shape out of this historical process of creating female jobs in the early twentieth century (fig. 10.1).

The offal and casings departments are examples of areas that had been transformed by new patterns of meat consumption. These departments prepared internal animal organs such as intestines, bladders, and hearts—colloquially known as the "pluck"—for later use in processed meat products. They only became female-dominated in the twentieth century. Until the 1890s slaughterhouses generally sold the pluck to small firms specializing in sausage manufacturing or simply discarded the organs. The minimal processing and packaging of the innards in preparation for sale had been a job performed by men on one side of the killing floors. As firms expanded production of lard, sausage, and animal by-products, work in offal and casings became far more detailed. Women hired in these areas actually performed very different tasks than the men they replaced, as they sorted, cleaned, and routed the animal organs to other areas of the plant, whereas men previously had simply packed the entire pluck for immediate sale or disposal. These expanded offal operations necessitated construction of new full-fledged departments, usually located directly underneath the killing floors. The dominant meatpacking union, the Amalgamated Meat Cutters, used the rationale that the jobs were "unfit for any women to perform" to object to female substitution but was too weak to prevent it at that time.[18]

While men in the nineteenth century had handled the pluck as a minor distraction from the production of fresh meat, female offal workers of the twentieth century performed the first step in the production of manufactured meat products. The subsidiary character of this labor is apparent: these women didn't manu-

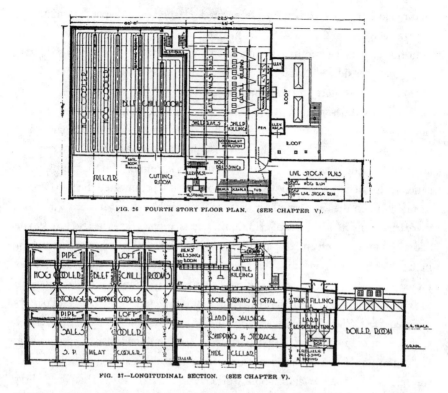

FIG. 26 FOURTH STORY FLOOR PLAN. (SEE CHAPTER V).

FIG. 27—LONGITUDINAL SECTION. (SEE CHAPTER V).

Figure 10.1. Diagrams from the 1910s of the relatively small Pfaelzer & Sons packinghouse in Chicago. The animals walk up a long ramp to the fourth floor, where the slaughtering process begins the production of meat. The fourth story plan shows how the killing floors for pigs, lamb, and cattle are immediately adjacent to each other, with the coolers that chill the carcasses in a separate, but connected building. Note that the cutting room (a man's department) is adjacent to the coolers and on the same level as the male-dominated killing departments. The longitudinal section shows the vertical arrangement of departments. The female departments are located directly below the kill floor, along with oleomargarine, which uses beef fat. One floor below are lard refining (pig fat) and sausage manufacturing. The diagram does not include the many chutes that used gravity to move meat from one department to another. As a firm like this one expanded into processed meat products after 1920 (relying heavily on female labor), additional separate buildings would be constructed, maintaining the physical separation of men's and women's departments. (H. Peter Henschien, *Packing House and Cold Storage Construction* [Chicago, 1916], pp. 40–43.)

facture anything themselves. They separated, cleaned, and dispatched materials for use elsewhere in the plant, while the main flow of meat from the killing floors went through male-dominated departments. By the 1930s, work in this department, once considered "unfit" for women workers, could be described by Tillie Olsen as an area where "men will not work."[19]

Olsen's vivid description of the casings and offal departments also was accurate. Under harsh conditions, women in both departments used their hands and small tools while men operated the only machine, a "splitting wheel" that crushed an animal's skull. Women in offal received the pluck from men in the killing areas, generally located one floor above, through a chute that connected the two departments. Women then separated fat that would later be refined into lard or oleomargarine, sent glands to the pharmaceutical area, prepared hearts and lungs to be added to sausage filler, and dispatched intestines to be further processed next door in the adjacent casings room. There, women cleaned, salted, and packed intestines for use as containers in sausage-making. Their hands were constantly immersed in the water used to clean animal organs and to flush the intestines of fat and partly-digested food. Water mixed with animal fat and other bodily fluids to create a slick surface on the floor of both departments, and combined with oppressive heat (exceeding 100°F in the summer) to generate unhealthy, steamy conditions. "It was terrible," recalled packinghouse worker Virginia Houston. "That was a dirty job!" The arduous work was poorly paid, subject to repeated layoffs, and provided little room for advancement. Women workers in casings typically suffered from debilitating conditions such as pneumonia, rheumatism, and arthritis.[20]

We have to link gender to patterns of racial and ethnic labor market segmentation, and to the family needs of female packinghouse workers, to explain why black women dominated these departments and young ethnic women held these jobs only in the absence of black women. The ethnic women in casings and offal were unreliable employees; generally unmarried, they quit once they had children or found a better job. Black women, in this sense, actually were more desirable employees for management, as they could be counted on to retain their jobs as among the best available to them. Packing firms "praised" black women to researcher Alma Herbst for their "loyalty" and "willingness to do the dirty work which soon became distasteful to foreign women." Black women also could be paid less than ethnics to perform what were known as "nigger jobs" because, as one packing executive explained to Herbst, their limited employment opportunities put them "on their haunches." Women in Sioux City's all-white offal departments received

$17.85 a week in 1929, while the black women who performed the same jobs in Chicago earned only $14.65, a 20 percent differential. By comparison, the weekly pay for men who operated the skull splitting machine was $26.13. While changes in the meat business may have opened jobs in casings and offal to women, patterns of racism and family needs determined what kind of women worked there.[21]

The women's jobs in pork trim offer a sharp contrast to the work in offal and casings. This was an entirely new department, created in the early twentieth century because of the increased production of processed meat products. Pork trim workers refined pieces of meat left over from the work of men who carved the pig into basic cuts for consumer use, such as chops and hams. The women used knives to trim fat or rind from small pieces of meat that later would be ground into sausages or canned. Although the best-paid jobs open to women, the work was performed under onerous conditions. The women worked on frozen meat in dingy, windowless rooms rarely warmer than 45°F. Water condensed on the walls—forming icicles when the temperature dropped below freezing—and combined with the fat trimmed from the meat to coat the floors with a permanent slippery layer that caused workers to slip and fall. The chilled meat numbed hands, and in this "ice hell," women kept warm by wrapping newspapers and gunny sacks around their legs and donning heavy sweaters under their work frocks. As women were paid by the piece (unlike the hourly work that prevailed in offal and casings), they worked at a furious rate to fill the buckets with the lean meat that determined their remuneration.[22]

Trimming pork requires considerable skill with a knife and entailed close to a year of experience to become proficient. Women's dominance in this area thus contradicts the argument of French scholar Noelie Vialles that meatpacking's main tool, the knife, was an intrinsically masculine instrument. As pork trim workers also had to move buckets of the trimmings, generally weighing 40–50 pounds, their actual daily tasks contradicted the rhetoric that women could not perform heavy jobs.[23]

Again, it is the ancillary character of pork trim work that suggests why men did not protest when women took these jobs. As in offal and casings, pork trim workers refined and routed pieces of the animal left over from the far more valuable products made by men. The department's structure symbolically encoded women's subordinate relationship to male workers; men in the pork cut department upstairs literally pushed their scraps down chutes which directed the material to the tables of the female trimmers. It is telling that despite considerable opportunities for advancement within the pork trim for adept women, they could not move into the butchering departments more central to the production of meat — ostensibly because they weren't strong enough. In fact, there were plenty of operations in the pork cut which the tough women butchers could have performed. The actual barrier was neither strength nor ability to wield a knife but the difference between a department which played a central role in the flow of fresh meat products — and thus capital — and one that was on a siding, a feeder to the packinghouse's manufacturing divisions.[24]

The skilled work and high pay in the pork trim also explain why it became dominated by middle-aged ethnic women who were either married or the main supporters of their families. Eighty percent of the pork trim workers surveyed in the late 1920s were or had been married, and 65 percent were over thirty. Skill protected them from layoffs, and piece-work payment systems brought them the highest earnings for women in meatpacking, a national average of $20.40 a week. Experienced pork trim workers could make almost $30 a week, pay comparable to that of male butchers. In need of dependable income for their families over an extended period of time, ethnic women sought these jobs through connections with fellow ethnics who were working in a packinghouse. At the same time, when foremen needed more workers, they would recruit the friends, neighbors, and relatives of dependable ethnic women already in their departments. The result was a reinforcing cycle of the ethnic family strategy meshing with local labor markets and the needs of lower-level management. The same cycle excluded black women, as they were not tied to the ethnic networks which overlapped work and community and helped white women dominate these jobs. As in casings and offal, explaining why certain women worked in pork trim requires consideration of more than gender.[25]

Sausage manufacturing represents a mixed picture of radically altered work methods and entirely new procedures in a department that had once been an all-male preserve. Sausage is made by filling animal intestines or artificial casings with meat scraps and various kinds of filler. Before 1900 this was a skilled job dominated by German craftsmen who chopped the meat, mixed in spices, and stuffed the casings by hand. Packing firms realized that they could make profitable use of meat scraps and internal organs that were, as one Swift & Co. executive claimed, "as wholesome as porterhouse steak, but not so palatable," by mass producing sausage. Firms mechanized the chopping of meat in the 1890s and introduced automatic stuffing machines in the early twentieth century. While craftsmen retained footholds in specialty products, the bulk of sausage production became routinized, with hundreds of pounds of meat and filler chopped in devices like the "Buffalo Silent Cutter" and funneled into stuffing machines which injected the material into casings. Packing firms hired women, generally recent immigrants, to fill jobs created by the new technology and division of labor. Although the skilled men objected strenuously to the introduction of women, the defeat of the Amalgamated's 1904 strike ended effective opposition to management's transformation of sausage making. By 1906, Upton Sinclair could describe a sausage department composed almost entirely of women.[26]

Despite the expansion of women's work in sausage, they held positions that clearly made them subordinate to male workers who retained deskilled versions of jobs they had dominated. In the modern departments men still controlled the stuffing of sausages—only now by operating the machines which chopped the meat and filled the casings, rather than through old craft techniques. As a result, male workers still regulated the tempo of sausage making. With little power over the pace of work, women dominated linking—separating the casings into the sections that later would become individual sausages—and other subsidiary tasks. Women "linked" most sausages by flipping the meat-filled intestines so as to twist the casings at regular intervals. While requiring no tools, it took women several months of training and experience to become proficient. An adept linker, Sinclair wrote, "worked so fast that the eye could literally not follow her, and there was only a mist of motion, and tangle after tangle of sausages appearing." While a far cleaner job than offal, casings, or pork trim, many "linkers" suffered from what we know today as carpal tunnel syndrome because of the feverish pace induced by piecework. Women could progress from unskilled jobs of moving and hanging the sausages (paying $16.91 a week) to linking (paying $18.47 weekly). Women could not obtain the job of operating the mechanical stuffing machines—which paid men $28.78 a week.[27]

Older ethnic women dominated sausage manufacturing operations because of dynamics similar to the pork trim—the work required skill, was less subject to employment fluctuations than offal and casings, and provided opportunity for advancement to relatively well-paid jobs. Desirous of stable, long-term employment, and recruited by foremen who wanted reliable help, the women workers were, as Sinclair noted, "apt to have a family to keep alive." As in pork trim, the dynamics of the family economy and supervisor's labor recruitment strategies brought ethnic women into the top female jobs. In the late 1920s, two-thirds of the sausage manufacturing workers were or had been married, and the same proportion was over twenty-five. And, again similar to pork trim, sausage generally was all-white.[28]

Efforts to increase processed meat sales also led packing firms to build sliced bacon departments, usually located in a separate building, in the 1910s and 1920s. Bacon is made from the cured belly of a pig; the development of a slicing machine in 1915 allowed firms to shift fabrication of bacon for consumer use from the butcher store to the packinghouse. To a greater degree than in other women's departments, conscious managerial strategy shaped the gender, racial, and ethnic characteristics of sliced bacon workers.

From their inception, sliced bacon departments had an unusual dual role: as a locus for the production of meat and a convenient vehicle for improving the industry's troubled public image. Some departments even had one glass wall to facilitate observation by tours. What impressed most visitors were the clean, well-lit rooms in which young white women, neatly dressed in immaculate white uniforms, performed their tasks while seated comfortably at long tables. Some plants even installed pipes carrying warm air to serve as footrests and to make the workers more comfortable without raising the room temperature. More critical observers also could have noticed the subordination of these women to men. While the few men performed the key step in the production of bacon, slicing the cured pig's belly by a machine, the women were limited to arranging the slices in rows, weighing, and wrapping.[29]

The dual role of sliced bacon departments shaped the composition of its female work force. Three-fifths of the workers were under the age of twenty-five, and 84 percent were native-born. The presence of young, white "American" women was well-suited to a positive public representation of meat production. As these jobs exacted little physical toll, white women with other options were willing to consider working in sliced bacon despite the undesirable status of packinghouse employment. At the same time, employers considered these women unreliable employees who could find other jobs and would not work after marriage. Hence it made sense to place them on jobs that they could master in a few hours.[30]

The same dynamics resulted in the complete exclusion of black women from sliced bacon. Management preferred to place black women on jobs that other workers eschewed, and feared public reaction to the sight of black women preparing bacon. As one management official explained to Alma Herbst "only white hands are fit to touch the meat." Placing black women in sliced bacon would have undermined management's labor market strategies and the self-portrait firms were painting for middle-class consumers of processed meat.[31]

One packinghouse publicist attributed the peculiar pattern of women's work in meatpacking to management's appreciation for the "passion for cleanliness, which seems to be inborn in many women, however slovenly their dress."[32] This justification, which struggles to balance gender stereotypes with racial and ethnic prejudice, is a typical example of a rationale which fails to explain the actual gendering of work in meatpacking. As we have seen, "cleanliness" is an apt term for neither the slimy conditions in the pork trim nor the stench of the offal department. What these women's tasks shared was not the particular qualities of each job but an ancillary relationship to the production of fresh meat that placed women in a subordinate relationship to male workers. The use of chutes to convey product from a male-dominated department to female workers below them aptly represented in physical form the relationship between men and women in a packinghouse. It also is telling that where there was machinery to operate, in the offal, sausage, and bacon rooms, men performed those tasks. All these women's jobs were linked to the manufacturing of finished, cured consumer products, rather than the fabrication of consumer cuts. They received the leftovers of male departments, and the amount of product and speed that it reached the women was regulated by men. Despite the lofty claims of promotional language, we can only make sense of where women worked in meatpacking by looking at the spatial arrangement, work process, and technology employed in the making of meat.

The Transformation of Women's Work in Meatpacking, 1955–1980

When unions finally secured collective bargaining in meatpacking in the 1940s, the sexual division of labor seemed so natural that it was not questioned by any definable group of workers. The local unions which created the United Packinghouse Workers of America (UPWA) in 1943 supported equal pay for equal work while enforcing the boundaries between male and female jobs. Through the 1940s and 1950s the UPWA maintained this posture and pushed aggressively for the elimination of the substantial gap between male and female starting pay. By the mid-1950s the union succeeded, eliminating the formal wage differential. At the same time, however, the UPWA encoded the sexual division labor into a rigid system of explicitly sex-defined jobs, and separate male-female seniority lists. While women's base rate rose to the same level as men, they had far fewer opportunities to climb into the higher pay brackets and could not move into male jobs.[33]

The union's acquiescence to the traditional sexual division of labor stood in marked contrast to its sustained challenge to entrenched patterns of racial and ethnic job allocation. Beginning in the early 1950s, local unions began a concerted campaign to integrate all-white departments by encouraging blacks to use the seniority provisions in UPWA contracts to transfer into lily-white areas. When whites in departments such as sliced bacon resisted the transfers, local union leaders usually responded by informing the whites, "you either work with them or you find another job," recalled Kansas City Armour steward Nevada Isom. "That was spelled out to them and they cooperated very nicely." In Fort Worth, white women in sliced bacon stopped work to prevent black women from entering the department. Local union president Mary Salinas bluntly informed them "they were going back to work and accept it as it was, and if they weren't satisfied they could just quit." Black hog kill workers struck in the Kansas City Wilson; Chicago Wilson; and Waterloo, Iowa, Rath plants to secure management's cooperation with the union's effort. Union victories were truly impressive: within five years black men and women broke through deeply entrenched racial barriers in dozens of plants. The UPWA's ability to alter the traditional racial division of labor among women and men indicates how shifting power relations between labor and capital, and among workers, had the potential to change patterns of work in meatpacking.[34]

The advances of black men and women occurred within the strict confines of the sexual division of labor, which was not appreciably affected. However, disruption of traditional ethnic and racial boundaries, and the mobilization of black women in meatpacking, inexorably brought the sexual division of labor into question. The gendering of packinghouse work in the early twentieth century had been intertwined with race and ethnicity; once those latter two categories disintegrated as rationales for job placement, gender stood uncomfortably on its own. Soon women who moved into jobs previously denied them because of race began to question why they were barred from men's jobs which they could perform. Indeed, the first calls for eliminating separate categories of men's and women's work came from black women. While a minority opinion in the union and among women workers in the 1950s, these were the first explicit arguments against the "axiomatic" sexual division of labor in meatpacking.[35]

Technological change in the late 1950s and 1960s dramatically reduced the employment in female departments and stimulated greater opposition among women to the traditional sexual division of labor. Motivating these innovations were efforts by firms to capitalize on expanding consumer demand for processed meat. Sausage manufacturing by packinghouses quadrupled between 1925 and 1963 to 3.7 billion pounds, while fresh meat production grew by a comparatively modest 150 percent. Sliced bacon, which did not even appear as a category in the census of manufactures before the war, was produced in equal volume to cured hams in 1963. Moreover, these products yielded more profits per pound than fresh meat. Upstart firms specializing in processed pork had profit rates that were double or triple those of the older Big Four companies who relied more on sales of fresh meat. While the volume of fresh meat production was still three to four times that of processed products, meatpacking firms could see where the money was going to be in the future. Hence it made considerable business sense to invest in ways to produce processed meat products more quickly—and with fewer workers.[36]

The new processing techniques disrupted established barriers between men's and women's work. "Oh yes, they were getting rid of the women," Velma Otterman Schrader complained when she described the dramatic changes in traditional work methods. Circular electric "wizard" knives in the offal and trim departments reduced the work force by doubling productivity. In sliced bacon, new technology increased production speeds and reduced labor needs by 40 to 60 percent. Especially massive job loss occurred in the sausage department, where machinery replaced labor at every step of the process. Power-driven grinders fed mixing machines which prepared sausage filler and automatically injected the mixture into long artificial casings, which then were twisted into links and hung onto curing racks by yet another machine. In the Waterloo, Iowa, Rath plant, employment fell from 300 to 12 in this department without any loss in the volume of production. Because of separate male-female jobs and seniority lists, it was routine for women to be laid off while men with considerably less seniority continued to work. The experiences of women had a profound effect on their attitudes toward the gendering of packinghouse work. By the mid-1960s female union members were clamoring for alteration of the union-enforced sexual division of labor in meatpacking.[37]

Changes in the physical structure of meat production added to the pressure on the traditional sexual division of labor. Between 1955 and 1965 packing firms closed dozens of old multistory plants in urban areas and replaced them with technologically advanced, single-story facilities in outlying suburbs or rural locations. The relationship between departments, as well as the work process within departments, underwent radical change. The offal and casings areas were no longer below the killing floors, and the female workers in pork trim now labored in rooms adjacent to the men in the pork cut. With the changing physical structure, new sliced bacon departments were no longer fashioned to serve as the public face of the industry. Single-floor packinghouses also eliminated the highly symbolic chute system of distributing meat by replacing them with powered conveyer belts. These changes seriously disrupted the manner in which gender had become encoded into the physical structure of meat production and thus facilitated challenges by women workers to the sexual division of labor.[38]

Significantly, male resistance to changes in the sexual division of labor in the 1960s relied on defending the right of men to have first claim on employment, rather than the intrinsic character of a job. "You heard all this stuff about the man's the breadwinner and therefore you shouldn't force him off his job or lay him off and let the women work," recalled the male chief steward in the Waterloo Rath plant. The restructuring of meat production, technological change, and union attacks on ethnic and racial discrimination had undermined old justifications for sex-typing of particular tasks. Nonetheless, the rhetoric of the family wage was sufficiently widespread among male union members that women, only 20 percent of the UPWA, were unable to alter union support for contractual barriers between men's and women's work.[39]

The 1964 Civil Rights Act changed the rules of the game by placing the allocation of jobs to women and men under federal supervision. Title VII's prohibition of employment discrimination by race and sex threatened, at least potentially, a system of job classification that clearly was having discriminatory consequences. In essence, federal legislation gave women workers the leverage to pressure their union and the meatpacking firms to fundamentally alter entrenched patterns of where women could work.

It is a measure of the discontent among female packinghouse workers, and the presence of strong women stewards and local leaders, that UPWA members were among the first women workers to file lawsuits against their union and their employers for violations of Title VII. Thoroughly embarrassed by the accusations, the UPWA established a national commission of female union leaders to develop proposals to correct the situation. This committee proposed a system that became, at least formally, a compromise accepted by all parties: alteration of job classifications into an "ABC" system consisting of "A" jobs open to men, "B" jobs open to women, and "C" jobs open to both. As union and management representatives painstakingly reclassified work inside packinghouses over the next few years, the old rigid system was significantly modified by many tasks becoming "C" jobs open to workers of both sexes. By the 1970s women were slowly breaking down the traditional sexual division of labor in meatpacking.[40]

The events of the 1950s and 1960s fundamentally changed the pattern of men and women's work in meatpacking. Among the factors producing this transformation, the pressure exerted by organized women workers stands out. Of course, women's efforts took place in a context which made success far more possible. Technological change provided a necessary stimulus for women workers to challenge the separate female space in meatpacking's "master machine," and their struggle was measurably aided by Title VII of the 1964 Civil Rights Act, itself a product of the black freedom struggle. But it took the initiatives of women and their female union leaders to pressure the Equal Employment Opportunities Commission to investigate the contractual separation of men and women's work and to make the UPWA take the problem seriously. With their power enhanced by federal legislation and the existence of collective bargaining agreements, female packinghouse workers forced their way onto jobs where men still believed women did not belong, and ended the limitation on women's employment opportunities to those places where "men will not work."

Epilogue

By the 1980s it was no longer the case that women generally held ancillary jobs that made them subordinate to men in the production process. However, persistent differences between men's and women's work remain in the contemporary meatpacking industry, still influenced by the dynamics of the early twentieth century. Fragmentary evidence suggests that women moved into many jobs that entailed operation of machines — as long as these were in processing departments, which had always been open to women. Women were able to move into new butchering jobs — as long as they were in newly created departments, such as the production of boxed beef, an operation developed in the 1960s. While there is no recent survey comparable to the 1932 Women's Bureau study, recent ethnographic studies of the contemporary meatpacking industry generally agree that the "view of meat cutting as essentially male employment seems to persist in the meat industry." While the information is suggestive, understanding the new patterns of gender in the meatpacking industry must await the assimilation of comparable data.[41]

Ironically, female packinghouse workers gained increased access to employment at the same time as working conditions declined because of a deterioration in union strength. Turmoil in the meatpacking industry in the 1980s destroyed national contracts, reduced the strength of labor organizations in national firms, and resulted in a 30 percent fall in real wages. This has measurably weakened the power of female packinghouse workers. Unions may have entrenched the sexual division of labor after World War II, but they also served as a vehicle through which women could eliminate glaring inequalities, improve the conditions under which they worked, and eventually erase the line between male and female jobs.[42]

The conditions under which women labor in the 1990s all too often recall Tillie Olsen's vivid account of the 1930s. Women operating bacon slicing machines have lost fingers because of the pace of work. Women in the butchering jobs now open to them routinely develop carpal tunnel syndrome. And in the new sexual division of labor that is taking shape, family needs still drive women to work. One packing firm executive told researchers that his firm preferred to hire single mothers "because they probably aren't as mobile" and need the work. His comment reflects how managerial hiring strategies, labor markets, and the relationship of women to the family still shape why and where women work in meatpacking.[43]

Consider this contemporary story from a 1990 study. A recently divorced mother of two in Garden City, Kansas, could only find work in meatpacking because she had no skills. For several years she performed a job that has been "women's work" since the turn of the century—removing brains from an animal's skull. She reached into the skull cavity and massaged the membrane around the brain, performing a "squishing" motion with her hands every two seconds—or 1,800 times an hour. Now she suffers from severe pains in her arms and can barely perform housework, much less pick up her children. Unable to work, she was fired and then denied workmen's compensation when the company successfully contested her claim.[44]

This woman's sad story of being forced to work because of family needs, restricted in her options because of limited geographic mobility and the structure of local labor markets, and relegated to a traditional female job attests to the centrality of gender in the organization of meatpacking's modern "master machines." This reflects a larger pattern: gender is a permanent, yet highly fluid, social relationship, not intrinsically tied to particular artifacts and practices while always present in them. Close attention to the process of making meat, as well as those who make the meat, opens up how gender shapes the things we consume and the way they are created.

NOTES

This paper benefited from comments by Arwen Mohun, Nina Lerman, John Staudenmeier, and three anonymous reviewers for *Technology and Culture*. Research was facilitated by grants from the National Endowment for the Humanities, the American Historical Association, and the University of Delaware.

1. Tillie Olsen, *Yonnondio of the 1930s* (New York, 1974), p. 134.

2. U.S. Department of Commerce and Labor, *Manufactures, 1905, Part I* (Washington, D.C., 1907), p. 17. U.S. Department of Commerce, *Fourteenth Census of the United States, Vol. 10: Manufactures* (Washington, D.C., 1923), p. 48. Mary Elizabeth Pigeon, *The Employment of Women in Slaughtering and Meat Packing*, Women's Bureau Bulletin No. 88 (Washington, D.C., 1932).

3. Biggs's term refers to the way industrial engineers planned the very structure of factories so that they would "work as predictably and obediently as a machine" by facilitating the most efficient flow of product through a plant. Meatpacking firms pioneered the development of the "rational factory" in the early twentieth century with influential builders like H. Peter Henschien disseminating architectural plans and construction standards widely emulated by the industry. Lindy B. Biggs, "Industry's Master Machine: Factory Planning and Design in the Age of Mass Production, 1900 to 1930," (Ph.D. diss., MIT, 1987), p. 94, and *The Rational Factory: Architecture, Technology, and Work in America's Age of Mass Production* (Baltimore, 1996). H. Peter Henschien, *Packing House and Cold Storage Construction* (Chicago, 1916).

4. Joan Scott, "Reply to Criticism," *International Labor and Working Class History* 32 (Fall 1987): 39–45.

5. Ruth Milkman, *Gender at Work: The Dynamics of Job Segregation by Sex during World War II* (Urbana, Ill., 1987), pp. 15–19. Carole Turbin, "Beyond Conventional Wisdom: Women's Wage Work, Household Economic Contribution, and Labor Activism in a Mid-Nineteenth-Century Working-Class Community," in *"To Toil the Livelong Day": America's Women At Work, 1780–1980*, ed. Carol Groneman and Mary Beth Norton (Ithaca, N.Y., 1987), pp. 47–67, and *Working Women of Collar City: Gender, Class, and Community in Troy, New York, 1864–86* (Urbana, Ill., 1992). Ava Baron, ed., *Work Engendered: Toward A New History of American Labor* (Ithaca, N.Y., 1991).

6. Ava Baron, "Contested Terrain Revisited: Technology and Gender Definitions of Work in the Printing Industry, 1850–1920," in *Women, Work, and Technology*, ed. Barbara Drygulski Wright et al. (Ann Arbor, Mich., 1987), pp. 58–83. Judith A. McGaw, "No Passive Victims, No Separate Spheres: A Feminist Perspective on Technology's History," in *In Context: History and the History of Technology*, ed. Stephen H. Cutliffe and Robert C. Post (Bethlehem, Pa., 1983), pp. 172–91. Angel Kwolek-Folland, *Engendering Business: Men and Women*

in the Corporate Office, 1870–1930 (Baltimore, 1994). Dolores Hayden, *The Grand Domestic Revolution* (Cambridge, Mass., 1981). Ruth Schwartz Cowan, *More Work for Mother: The Ironies of Household Technology from the Open Hearth to the Microwave* (New York, 1983). Patricia A. Cooper, *Once a Cigar Maker: Men, Women, and Work Culture in American Cigar Factories, 1900–1919* (Urbana, Ill., 1987). David R. Roediger, *The Wages of Whiteness* (New York, 1991). Robin D. G. Kelley, "'We Are Not What We Seem': Rethinking Black Working Class Opposition in the Jim Crow South," *Journal of American History* 80, no. 1 (June 1993), pp. 75–112. Harry Braverman, *Labor and Monopoly Capital* (New York, 1974). Andrew Sayer and Richard Walker, *The New Social Economy: Reworking the Division of Labor* (Cambridge, Mass., 1992).

7. The most forceful argument for privileging language over experience in historical analysis is in Joan W. Scott, "The Evidence of Experience," *Critical Inquiry* 17 (Summer 1991): 773–97. For a sharp critique of this perspective, see Charles Tilly, "Softcore Solipsism," *Labour/Le Travail* 34 (Fall 1994): 259–68.

8. U.S. Bureau of Corporations, *Report on the Beef Industry* (Washington, D.C., 1905), pp. 1–8. *Report of the Federal Trade Commission on the Meat-Packing Industry* (Washington, D.C., 1919), Summary and Part I, p. 127. FTC Part I, map and table opposite 394, map and table opposite 397. See also Margaret Walsh, *The Rise of the Midwestern Meat Industry* (Lexington, Ky., 1982).

9. U.S. Department of Commerce, *Fourteenth Census of the United States, Volume 10: Manufactures* (Washington, D.C., 1923), p. 56. U.S. Department of Commerce, *Fifteenth Census of the United States, Volume 2: Manufactures* (Washington, D.C., 1933), p. 176.

10. For the development of refrigeration in meatpacking, see "The 'Significant Sixty': A Historical Report on the Progress and Development of the Meat Packing Industry, 1891–1951," *National Provisioner,* January 26, 1952, pp. 197–216, 360; and Alfred D. Chandler Jr., *The Visible Hand: The Managerial Revolution in American Business* (Cambridge, Mass., 1977), pp. 299–301.

11. James Barrett, *Work and Community in the Jungle: Chicago's Packinghouse Workers, 1894–1922* (Urbana, Ill., 1987), p. 22. John R. Commons, "Labor Conditions in Meat Packing and the Recent Strike," *Quarterly Journal of Economics,* 19 (November 1904): 1–32.

12. William Raspberry, personal interview by Roger Horowitz, December 10, 1987. Chicagoan Dempsey Travis, a black soldier, conveys the same idea in his description of his father, who was a stockyard worker. "He had a walk and a proudness and an importance that you couldn't believe." Studs Terkel, *"The Good War": An Oral History of World War Two* (New York, 1984), p. 151.

13. James R. Barrett, "Women's Work, Family Economy and Labor Militancy: The Case of Chicago's Packinghouse Workers, 1900–1922," in *Labor Divided: Race and Ethnicity in United States Labor Struggles, 1835–1960,* ed. Robert Asher and Charles Stephenson (Albany, N.Y., 1990), pp. 249–66. See also Sonya O. Rose, "Gender Segregation in the Transition to the Factory: The English Hosiery Industry, 1850–1910," *Feminist Studies* 13 (Spring 1987): 163–84. The one job that Women's Bureau investigators noticed that women held in the killing floors was stamping the animals who had been inspected and approved by government inspectors. This was an entirely new job created after the 1906 Meat Inspection Act.

In this way, women who did work in the killing floor area only were in jobs that had not previously been performed by men. Mary Elizabeth Pidgeon, *The Employment of Women in Slaughtering and Meat Packing* (Washington, D.C., 1932), p. 34. Tillie Olsen also identifies this as a women's job: "Hogs dangling, dancing along the convey, 300, 350 an hour; Mary running along the rickety platform to keep up, stamping, stamping the hides." Olsen (n. 1 above), p. 133.

14. Estelle Zabritsky interview in *First Person America*, ed. Ann Banks (New York, 1980), pp. 56–59 (quote p. 56).

15. Eric Brian Halpern, "'Black and White, Unite and Fight': Race and Labor in Meatpacking, 1904–1948," (Ph.D. diss., University of Pennsylvania, 1989), p. 233. Sophie Kosciowlowski interview by Les Orear, Chicago Stock Yards Interview Series, 1971, and by Elizabeth Butters, January 15, 1971, both deposited with the Oral History Project in Labor and Immigration History, Roosevelt University. Carole Turbin, "Beyond Conventional Wisdom," and *Working Women of Collar City* (n. 5 above). Quote from *Kansas City Star,* October 19, 1913.

16. Quote from *Kansas City Star,* October 19, 1913. For example, in Kansas City, Missouri in the 1910s, 63 percent of the area's black women worked for wages; 83 percent of them made less than $4 per week as cooks, housekeepers, or laundresses. Clifford Naismith, "History of the Negro Population of Kansas City, Missouri 1870–1930," (manuscript, n.d.), pp. 202–238, in A. Theodore Brown Collection, folder 5, Western Historical Manuscripts Collection, University of Missouri, Kansas City. Asa Martin, *Our Negro Population* (Kansas City, Mo., 1913), pp. 50–51, 56–57.

17. Joanne J. Meyerowitz, *Women Adrift: Independent Wage Earners in Chicago, 1880–1930* (Chicago, 1988), pp. 35–36. Quote from *Employment of Women in Slaughtering and Meatpacking,* (n. 13 above), p. 39; Alma Herbst, *The Negro in the Slaughtering and Meat-Packing Industry* (1932; reprint, New York, 1971), p. 76.

18. Quote from Bruce R. Fehn, "Striking Women: Gender, Race and Class in the United Packinghouse Workers of America (UPWA), 1938–1968" (Ph.D. diss., University of Wisconsin, Madison, 1991), p. 37. "The 'Significant Sixty'" (n. 10 above), pp. 276–78.

19 . Pidgeon (n. 13 above), pp. 34–35.

20. Herbst, pp. 169–71. Pidgeon, pp. 20–23, 34–35. C. V. Whalin, "By-Products of the Slaughtering and Meat Packing Industry," in Federal Trade Commission, *Report on the Meat-Packing Industry,* Vol. 1, (Washington, D.C., 1919), pp. 547–57. Dominic Anthony Pacyga, "Villages of Packinghouses and Steel Mills: The Polish Worker on Chicago's South Side, 1880 to 1921," (Ph.D. diss., University of Illinois, 1981), pp. 68–69. Virginia Houston interview, United Packinghouse Workers of America Oral History Project (principal interviewers Rick Halpern and Roger Horowitz), State Historical Society of Wisconsin (hereafter cited as UPWAOHP). Eunetta Pierce, Leona Tarnowski, Nevada Isom interviews, Jeanette Haymond and Louise Townshend interview, UPWAOHP.

21. Quotes from Herbst, pp. 75–77, except for reference to "nigger work," which is from Homer Early interview, UPWAOHP. Pidgeon, pp. 144–45. U.S. Department of Labor, *Wages and Hours of Labor in the Slaughtering and Meat-Packing Industry, 1929,* Bureau of Labor Statistics Bulletin No. 535 (Washington, D.C., 1931), p. 10.

22. Herbst, p. 171. Pidgeon, pp. 23–24, 36–37, 42–45. Hattie Jones interview, Jeanette Haymond and Louise Townshend interview, UPWAOHP.

23. Noelie Vialles, *Animal to Edible* (New York, 1994), pp. 94–110. Pidgeon (n. 13 above), p. 24. Pacyga, p. 60.

24. Women in the pork trim, paid according the amount of lean meat in their bucket, frequently squabbled over pieces of meat. "One [would] reach over here for a big piece of meat that had a lot of lean in it," recalled former Chicago Swift worker Philip Weightman, "another one would walk over and start fighting with the hook to get it." Philip Weightman interview, UPWAOHP. Similar comments appear in interviews with Walt Mason and Darrel Poe, UPWAOHP.

25. Pidgeon, pp. 141–42, 144–45. Herbst (n. 17 above), pp. 91, 123.

26. F. W. Wilder, *The Modern Packing House* (Chicago, 1905), pp. 344–410, quote on p. 341. Upton Sinclair, *The Jungle* (1905; reprint, Memphis and Atlanta, 1988), pp. 119–20. "The 'Significant Sixty'" (n. 10 above), p. 276. *Kansas City Star*, December 5, 1915. Fehn (n. 18 above), pp. 67–71.

27. Sinclair, p. 120. Betty Watson, Helen Zrudsky interviews, UPWAOHP. U.S. Department of Labor, *Wages and Hours of Labor* (n. 21 above), pp. 36–39. Herbst, p. 117. Pidgeon, pp. 22–27. While female linkers complained to investigators in the late 1920s of numbness and weakness in their arms, a 1943 report on work hazards in meatpacking neither reported nor investigated the extent of this occupational disorder. U.S. Department of Labor, *Injuries and Accident Causes in the Slaughtering and Meat-Packing Industry, 1943*, Bulletin No. 855 (Washington D.C., 1943), esp. p. 10.

28. Pidgeon, pp. 141–42. Sinclair, p. 120. Offal and casings experienced greater employment fluctuation because these departments were linked to the slaughtering departments, where production peaked between September and April. Sausage workers tended to have a more steady supply of meat with which to work and hence experienced fewer layoffs.

29. "The 'Significant Sixty,'" p. 278. Pidgeon (n. 13 above), pp. 7, 28–29, 40, 140–41.

30. Pidgeon, pp. 140–41.

31. Herbst (n. 17 above), pp. 77–78.

32. Quote from *Kansas City Star*, December 5, 1915.

33. The union's treatment of women, equal pay for equal work, and acceptance of separate job categories is discussed in Roger Horowitz, *"Negro and White, Unite and Fight!" A Social History of Industrial Unionism in Meatpacking, 1930–1990* (Urbana, Ill., 1997), esp. chap. 9. A good survey of wage patterns and brackets can be found in U.S. Department of Labor, *Wage Chronology, Armour and Co., 1941–63*, Bureau of Labor Statistics Bulletin No. 187 (Washington, D.C., 1965). For the experience of other unions, see Nancy Gabin, *Feminism in the Labor Movement: Women and the United Auto Workers, 1935–1975* (Ithaca, N.Y., 1990), and Lisa Kannenberg, "The Impact of the Cold War on Women's Trade Union Activism: The UE Experience," *Labor History* 34 (1993): pp. 309–23.

34. "Departments Practicing Discrimination," "Some of the Areas Where Discrimination is at its Worst," October 30–31 and November 1, 1953, in personal papers of Herbert Hill. Interviews: Nevada Isom, Mary Salinas, Sam Parks, Walter Bailey, and Robert Burt, UPWAOHP. "Program Department Report," January 12, 1954, UPWA papers, State Histori-

cal Society of Wisconsin, box 359, folder 7. "Convention report of UPWA Anti-Discrimination and Program Departments," June 18, 1956, UPWA papers, box 347, folder 13.

35. Interviews: Marion Simmons, Rowena Moore, and Addie Wyatt, UPWAOHP. See also Fehn (n. 18 above), pp. 264–66. For "axiomatic" quote, see Joan Scott, "Reply to Criticism" (n. 4 above).

36. U.S. Department of Commerce, *1963 Census of Manufactures, Vol. II,* Part 1 (Washington, D.C., 1966), pp. 20A-14, 20A-15. James Franklin Crawford, "Wage Pattern Following in the Meat Packing Industry," (Ph.D. diss., University of Wisconsin, Madison, 1957), p. 79. Marvin Hayenga et al., *The U.S. Pork Sector: Changing Structure and Organization* (Ames, Iowa, 1985), p. 128.

37. UPWAOHP interviews: Lucille Bremer, Robert Burt (July 30, 1986), Betty Watson, Charles Mueller (July 30, 1986), Bill Webster, Mary and Alvin Edwards, Eugene Crowley and Marjorie Carter, Max Graham, Virginia Houston, and Walter Bailey. "How Mechanization and Plant Closings Are Holding Down Packinghouse Employment," 1957, UPWA papers, box 382, folder 5. Robert Lubar, "Armour Sees Fat Years Ahead," *Fortune,* October 1959, p. 25.

38. "Meat Packing Becomes Decentralized," *Business Conditions,* November 1959, pp. 4–10. Willard F. Williams, "Structural Changes in the Meat Wholesaling Industry," *Journal of Farm Economics* 40 (May 1958): 315–29. Richard J. Arnould, "Changing Patterns of Concentration in American Meat Packing, 1880–1963," *Business History Review* 45 (Spring 1971): 18–34. *National Provisioner,* July 4, 1981, special two-volume issue titled "Meat for the Multitudes," vol. 2, pp. 86–91. U. S. Department of Labor, *Industry Wage Survey: Meat Products* (Washington, D.C., November 1963), p. 12. U.S. Department of Labor, *Industry Wage Survey: Meat Products* (Washington, D.C., June 1984), p. 6.

39. Quote from Charles Mueller interview, July 30, 1986, UPWAOHP. For an example of resentment by younger workers, see Max Graham and Herb Cassano interviews, UPWAOHP.

40. *Packinghouse Worker,* March 1965. Betty Watson interview, UPWAOHP. The idea of an "ABC" system had been circulating in the organization since 1956, when Marian Simmons raised an idea for a three-way division of the jobs at the UPWA convention. See Fehn (n. 18 above), pp. 273–74, for Simmons's views, and pp. 256–306, for a full discussion of these dynamics. See also Bruce Fehn, "Chickens Come Home to Roost: Industrial Organization, Seniority, and Gender Conflict in the United Packinghouse Workers of America, 1955–1966," *Labor History* 34 (1993): 324–41, and Dennis A. Deslippe, "'We Had an Awful Time with Our Women': Iowa's United Packinghouse Workers of America, 1945–75," *Journal of Women's History* 5 (Spring 1993): 10–32.

41. Information on women in the contemporary meatpacking industry is drawn from the following: Christopher Drew, "Regulators Slow Down as Packers Speed Up," *Chicago Tribune,* October 26, 1988. U. S. House of Representatives, *Here's the Beef: Underreporting of Injuries, OSHA's Policy of Exempting Companies from Programmed Inspections Based on Injury Records, and Unsafe Conditions in the Meatpacking Industry,* 100th Cong., 1st sess., 1988, H. Rept. 100-542. William Glaberson, "Misery on the Meatpacking Line," *New York Times,* June 14, 1987. House Committee on Government Operations, *Underreporting*

of Occupational Injuries and Its Impact on Workers' Safety: Hearings before the Subcommittee of the Committee on Government Operations of the House of Representatives, 100th Congress, 1st sess., March 1987. Quote from Janet E. Benson, "The Effects of Packinghouse Work on Southeast Asian Refugee Families," in *Newcomers in the Workplace: Immigrants and the Restructuring of the U.S. Economy,* ed. Louise Lamphere et al. (Philadelphia, 1994), p. 111. On contemporary men's work in meatpacking, see also Donald D. Stull, "Knock 'Em Dead: Work on the Killfloor of a Modern Beefpacking Plant," in Lamphere, pp. 44–77, and Ken C. Erickson, "Guys in White Hats: Short-Term Participant Observation among Beef-Processing Workers and Managers," in Lamphere, pp. 78–98.

42. American Meat Institute, *Meat Facts 1991* (Washington, DC, 1992), p. 34. U.S. Department of Commerce, *Statistical Abstract of the United States, 1991* (Washington D.C., 1991), p. 474. This argument is developed more fully in Horowitz, *"Negro and White, Unite and Fight!"* (n. 33 above), chap. 10.

43. Quote from Donald D. Stull et al., "The Price of a Good Steak: Beef Packing and Its Consequences for Garden City, Kansas," in *Structuring Diversity: Ethnographic Perspectives on the New Immigration,* ed. Louise Lamphere (Chicago, 1992), p. 51.

44. Donald D. Stull et al., "Changing Relations: Newcomers and Established Residents in Garden City, Kansas," Institute for Public Policy and Business Research, University of Kansas, Report No. 172, February 5, 1990, p. 59.

ELEVEN

Programming

JENNIFER LIGHT

Among the jobs women took during World War II was that of "computer"—computing ballistics trajectories and solving complex mathematical equations of wartime research. Although their work required advanced mathematics training, the job was labeled as subprofessional, a kind of clerical work. The calculations, which could take days or weeks to complete, were the bottleneck the electronic computer ENIAC was designed to automate. When ENIAC was ready in 1945, some of the human computers were converted to machine "operators"—the first computer programmers, a task which at the time required learning the circuitry of the machine.

On one level this is a story of workplace mechanization, but, as Jennifer Light argues, it is also a story of rendering women's technological contributions invisible: standard accounts of the history of computing barely mention the "ENIAC girls," and postwar publicity about the project skimmed over their presence as well. Yet ENIAC was a pile of circuitry without them. How did gender shape the assignments of technological work? The portrayal of technological knowledge? Did the content of the women's work match or contradict wartime gender expectations? Are the gender ideologies portrayed here similar to or different from those in Paul Edwards's article on subsequent generations of military computers?

"When Computers Were Women," *Technology and Culture* 40, no. 3 (1999): 455–83. Reprinted by permission of the Society for the History of Technology.

J. Presper Eckert and John W. Mauchly, household names in the history of computing, developed America's first electronic computer, ENIAC, to automate ballistics computations during World War II. These two talented engineers dominate the story as it is usually told, but they hardly worked alone. Nearly two hundred young women, both civilian and military, worked on the project as human "computers," performing ballistics computations during the war. Six of them were selected to program a machine that, ironically, would take their name and replace them, a machine whose technical expertise would become vastly more celebrated than their own.[1]

The omission of women from the history of computer science perpetuates misconceptions of women as uninterested or incapable in the field. This article retells the history of ENIAC's "invention" with special focus on the female technicians whom existing computer histories have rendered invisible. In particular, it examines how the job of programmer, perceived in recent years as masculine work, originated as feminized clerical labor. The story presents an apparent paradox. It suggests that women were somehow hidden during this stage of computer history while the wartime popular press trumpeted just the opposite — that women were breaking into traditionally male occupations within science, technology, and engineering. A closer look at this literature explicates the paradox by revealing widespread ambivalence about women's work. While celebrating women's presence, wartime writing minimized the complexities of their actual work. While describing the difficulty of their tasks, it classified their occupations as subprofessional. While showcasing them in formerly male occupations, it celebrated their work for its femininity. Despite the complexities — and often pathbreaking aspects — of the work women performed, they rarely received credit for innovation or invention.

The story of ENIAC's female computers supports Ruth Milkman's thesis of an "idiom of sex-typing" during World War II — that the rationale explaining why women performed certain jobs contradicted the actual sexual division of labor.[2] Following her lead, I will compare the actual contributions of these women with their media image. Prewar labor patterns in scientific and clerical occupations significantly influenced the way women with mathematical training were assigned to jobs, what kinds of work they did, and how contemporary media regarded (or failed to regard) this work. This article suggests why previous accounts of computer history did not portray women as significant and argues for a reappraisal of their contributions.[3]

Women in Wartime

Wartime literature characterized World War II as a momentous event in the history of women's employment. In 1943 *Wartime Opportunities for Women* proclaimed, "It's a Woman's World!"[4] Such accounts hailed unprecedented employment opportunities as men were recruited for combat positions. New military and civilian women's organizations such as the Army's Women's Auxiliary Army Corps (WAAC, converted to full military status in 1943 and renamed the Women's Army Corps [WAC]), the Navy's Women Accepted for Volunteer Emergency Service (WAVES), and the American Women's Voluntary Services (AWVS) channeled women into a variety of jobs. The press emphasized the role of machines in war and urged women with mechanical knowledge to "make use of it to the best possible purpose."[5] *Wartime Opportunities for Women* urged: "In this most technical of all wars, science in action is a prime necessity. Engineering is science in action. It takes what the creative mind behind pure science has to offer and builds toward a new engine, product or process."[6] According to the U.S. Department of Labor's Women's Bureau: "The need for women engineers and scientists is growing both in industry and government. . . . Women are being offered scientific and engineering jobs where formerly men were preferred. Now is the time to consider your job in science and engineering. There are no limitations on your opportunities. . . . In looking at the war job opportunities in science and engineering, you will find that the slogan there as elsewhere is 'WOMEN WANTED!'"[7]

A multiplicity of books and pamphlets published by the U.S. War Department and the Department of Labor, with such titles as *Women in War, American Women in Uniform, Back of the Fighting Front,* and *Wartime Opportunities for Women,* echoed this sentiment. Before World War II, women with college degrees in mathematics generally taught primary or secondary school. Occasionally they worked in clerical services as statistical clerks or human computers. The war changed job demands, and one women's college reported that every mathematics major had her choice of twenty-five jobs in industry or government.[8]

Yet, as Milkman suggests, more women in the labor market did not necessarily mean more equality with men. Sexual divisions of labor persisted during wartime.

The geography of women's work settings changed, but the new technical positions did not extend up the job ladder. A widely held belief that female workers would be dismissed once male veterans returned from the war helps to explain the Women's Bureau acknowledgement that "except for Ph.D.'s, women trained in mathematics tend to be employed at the assistant level."[9] The War Department and the Department of Labor actively promoted women's breadth of opportunity yet in some areas explicitly defined which jobs were "open to women." Classified advertisements ran separate listings for "female help wanted" and "male help wanted."

Women's Ambiguous Entry into Computing

Women's role in the development of ENIAC offers an account of the feminization of one occupation, "ballistics computer," and both the creation of and gendering of another, "operator" (what we would now call programmer). Ballistics computation and programming lay at the intersection of scientific and clerical labor. Each required advanced mathematical training, yet each was categorized as clerical work. Such gendering of occupations had precedent. Since the late nineteenth century feminized jobs had developed in a number of sciences where women worked alongside men. Margaret Rossiter identifies several conditions that facilitated the growth of "women's work."[10] These include the rise of big science research projects, low budgets, an available pool of educated women, a lack of men, a woman who could act as an intermediary (such as a male scientist's wife), and a somewhat enlightened employer in a climate generally resistant to female employees entering traditionally male domains. Craving opportunities to use their skills, some women colluded with this sexual division of labor. Many did not aspire to professional employment at higher levels.[11]

Occupational feminization in the sciences fostered long-term invisibility. For example, beginning in the 1940s, laboratories hired women to examine the nuclear and particle tracks on photographic emulsions.[12] Until the 1950s, published copies of photographs that each woman scanned bore her name. Yet eventually the status of these women's work eroded. Later publications were subsumed under the name of the lab leader, inevitably a man, and publicity photographs rarely showcased women's contributions. Physicist Cecil Powell's request for "three more microscopes and three girls" suggests how invisibility and interchangeability went hand in hand.[13] In a number of laboratories, scientists described women not as individuals but rather as a collective, defined by their lab leader ("Cecil's Beauty Chorus") or by their machines ("scanner girls"). Likewise in the ENIAC project, female operators are referred to as "[John] Holberton's group" or as "ENIAC girls." Technicians generally did not author papers or technical manuals. Nor did they acquire the coveted status symbols of scientists and engineers: publications, lectures, and membership in professional societies. Ultimately these women never got a public opportunity to display their technical knowledge, crucial for personal recognition and career advancement.

Wartime labor shortages stimulated women's entry into new occupations, and computing was no exception.[14] Ballistics computing, a man's job during World War I, was feminized by World War II. A memorandum from the Computing Group Organization and Practices at the National Advisory Committee for Aeronautics (NACA), dated 27 April 1942, explains how the NACA conceived the role of computers: "It is felt that enough greater return is obtained by freeing the engineers from calculating detail to overcome any increased expenses in the computers' salaries. The engineers admit themselves that the girl computers do the work more rapidly and accurately than they would. This is due in large measure to the feeling among the engineers that their college and industrial experience is being wasted and thwarted by mere repetitive calculation."[15]

Patterns of occupational segregation developed in selected industries and job categories newly opened to women.[16] Women hired as computers and clerks generally assisted men. Captain Herman Goldstine, an ENIAC project leader, served as liaison from the U.S. Army's Ballistic Research Laboratory (BRL) to the Moore School of Electrical Engineering at the University of Pennsylvania, which produced ENIAC, and director of computer training for BRL. He recalls that by World War II "there were a few men [computers] but only a few. Any able-bodied man was going to get taken up into the armed forces."[17] With feminization came a loss of technical status, since other men doing more "important" technical and classified work remained in noncombatant positions. Thus, the meaning of "wartime labor shortage" was circumscribed even as it came into being. While college-educated engineers considered the task of computing too tedious for themselves, it was not too tedious for the college-educated women who made up the majority of computers.[18] These were not simply cases of women taking on men's tasks but rather of the emergence of new job definitions in light of the female workforce.[19] Celebrations of women's wartime contributions thus rarely challenged gender roles.

Rather, popular accounts portrayed civilian jobs for women as appropriately feminine "domestic" work for the nation—despite the fact they were formerly done by
men.[20]

The introduction of technology also facilitated women's entry into paid labor.
Machines stimulated the reorganization of work processes, often leading to the
creation of new occupations and the culling of older ones. In both clerical and factory work, introducing technology changed some jobs so that women performed
slightly different tasks rather than substituting directly for men. Women's entry
into the workforce was greatest in new occupations where they did not displace
men.[21] Once a particular job was feminized this classification gathered momentum, often broadening to include other occupations.[22] By World War II, computing
was feminized across a variety of fields, including engineering, architecture, ballistics, and the aircraft industry. The new machines, capable of replacing hundreds
of human computers, required human intervention to set up mathematical problems. Without a gendered precedent, the job of computer operator, like the newly
created jobs of "stenographer typist" and "scanning girl," became women's work.
There is, of course, a fundamental difference between the human computer and
the programmer who transfers this skill to an automated process. In the 1940s, the
skill of transferring this information—what we now call programming—fit easily
with notions about women's work. As an extension of the job of a human computer, this clerical task offered slightly higher status and higher pay than other
kinds of clerical labor.[23]

Female Computers and ENIAC Girls

Like much of scientific research and development during World War II, the
ENIAC was the offspring of a wartime alliance between a university (the University of Pennsylvania, specifically the Moore School of Electrical Engineering)
and the U.S. armed forces, in this case the Army Proving Ground (APG) in Aberdeen, Maryland. The APG housed the army's Ballistic Research Laboratory (BRL),
which produced range tables for gunners. During the war, BRL recruited approximately two hundred women to work as computers, hand-calculating firing tables
for rockets and artillery shells. In 1940, when President Franklin D. Roosevelt declared a national emergency, BRL commandeered the Moore School's differential
analyzer and began to move some of its work to the university.[24]

One of the first women the army hired to work at the Moore School was twenty-two-year-old Kathleen McNulty. She had graduated in 1942 from Chestnut Hill College, in Philadelphia, with one of the three math degrees awarded in her class. McNulty and her friend Frances Bilas answered an advertisement in a local paper that said Aberdeen was hiring mathematicians:

> I never heard of numerical integration. We had never done anything like that. Numerical integration is where you take, in this particular case . . . [the] path of a bullet from the time it leaves the muzzle of the gun until it reaches the ground. It is a very complex equation; it has about fifteen multiplications and a square root and I don't know what else. You have to find out where the bullet is every tenth of a second from the time it leaves the muzzle of the gun, and you have to take into account all the things that are going to affect the path of the bullet. The very first things that affect the path of the bullet [are] the speed at which it shoots out of the gun [the muzzle velocity], the angle at which it is shot out of the gun, and the size. That's all incorporated in a function which they give you—a [ballistic] coefficient.
>
> As the bullet travels through the air, before it reaches its highest point, it is constantly being pressed down by gravity. It is also being acted upon by air pressure, even by the temperature. As the bullet reached a certain muzzle velocity—usually a declining muzzle velocity, because a typical muzzle velocity would be 2,800 feet per second [fps]—when it got down to the point of 1,110 fps, the speed of sound, then it wobbled terribly. . . . So instead of computing now at a tenth of a second, you might have broken this down to one one hundredth of a second to very carefully calculate this path as it went through there. Then what you had to do, when you finished the whole calculation, you interpolated the values to find out what was the very highest point and where it hit the ground.[25]

The work required a high level of mathematical skill, which included solving nonlinear differential equations in several variables: "Every four lines we had to check our computations by something called Simpson's rule to prove that we were performing the functions correctly. All of it was done using numbers so that you kept constantly finding differences and correcting back."[26] Depending upon their method, the computers could calculate a trajectory in somewhere between twenty minutes and several days, using the differential analyzer, slide rules, and desktop commercial calculators.[27] Despite the complexities of preparing firing tables, in this feminized job category McNulty's appointment was rated at a subprofessional grade. The BRL also categorized women like Lila Todd, a computer supervisor when McNulty started work at the Moore School, as subprofessional.[28]

Herman Goldstine recalls that BRL hired female computers almost exclusively. At first, most women were recent college graduates in the Baltimore and Philadelphia area. Adele Goldstine, his wife and a senior computer, expanded recruiting to include colleges across the Northeast, but the project still needed more personnel.[29] In a short time, recalls Goldstine, "We used up all of the civil service women we could get our hands on."[30] A memo to University of Pennsylvania provost George McClelland from Harold Pender, dean of the Moore School, explained how BRL sought to remedy the situation: "Colonel Simon, Chief of the Ballistic Research Laboratory, has had a specially selected group of WACs assigned to the Laboratory. Although these women have been individually picked they are for the most part ready for training and are not trained persons who can enter fully into the Laboratory's work. . . . By consulting appropriate persons on the campus it appears that this can be carried out without interfering with any of the University's regular work. . . . Under the above circumstances it appears that the University's regular work will not be disturbed and at the same time we will have the opportunity to do a rather important service."[31] Pender's memo embodies a more widespread ambivalence about women's wartime contributions, particularly as members of the military. While "specially selected" for a "rather important" task, these women were simultaneously "not trained persons" and could not enter "fully" into the BRL's work.

Colonel Simon assigned two groups of WACs to work as computers. One used desk calculators and the differential analyzer for practical work at the BRL, while the other studied mathematics for ballistics computations at the University of Pennsylvania. These two groups alternated monthly for eight months. The first WAC course started on 9 August 1943. According to reports in the *Daily Pennsylvanian,* the university's student newspaper, these women assimilated smoothly into campus life:

> The WACs at present stationed on the University campus are members of two groups alternating in a special course at the Moore School of Electrical Engineering, and were detached from the unit at Aberdeen Proving Ground, Maryland. At Aberdeen most of them were assigned as computers. The two sections, each of which numbers approximately thirty women, are commanded by second lieutenants and corporals. They are taking courses that are equivalent to the work of a college mathematics major. The results of these studies will later be used in ballistic work at the Ballistic Research Laboratory of the Army Ordnance Department. They are stationed at the

Moore School of Electrical Engineering rather than at any other University school because of the large amount of work that the Moore School has done in collaboration with the Ballistic Research Laboratory. They are quartered in the fraternity house [Phi Kappa Sigma], messed in Sergeant Hall, and receive physical training at Bennett Hall. They are required to police their own rooms and be in bed at eleven forty-five P.M., with the exception of weekends. Reveille must be answered at 7:10 A.M.[32]

In this straightforward report, the student reporter neglects to mention the concurrent and widespread tensions surrounding WACs. Only a month earlier, on 1 July 1943, President Roosevelt had signed legislation converting the Women's Auxiliary Army Corps to full military status as the WAC. The conversion was scheduled for implementation by 1 October. According to WAC historian Mattie Treadwell, "The following ninety days of the summer of 1943, initially called The Conversion, were perhaps the busiest in the history of the Corps."[33]

While the article quoted several WACs commenting about their campus lives in a quite positive tone, Adele Goldstine, in an undated letter to a correspondent, reported, "Rumor hath it that the WACs (Sec. I) have been told that they're unloved by everybody including the ES&MWTesses. If it's true, I'm sorry to hear it because I'm afraid it will make our uphill fight steeper."[34] Her letter suggests that the women's presence on campus had become the "interference" and "disturbance" intimated by Simon's memo. Indeed, ambivalence about The Conversion had triggered slander campaigns against WACs from 1943. The cold reception of WAC volunteers was a product not only of news media but also of local gossip: "Resentment was expressed in towns where WACs were quartered, to the effect that they were spoiling the character of the town."[35] The WACs in Philadelphia may have experienced some of the more widespread hostility towards enlisted women.

Separated by skill level into two groups, the WACs at the Moore school had forty hours of classroom instruction per week. According to the syllabus, the course was designed to treat "in succinct form the mathematics which a person should have to work on physical problems such as those likely to be met in the Ballistic Research Laboratory."[36] The mathematics ranged from elementary algebra to simple differential equations. In addition, a unit on the use of calculating machines covered computation- and calculation-machine techniques, handling numerical data, organizing work for machine calculation, and using slide rules.

The instructors included three men (a Dr. Sohon, a Mr. Charp, and a Mr. Fliess) and nine women (Adele Goldstine, Mary Mauchly, Mildred Kramer, Alice Burks, a Mrs. Harris, a Miss Mott, a Miss Greene, a Mrs. Seeley, and a Mrs. Pritkin). Accounts of ENIAC that discuss the WAC course, such as Goldstine's book and the civilian women's own reflections, mention as instructors only three married women: Adele Goldstine; Mary Mauchly, wife of John Mauchly of the Moore School; and Mildred Kramer, wife of Samuel Noah Kramer, a professor of Assyriology at the University of Pennsylvania. Yet archival records show that this is not the full story.[37] Perhaps this oversight is consistent with a different trend Rossiter discusses — that more prominent women in science were often married to notable men, also often scientists. It is unclear whether Goldstine, Mauchly, and Kramer became "visible" because their husbands' visibility accorded them extra attention, because these men somehow facilitated their wives' careers, or because the women themselves campaigned for recognition.

"Thanks for the Memory," a song presumably written by several WACs, offers a playful account of their time at the Moore School:

Of days way back when school
Was just the daily rule
When we just studied theories
For fun and not as tools — thank you so much.

Of lectures running late
Of Math that's mixed with paint
Of dainty slips that ride up hips
And hair-do-ups that ain't — thank you so much.

Many's the time that we fretted
And many's the time that we sweated
Over problems of Simpson and Weddle
But we didn't care — for c'est la guerre!

That Saturday always came
And teach ran for her train
If she didn't lam — like Mary's lamb
Her pets to Moore School came — thank you so much.

Machines that dance and dive
Of numbers that can jive
Of series that do leaps and bounds
Until you lose the five — thank you so much.

Of half-hour luncheon treks
How we waited for our checks!
Of assets, liabilities—
Till all of us were wrecks—thank you too much.

We squared and we cubed and we plotted
And many lines drew and some dotted
We've all developed a complex
Over wine, sex, and f(x)

Of private tete-a-tetes
And talk about our dates
And how we wish that teacher would oblige
By coming late—thank you so much.

And so on through the night.[38]

Even as the WAC courses went on, Moore School engineers were designing a machine to automate the production of the same firing and bombing tables calculated by the human computers: the ENIAC. Engineers wanted answers faster than women could supply them using available technologies. Yet ENIAC couldn't do everything itself. Programming equations into the machine required human labor.[39] The eventual transfer of computing from human to machine led to shifting job definitions. A "computer" was a human being until approximately 1945. After that date the term referred to a machine, and the former human computers became "operators."[40]

Herman Goldstine recounts selecting the operators. At BRL, one group of women used desk calculators and another the differential analyzer. Selecting a subgroup from each, Goldstine "assigned six of the best computers to learn how to program the ENIAC and report to [John] Holberton," employed by the Army Ordnance Department to supervise civilians.[41] With no precedents from either sex, the creation and gendering of "computer operator" offers insight into how sexual divisions of labor gather momentum. Computing was a female job, and other female clerical workers operated business machines. So it was not unusual that in July 1945 women would migrate to a similar but new occupation. The six women—Kathleen McNulty, Frances Bilas, Betty Jean Jennings, Ruth Lichterman, Elizabeth Snyder, and Marlyn Wescoff—reported to the Moore School to learn to program the ENIAC.

The ENIAC project made a fundamental distinction between hardware and software: designing hardware was a man's job; programming was a woman's job. Each of these gendered parts of the project had its own clear status classification. Software, a secondary, clerical task, did not match the importance of constructing the ENIAC and getting it to work.[42] The female programmers carried out orders from male engineers and army officers. It was these engineers and officers, the theoreticians and managers, who received credit for invention. The U.S. Army's social caste system is historically based on European gentlemen's social codes.[43] As civil servants, the six women computers chosen to operate the ENIAC stood outside this system.

Yet if engineers originally conceived of the task of programming as merely clerical, it proved more complex. Under the direction of Herman and Adele Goldstine, the ENIAC operators studied the machine's circuitry, logic, physical structure, and operation. Kathleen McNulty described how their work overlapped with the construction of the ENIAC: "Somebody gave us a whole stack of blueprints, and these were the wiring diagrams for all the panels, and they said 'Here, figure out how the machine works and then figure out how to program it.' This was a little bit hard to do. So Dr. Burks at that time was one of the people assigned to explain to us how the various parts of the computer worked, how an accumulator worked. Well once you knew how an accumulator worked, you could pretty well be able to trace the other circuits for yourself and figure this thing out."[44]

Understanding the hardware was a process of learning by doing. By crawling around inside the massive frame, the women located burnt-out vacuum tubes, shorted connections, and other nonclerical bugs.[45] Betty Jean Jennings's description confirms the ingenuity required to program at the machine level and the kinds of tacit knowledge involved:

We spent much of our time at APG learning how to wire the control board for the various punch card machines: tabulator, sorter, reader, reproducer, and punch. As part of our training, we took apart and attempted to fully understand a fourth-order difference board that the APG people had developed for the tabulator. . . . Occasionally, the six of us programmers all got together to discuss how we thought the machine worked. If this sounds haphazard, it was. The biggest advantage of learning the ENIAC from the diagrams was that we began to understand what it could and what it could not do. As a result we could diagnose troubles almost down to the individual vacuum tube. Since we knew both the application and the machine, we learned to diagnose troubles as well as, if not better than, the engineer.[46]

Framing the ENIAC story as a case study of the mechanization of female labor, it would be hard to argue that de-skilling accompanied mechanization.[47] The idiom of sex-typing, which justified assigning women to software, contradicted the actual job, which required sophisticated familiarity with hardware. The six ENIAC operators understood not only the mathematics of computing but the machine itself. That project leaders and historians did not value their technical knowledge fits the scholarly perception of a contradiction between the work actually performed by women and the way others evaluate that work. In the words of Nina Lerman, "Gender plays a role in defining which activities can readily be labeled 'technological.'"[48]

Meanwhile, at the Los Alamos Scientific Laboratory in New Mexico, scientists were preparing a new thermonuclear weapon, the Super. Stanley Frankel and Nicholas Metropolis, two Los Alamos physicists, were working on a mathematical model that might help to determine the possibility of a thermonuclear explosion. John Von Neumann, a technical consultant, suggested that Los Alamos use ENIAC to calculate the Super's feasibility. Once Von Neumann told Herman Goldstine about this possible use, Herman and Adele invited Frankel and Metropolis to Philadelphia and offered them training on the ENIAC. When the two physicists arrived in Philadelphia in the summer of 1945, Adele Goldstine and the women operators explained how to use the machine. McNulty recalled that "We had barely begun to think that we had enough knowledge of the machine to program a trajectory, when we were told that two people were coming from Los Alamos to put a problem on the machine."[49] Despite such self-effacing comments, the operators demonstrated impressive mastery of the ENIAC during the collaboration with the Los Alamos physicists. By October, the two theoretical physicists had programmed their elaborate problem on huge sheets of paper. Then, the women programmed it into the machine, which no one had formally tested. As McNulty explained, "No one knew how many bad joints there were, and how many bad tubes there were, and so on."[50] The cooperative endeavor furthered the operators' intimate understanding of ENIAC as they pushed it to a new level of performance. Programming for Frankel and Metropolis took one million IBM punch cards, and the machine's limited memory forced the women to print out intermediate results before re-punching new cards and submitting them to the machine. Within a month, the Los Alamos scientists had their answer—that there were several design flaws.[51]

The "ENIAC girls" turned their attention back to shell trajectory calculations and were still engaged on that project when the war ended. The ENIAC, designed and constructed in military secrecy, was prepared for public unveiling in early 1946. A press conference on 1 February and a formal dedication on 15 February each featured demonstrations of the machine's capabilities. According to Herman Goldstine, "The actual preparation of the problems put on at the demonstration was done by Adele Goldstine and me with some help on the simpler problems from John Holberton and his girls."[52] Indeed, Elizabeth Snyder and Betty Jean Jennings developed the demonstration trajectory program.[53] Although women played a key role in preparing the demonstrations, both for the press and for visitors to the laboratory, this information does not appear in official accounts of what took place.

Contemporary Accounts of ENIAC

Social constructionist historians and sociologists of science take the position that scientists describing their experimental work do not characterize events as they actually happened.[54] Publicity for technical demonstrations is not so different. In presenting ENIAC to the public, engineers staged a well-rehearsed event. They cooperated with the War Department, which controlled representations of the project through frequent press releases to radio and newspapers.

It is a curious paradox that while the War Department urged women into military and civil service and fed the media uplifting stories about women's achievements during the war, its press releases about a critical project like the ENIAC do not mention the women who helped to make the machine run. War Department press releases characterize ENIAC as "designed and constructed for the Ordnance Department at the Moore School of Electrical Engineering of the University of Pennsylvania by a pioneering group of Moore School experts."[55] They list three individuals as "primarily responsible for the extremely difficult technical phases of work . . . Eckert—engineering and design; Mauchly—fundamental ideas, physics; Goldstine—mathematics, technical liaison."[56] The War Department's selective press releases highlighted certain individuals involved in the ENIAC project while omitting others, specifically the women operators. Because of these omissions the operators were neither interviewed nor offered the opportunity to participate in telling the ENIAC story. Newspaper accounts characterize ENIAC's ability to perform tasks as "intelligent," but the women doing the same computing tasks did not receive similar acclaim.[57] While the media publicly hailed hardware designers as having "fathered" the machine, they did not mention women's contributions. The difference in status between hardware and software illustrates another chapter in the story of women in the history of science and technology. The unmentioned computer technicians are reminiscent of Robert Boyle's "host of 'laborants,' 'operators,' 'assistants' and 'chemical servants'" whom Steven Shapin described as "invisible actors." Working three centuries earlier, their fate was the same: they "made the machines work, but they could not make knowledge."[58]

The *New York Times* of 15 February 1946 described Arthur Burks's public demonstration: "The ENIAC was then told to solve a difficult problem that would have required several weeks' work by a trained man. The ENIAC did it in exactly 15 seconds."[59] The "15 seconds" claim ignores the time women spent setting up each problem on the machine. Accompanying photographs of Eckert and Mauchly, the article reported that "the Eniac was invented and perfected by two young scientists of the [Moore] school, Dr. John William Mauchly, 38, a physicist and amateur meteorologist, and his associate, J. Presper Eckert Jr., 26, chief engineer on the project. Assistance was also given by many others at the school. . . . [The machine is] doing easily what had been done laboriously by many trained men. . . . Had it not been available the job would have kept busy 100 trained men for a whole year."[60] While this account alludes to the participation of many individuals other than Eckert and Mauchly, the hypothetical hundred are described as men. Why didn't the article report that the machine easily did calculations that would have kept one hundred trained women busy, since BRL and the Moore School hired women almost exclusively as computers? Even in an era when language defaulted to "he" in general descriptions, this omission is surprising, since the job of computer was widely regarded as women's work.[61] Women seem to have vanished from the ENIAC story, both in text and in photographs. One photograph accompanying the *New York Times* story foregrounds a man in uniform plugging wires into a machine. While the caption describes the "attendants preparing the machine to solve a hydrodynamical problem," the figures of two women in the background can be seen only by close scrutiny (fig. 11.1). Thus, the press conference and follow-up coverage rendered invisible both the skilled labor required to set up the demonstration and the gender of the skilled workers who did it.

The role of the War Department and media in shaping public discourse about the machine and its meaning is significant. Several potential opportunities for the women operators to get some public attention and credit for their work never materialized. For example, the publicity photograph of the ENIAC printed in the *New York Times* was among the most widely disseminated images of the machine.

Figure 11.1. One of the most widely reprinted photographs of ENIAC, from the *New York Times,* 15 February 1946. (Courtesy of New York Times Pictures.)

When it was published as an army recruitment advertisement (fig.11.2), the women were cropped out.[62] This action is understandable, at one level, since the operators were all civilians. Yet given the important participation of WACs in closely related wartime work, it constituted another missed opportunity to give the women their due.

Archival records show that photographers came in to record the ENIAC and its engineers and operators at least twice. Neither visit resulted in any publicity for the women. On the first occasion, an anonymous photographer's pictures of the ENIAC group turned out poorly. Herman Goldstine wrote apologetically to Captain J. J. Power, Office of the Chief of Ordnance: "Dear John, I am returning herewith the photographs with sheets of suggested captions. As you can see from looking at these photographs, many of them are exceedingly poor, and, I think, unsuitable for publication."[63] Nonetheless, the captions for these unsuitable photographs are instructive:

VIEW OF ONE SIDE OF THE ENIAC: Miss Frances Bilas (Philadelphia, Pa.) and Pfc. Homer W. Spence (Grand Rapids, Mich.) are setting program switches. Miss Bilas is an ENIAC operator in the employ of the Ballistic Research Laboratory, Aberdeen Proving Ground, Md., and Pfc. Spence is a maintenance engineer. . . .

SETTING UP A PROBLEM ON THE ENIAC: Reading from left to right, Miss Akrevoe Kondopria (Philadelphia, Pa.) at an accumulator, Miss Betty Jennings (Stanbury, Mo.), Cpl. Irwin Goldstein (Brooklyn, NY) and Miss Ruth Lichterman (Rockaway, NY) standing at function tables. Miss Kondopria is a Moore School employee on the ENIAC project; Miss Jennings and Miss Lichterman are ENIAC operators em-

Figure 11.2. This advertisement appeared in *Popular Science Monthly,* October 1946. (Army materials courtesy of the U.S. Government, as represented by the Secretary of the Army.)

ployed by the Ballistic Research Laboratory, Aberdeen Proving Ground, Md., and Cpl. Goldstein is a maintenance engineer. . . .

SETTING UP A PROBLEM ON THE ENIAC: Reading from left to right, Miss Betty Snyder (Narberth, Pa.), Miss Betty Jennings (Stanbury, Mo.), Miss Marlyn Wescoff (Philadelphia, Pa.) and Miss Ruth Lichterman (Rockaway, NY). Miss Snyder is setting program switches on an accumulator; Miss Jennings is setting up numbers to be remembered in the function table . . . Miss Wescoff and Miss Lichterman are working at the printer. . . . The function table which stores numerical data set up on its switches is seen at the right and its two control panels are behind Miss Frances Bilas (Philadelphia, Pa.) who is plugging a program cable in the master programmer. Miss Bilas is an ENIAC operator in the employ of the Ballistic Research Laboratory, Aberdeen Proving Ground, Maryland.[64]

"Setting switches," "plugging cables," and "standing at function tables" — such captions understate the complexities of women's work. While two men appear alongside the operators, they are "maintenance engineers," occupational titles suggesting technical expertise.

The second photographer was Horace K. Woodward Jr., who wrote an article about ENIAC for *Science*. He wrote to Adele Goldstine: "Dear Mrs. Goldstine and other mENIACS, You will be perturbed to hear that the color flesh shots (oops, flAsh shots) that I was taking 1 feb 46 turned out nicely. I hadn't intended them for publication but thought you folks might like them."[65] His article in *Science* carried no photographs of the women and made no reference to their existence.

More surprising still, the media reports did not highlight Adele Goldstine, despite her leadership position and her expertise in a technical realm that had not earlier existed for either sex.[66] An affidavit Adele Goldstine submitted as testimony in *Sperry Rand v. Bell Labs* explains how she saw her own role: "I did much of the programming and the setting up of the ENIAC for the various problems performed on it while I was at the Moore School. I also assisted my husband in training Mr. Holberton and a group of girls to set up problems on the ENIAC. . . . I worked with Mr. Holberton and his group to program each problem which they put on the ENIAC up to and including the demonstration problems for the ENIAC dedication exercises."[67] Adele Goldstine and Moore School professor Harry Huskey were charged with producing an ENIAC operating manual, a complete technical report, and a maintenance manual.[68] Herman Goldstine explains: "The only persons who really had a completely detailed knowledge of how to program the

ENIAC were my wife and me. Indeed, Adele Goldstine wrote the only manual on the operation of the machine. This book was the only thing available which contained all the material necessary to know how to program the ENIAC and indeed was its purpose."[69] In addition, he reports that his wife contributed heavily to a 1947 paper he coauthored with John Von Neumann, *Planning and Coding Problems for an Electronic Computing Instrument*.[70]

It is an overstatement to say that female computers and operators were never covered in any media. A few articles mention them, as in this example:

> An initial group, consisting primarily of women college graduates, especially trained for work by the Moore School, began the work in ground gunfire, bombing and related ballistics studies immediately after Pearl Harbor, when the Aberdeen Proving Ground's Ballistic Research Laboratory broadened its program at the University.
>
> Forerunners of a group eventually numbering more than 100, they made use of the Moore School's differential analyzer, which is equally useful in the realm of ballistics and the solution of peacetime mathematical problems.
>
> Two other groups were organized later, under separate contracts, one of which was devoted to analysis of experimental rocket firing at Aberdeen, while the other assisted in the proving ground development of new shells and bombs.[71]

This recognition is quite different from the publicity accorded to male officers and engineers associated with the project.[72] The article cited here portrays the women as interchangeable. Even if it were too space-consuming to name each human computer, it is still notable that the article describes the women as being trained for work "by the Moore School" as opposed to "by Adele Goldstine" or by her many female colleagues.[73] That ENIAC's 1946 demonstration doubled as a vanishing act for its female participants fits neatly with postwar propaganda that as early as 1944 began redirecting women into more traditional female occupations or out of the paid labor force entirely.[74]

And what of the several years after World War II? While the Department of Labor acknowledged women's desire to stay on in paid employment, its publications were not so optimistic.[75] An avalanche of materials urged women to leave work. A 1948 *Women's Bureau Bulletin* reported on the situation for women with mathematics education who sought paid work:

> Although, during the war, production firms and Government projects were important outlets for women trained in mathematics, the emphasis, following the end of

hostilities, shifted back to the more usual channels. Teaching and employment with insurance and other business firms became the principal outlets for women college graduates with mathematical training. . . . Most of the wartime research projects sponsored by the Government were dropped after V-J day. In the few that continued, the small number of mathematical jobs were filled by the staffs of the institutions at which the research was being done and by men with mathematical skills who were being released from military service. The women's military services, which utilized women with mathematical training during the war, were reduced to very small staffs. . . . As women leave, men will be hired to replace them. . . . Although many women are continuing on their wartime mathematical jobs, it is difficult to say how much of the gain will be in terms of permanent opportunities for women.[76]

The Federal Bureau of Investigation dropped many of the women it had hired as cryptographers during the war. By 1946, the National Bureau of Standards had filled most of the vacancies on its computing staff with male veterans.[77] At the Ballistics Research Laboratory, an army memorandum detailed criteria for how individuals would be let go, with separate instructions for male officers and for WAC officers.[78] With this in mind, the absence of women from an October 1946 army recruitment ad makes sense. The "propaganda machine," as Herzenberg and Howes call it, that during the war had so successfully called women out of their homes, made a 180-degree turn, pushing many women back towards full-time domesticity.[79]

In the 1950s, new opportunities developed alongside continuing ambivalences about women's occupational roles. A 1956 U.S. Department of Labor report on employment opportunities for women mathematicians and statisticians is replete with examples of women's mathematical work — and the future need for women mathematicians — in a variety of fields including programming. Four "findings" appear as an executive summary:

1. More women mathematicians and statisticians are currently needed, and interesting jobs await those trained at the bachelor's degree as well as graduate levels.

2. Young women in high school should be encouraged to try mathematics and if they have the qualifications for success in mathematics and statistics should be encouraged to prepare for those fields; anticipated shortages make the long-run outlook exceptionally favorable.

3. Young women who combine the qualifications for teaching with ability in

mathematics should be encouraged to teach, at least part time, since in teaching they can magnify their contribution to the Nation's progress.

4. Mature college women who have majored in mathematics, possess the personal qualifications for teaching, and have time available to work, should prepare themselves through refresher courses in mathematics and education for teaching positions, if they live in one of the many communities experiencing or anticipating a shortage of mathematics teachers.[80]

The report explores a wide range of career options, including programming and actuarial work. Yet as the patriotic rhetoric of service "to the Nation's progress" makes clear, the Department of Labor prioritized teaching as a career choice. Science and engineering had won the war, and now the developing baby boom predicted a growing demand for math teachers.

Despite such exhortations, some women never left computer programming. Fran Bilas, Kay McNulty, and Betty Snyder continued briefly with ENIAC when it moved to BRL in 1947; Ruth Lichterman stayed on for two years.[81] Other women joined the ENIAC at BRL following the war. Betty Snyder Holberton went on to program UNIVAC and to write the first major software routine ever developed for automatic programming. She also collaborated on writing COBOL and FORTRAN with Grace Hopper, a key programmer of the Mark I. Hopper left active duty with the U.S. Navy as a lieutenant in 1946 but remained with the Navy reserves until 1966. From 1946 until it started running programs around 1951, the Electronic Computer Project at Princeton's Institute for Advanced Study employed mostly female programmers, who included Thelma Estrin, Hedi Selberg, Sonia Bargmann, and Margaret Lamb. Their accomplishments are future chapters for a history of computer programming.

Conclusion

The ENIAC story highlights several issues in the history and historiography of gender, technology, and labor. Major wars have unmistakable influences on gender relations and work, and those effects can be elusive and complex. Conflicts among representations of women's work in computing ensure work for the historian in distinguishing seeming gender changes from real ones. These conflicts and sometime contradictions lie at the heart of women's historical invisibility.

First, the variance between effusive wartime recruiting literature and historians' evaluations of women's actual opportunities is striking. Disputing the claims of propaganda, historians generally agree that during wartime women may have made some progress in expanding the varieties of work they could do. Yet rather than move up the ladder of success women's work appears to have added more rungs at the bottom. The narrative histories of the ENIAC since 1946 echo this finding. With few exceptions, they make the implicit or explicit assumption that, while women were involved, their participation was not sufficiently important to merit explication. Thus, this episode in the history and historiography of computing confirms Rossiter's "Matilda effect": individuals at the top of professional hierarchies receive repeated publicity and become part of historical records, while subordinates do not and quickly drop from historical memory.[82]

A second conflicting representation concerns the actual work performed by women contrasted with how employers categorized this work. As this article shows, the evidence of ENIAC challenges the implicit assumption of computing historians that the low-status occupations of women meant that their work could not be innovative. Wartime propaganda proclaimed "no limitations on your opportunities," yet only certain jobs were open to women. However, it was within the confines of precisely such low-status occupational classifications that women engaged in unprecedented work. Looking behind media accounts and later narratives of the development of ENIAC to consider primary source accounts of the work women actually performed reveals how its low-status categorization clashed with the kinds of knowledge required. Finding this mismatch offers the possibility that, in their work as operators, women moving into stereotypical male domains played a subversive role, challenging the gender status quo before the war. According to this view, women's invisibility reflects deep-rooted ambivalences about the roles women professionals began to occupy in the labor force. These ambivalences permeated both power relationships in the workplace and media portrayals of women's contributions.

Third, portrayals of women's postwar fate continue the ambivalence that characterized their wartime work. Women were seen as meeting a crisis—but only a temporary one. One 1943 guide to managers explained: "Women can be trained to do any job you've got—but remember 'a woman is not a man;' A woman is a substitute—like plastic instead of metal."[83] Both postwar propaganda and historians characterize women as retreating to teaching and homemaking after the war, abandoning their gains. Yet a fair number did not leave the workforce, a fact that the Department of Labor acknowledged even as it urged women toward teaching.[84]

The revised history of ENIAC presented here reveals that many of historians' questions about the history of computing reflect the unintentionally "male-centered terms" of history.[85] The result is a distorted history of technological development that has rendered women's contributions invisible and promoted a diminished view of women's capabilities in this field. These incomplete stories emphasize the notion that programming and coding are, and were, masculine activities. As computers saturate daily life, it becomes critical to write women back into the history they were always a part of, in action if not in memory.

NOTES

The author thanks Peter Buck, Herman Goldstine, Rachel Prentice, Sherry Turkle, and John Staudenmaier for their contributions to this article. An early version of the article was presented at "Gender, 'Race,' and Science," a conference at Queen's University, Kingston, Ontario, 12–15 October 1995.

1. History has valued hardware over programming to such an extent that even the *IEEE Annals of the History of Computing* issue devoted to ENIAC's fiftieth anniversary barely mentioned these women's roles. See *IEEE Annals of the History of Computing* 18, no. 1 (1996). Instead, they were featured two issues later in a special issue on women in computing.

2. Ruth Milkman, *Gender at Work: The Dynamics of Job Segregation by Sex During World War II* (Chicago, 1987).

3. Two books currently offer some information on the participation of women in computer history: see Autumn Stanley, *Mothers and Daughters of Invention: Notes for a Revised History of Technology* (Metuchen, N.J., 1993), and Herman Goldstine, *The Computer from Pascal to Von Neumann* (Princeton, 1972). For recollections from women who worked on the ENIAC, see W. Barkley Fritz, "The Women of ENIAC," *IEEE Annals of the History of Computing* 18, no. 3 (1996): 13–28. Other histories tend to make passing references to the women and to show photographs of them without identifying them by name.

4. Evelyn Steele, *Wartime Opportunities for Women* (New York, 1943), preface. For an analysis of American mobilization propaganda directed at women, see Leila Rupp, *Mobilizing Women for War: German and American Propaganda, 1939–1945* (Princeton, 1978).

5. Keith Ayling, *Calling All Women* (New York, 1942), 129.

6. Steele, 101.

7. Ibid., 99–100.

8. According to a *Women's Bureau Bulletin*, "A coeducational university, which before the war had few outlets for mathematics majors except in routine calculating jobs, found many attractive jobs available to mathematics majors during the war, mostly in Government-sponsored research. . . . There was a definite shift from the usual type of employment for mathematics majors in teaching and in clerical jobs in business firms to computing work in industry and on Government war projects." See United States Department of Labor, "The Outlook for Women in Mathematics and Statistics," *Women's Bureau Bulletin* 223–24

(1948): 3. According to this report, women comprised the majority of high-school mathematics teachers.

9. Ibid., 8. Margaret Rossiter, *Women Scientists in America: Before Affirmative Action, 1940–1972* (Baltimore, 1995), 13, confirms this practice more widely in the sciences. The few women who worked in supervisory roles generally supervised other women, a much less prestigious managerial role than supervising men. However, at the Work Project Administration's Mathematical Tables Project, women supervised male computers. See Denise W. Gürer, "Women's Contributions to Early Computing at the National Bureau of Standards," *IEEE Annals of the History of Computing* 18, no. 3 (1996): 29–35. The War Department in 1942 classified all military occupational specialties as either suitable or unsuitable for women; all jobs involving supervision over men were automatically declared unsuitable. Public Law 110 also made explicit that women could not command men without intervention from the secretary of war; see Bettie Morden, *The Women's Army Corps, 1945–1978* (Washington, D.C., 1990), 14.

10. See Margaret Rossiter, *Women Scientists in America: Struggles and Strategies to 1940* (Baltimore, 1982), also *Women Scientists in America: Before Affirmative Action, 1940–1972.* In the 1982 volume, p. 55, Rossiter describes the late-nineteenth-century star counters in astronomical laboratories who performed computer work for male astronomers. The famed astronomer Maria Mitchell was employed as a computer for the U.S. Coast and Geodetic Survey in the late 1860s. The term *computer,* meaning "one who computes," originally referred to the human who was assigned various mathematical calculations. Ute Hoffman dates the use of *computer* to the seventeenth century, when it was used in reference to men who tracked the course of time in their calendars. For decades the terms *computer* and *calculator* were interchangeable. In fact, early computers such as the ENIAC and Mark I were called electronic calculators. See Ute Hoffmann, "Opfer und Täterinnen: Frauen in der Computergeschichte," in *Micro Sisters: Digitalisierung des Alltags, Frauen und Computer,* ed. Ingrid Schöll and Ina Küller (Berlin, 1988). A number of other historians have documented women's work in other sciences. For example, Peter Galison, *Image and Logic: A Material Culture of Microphysics* (Chicago, 1997), discusses the work of women in high-energy physics laboratories, both those who counted flashes on the scintillator in Rutherford's laboratory and those who scanned the photographs from bubble-chamber experiments. Caroline Herzenberg and Ruth Howes, "Women of the Manhattan Project," *Technology Review* 8 (1993): 37, describe the work of women at Los Alamos, "some with degrees in mathematics and others with little technical background," who performed mathematical calculations for the design of the bomb. Amy Sue Bix, "Experiences and Voices of Eugenics Field-Workers: 'Women's Work' in Biology," *Social Studies of Science* 27 (1997): 625–68, reports the work of female field-workers at the Eugenics Record Office, who gathered data on individuals and families. In every case the work was subordinate to men's. See also Jane S. Wilson and Charlotte Serber, eds., *Standing By and Making Do: Women of Wartime Los Alamos* (Los Alamos, N.M., 1988).

11. See Rossiter, *Women Scientists in America* (both volumes). According to Herman Goldstine, it was the fact that women were not seeking career advancement that made them ideal workers: "In general women didn't get Ph.D.'s. You got awfully good women because

they weren't breaking their backs to be smarter than the next guy." Herman Goldstine, interview by author, Philadelphia, 16 November 1994. Goldstine also noted that the few men he encountered working on programming rarely conceived of their jobs as permanent. Rather, they were steps on the way to something better. These jobs were "never careers for them, but a way of making money for a short time." Consequently, Goldstine observes, "Men in general were lousy—the brighter the man the less likely he was to be a good programmer. . . . The men we employed were almost all men who wanted Ph.D.s in math or physics. This [hands-on work] was a bit distasteful. I think they viewed what they were doing as something they were not going to be doing for a career. If you take a woman like Hedi Selberg [a programmer at the Institute for Advanced Study Electronic Computer Project] she probably didn't want to sit around with the baby all the time."

12. Galison cites the invention and popularization of the term "scanner girl."

13. Ibid., 176.

14. For further discussion of prewar trends in hiring practices, see Lisa Fine, *The Souls of the Skyscraper: Female Clerical Workers in Chicago, 1870–1930* (Philadelphia, 1990), and Margery Davies, *Women's Place Is at the Typewriter: Office Work and Office Workers, 1870–1930* (Philadelphia, 1982). See also Milkman (n. 2 above), chaps. 1–3.

15. Paul Ceruzzi, "When Computers Were Human," *Annals of the History of Computing* 13 (1991): 242.

16. Cf. Milkman, 49: "The boundaries between 'women's' and 'men's' work changed location, rather than being eliminated. . . . Rather than hiring women workers to fill openings as vacancies occurred, managers explicitly defined some war jobs as 'suitable' for women, and others as 'unsuitable,' guided by a hastily revised idiom of sex-typing that adapted prewar traditions to the special demands of the war emergency." Both Milkman and Fine discuss how gender-specific advertisements reflect the feminization of specific occupations. Fine offers an analysis of the shifting gender imagery of some clerical occupations. On this point, however, note that focusing on the industry's language about women (in this case, the stories about the biological capacities and natural implications of womanhood— or, by extension, on the advertising techniques used to create a gendered labor force) can confuse industry ideals with women's actual practice. As Milkman's notion of the idiom of sex-typing suggests, there is indeed a disjuncture between women's prescribed place and what women actually did. This disjuncture is central to women's invisibility in technological history.

17. Goldstine interview (n. 11 above). The domain's masculinity appears in the preface of a textbook on exterior ballistics: Office of the Chief of Ordnance, *The Method of Numerical Integration in Exterior Ballistics: Ordnance Textbook* (Washington, D.C., 1921). "The names of the men who have contributed most to its [the text's] development, particularly Major Moulton and Professor Bliss, are mentioned in various places in the text, and to whom the writer might appropriately make personal acknowledgement, would amount practically to an enumeration of all the officers, civilian investigators, and computers who have been connected with the work in ballistics in Washington and at the Aberdeen Proving Ground."

18. The heads of the computing groups were all college graduates, as were the majority of computers.

19. "The title 'engineering computer' was created for these women, since such work before the war was done by young, junior engineers as part of their induction training following graduation from an engineering college." U.S. Department of Labor, "Women in Architecture and Engineering," *Women's Bureau Bulletin* 223–25 (1948): 56. See Sharon Hartmann Strom, *Beyond the Typewriter: Gender, Class, and the Origins of Modern American Office Work, 1900–1930* (Urbana, Ill., 1992), for a discussion of similar circumstances within American businesses. To call a particular job "feminized" does not restrict it to women. Certainly there were some male computers and programmers. For a review of literature on gender and technology, see Nina Lerman, Arwen Palmer Mohun, and Ruth Oldenziel, "Versatile Tools: Gender Analysis and the History of Technology," *Technology and Culture* 38 (1997): 1–30.

20. The idiom of sex-typing made the sexual division of labor seem natural; differences in work capacity were considered biologically based. Evelyn Steele, editorial director of *Vocational Guidance Research,* writes, "It is generally agreed that women do well at painstaking, tedious work requiring patience and dexterity of the hands. The actual fact that women's fingers are more slender than men's makes a difference. Also, women adapt themselves to repetitive jobs requiring constant alertness, nimble fingers and tireless wrists. They have the ability to work to precise tolerances, can detect variations of ten-thousandths of an inch, [and] can make careful adjustments at high speed with great accuracy"; Steele (n. 4 above), 46. Women's strengths thus lay in performing repetitive, detailed, unskilled tasks. Such statements were not new. Arguments made in favor of women working as telephone operators were similar: "The work of successful telephone operating demanded just that particular dexterity, patience and forbearance possessed by the average woman in a degree superior to that of the opposite sex." Brenda Maddox, "Women and the Switchboard," in *The Social Impact of the Telephone,* ed. Ithiel de Sola Pool (Cambridge, Mass., 1977), 266. See also Fine (n. 14 above), chap. 4, "The Discourse on Fitness: Science and Symbols." For a discussion of women's wartime labor as portrayed in literature and advertising, see Charles Hannon, "'The Ballad of the Sad Cafe' and Other Stories of Women's Wartime Labor," *Genders* 23 (1996): 97–119.

21. For a further discussion of the prewar situation and the complex interaction between new technologies and the sexual division of labor, see Fine, also Davies (n. 14 above). Jobs with a more established tradition of male employment were less likely to become feminized before World War II. For example, while "clerk" and "bookkeeper" stayed largely male, feminization was more widespread in stenography because it had not been defined as male. See Milkman (n. 2 above), chap. 4. For further discussion of how new jobs were gendered, see Heidi Hartmann, Robert Kraut, and Louise Tilly, eds., *Computer Chips and Paper Clips: Technology and Women's Employment,* 2 vols. (Washington, D.C., 1986), vol. 1, chap. 2.

22. See Rossiter, *Women Scientists in America: Struggles and Strategies to 1940* (n. 10 above), and Milkman.

23. At the time, women were concentrated in clerical roles more than in any other occupation; they comprised 54 percent of all clerical workers in 1940 and 62 percent in 1950. U.S. Department of Labor, "Changes in Women's Occupations 1940–1950," *Women's Bureau Bulletin* 253 (1954): 37. Clerical work encompasses a broad range of jobs, including office

machine operators. *The Employment and Training Administration and U.S. Employment Service's Dictionary of Occupational Titles* (Washington, D.C., 1939–41) classified computing-machine operator and calculating-machine operator as entry-level clerical occupations. For further discussion of the wide range of clerical jobs, see Strom (n. 19 above) and Fine. See also David Alan Grier, "The ENIAC, the Verb 'to program' and the Emergence of Digital Computers," *IEEE Annals of the History of Computing* 18, no. 1 (1996): 51–55.

24. It was part of a prior agreement with the Moore School that in times of national emergency the Aberdeen Proving Ground could commandeer the school's differential analyzer. Lydia Messer, oral history, interview by Cornelius Weygandt, 22 March 1988, University of Pennsylvania Archives, Philadelphia. Joel Shurkin, *Engines of the Mind* (New York, 1984), 119. BRL had apparently organized previous cooperative projects during World War I with the University of Pennsylvania. The U.S. Army Ordnance Department's *Course in Exterior Ballistics: Ordnance Textbook* (Washington, D.C., 1921) credits H. H. Mitchell of the University of Pennsylvania as "Master Computer, who organized the range table computation work at Aberdeen." Before 1941, the Moore School also provided computers for BRL. Nancy Stern, *From ENIAC to UNIVAC: An Appraisal of the Eckert-Mauchly Computers* (Bedford, Mass., 1981), 10.

25. Shurkin, 128.

26. Ibid., 127–28.

27. Stern, 13–14.

28. Not all women's jobs ranked lower or earned less than men's, but the history of female employment shows a persistent pattern into which the BRL's policies fit. For example, see Sharon Hartmann Strom, "'Machines Instead of Clerks': Technology and the Feminization of Bookkeeping, 1910–1950," in Hartmann, Kraut, and Tilly (n. 21 above), 2: 63–97. See Fritz (n. 3 above) for women's accounts of the work they performed and H. Polachek, "Before the ENIAC," *IEEE Annals of the History of Computing* 19, no. 2 (1997): 25–30, for the complexities of computations for preparing firing tables.

29. Adele Goldstine received her bachelor's degree from Hunter College in 1941, then a master's from the University of Michigan in 1942. In 1942 she taught mathematics in the public school system in Philadelphia. From late 1943 to March 1946 she worked for the ENIAC project at the Moore School and spent part of 1944 at the Aberdeen Proving Ground. In 1948, she resumed graduate study at New York University. She became a consultant to the Atomic Energy Commission project effective 7 June 1947, working on making the ENIAC into a stored-program computer. Herman Goldstine recalls that "Los Alamos was the major user of the ENIAC so it was [John] Von Neumann [who was using it]. Adele was his assistant. I was also a consultant but she was doing the major part." Goldstine interview (n. 11 above).

30. Ibid.

31. Harold Pender to George McCelland, 23 July 1943, Information Files: World War II: WAC Training: Miscellaneous, University of Pennsylvania Archives.

32. *Daily Pennsylvanian,* 29 September 1943, untitled clipping in Information Files: World War II: WAC Training: Miscellaneous, University of Pennsylvania Archives. While women received instructions from civilians (not an unusual practice in the armed services),

they were commanded by military second lieutenants and corporals. The WAC officer in charge of the detachment on campus was Lt. Mildred Fleming.

33. Mattie Treadwell, *United States Army in World War Two Special Series: The Women's Army Corps* (Washington, D.C., 1954), 221.

34. Adele Goldstine to J. G. Brainerd, n.d., "Monday Night," Information Files: World War II: WAC Training: Miscellaneous, University of Pennsylvania Archives. The ES&MWTesses were the women involved in the Engineering, Science, and Management War Training courses. J. G. Brainerd was a professor at the Moore School and liaison with U.S. Army Ordnance.

35. Helen Rogan, *Mixed Company: Women in the Modern Army* (New York, 1981), 41; Treadwell, chap. 4. Building on the work of historians such as Milkman (n. 2 above) and Fine (n. 14 above), who have analyzed the need for women in men's jobs to maintain femininity, Leisa Meyer has described the sexual politics of women's entrance into military service; see "Creating G.I. Jane: The Regulation of Sexuality and Sexual Behavior in the Women's Army Corps During World War Two," *Feminist Studies* 18 (1992) 581–601, and *Creating G.I. Jane: Sexuality and Power in the Women's Army Corps during World War Two* (New York, 1996).

36. "Topics Included in the Engineering, Science, and Management War Training Courses for Members of the W.A.C. from Aberdeen Proving Ground," Information Files: World War II: WAC Training: Miscellaneous, University of Pennsylvania Archives. There was a second training course in 1945; Herman Goldstine Papers, American Philosophical Society Library, Philadelphia (hereinafter Goldstine Papers).

37. Goldstine, *The Computer from Pascal to Von Neumann* (n. 3 above), 134; Fritz (n. 2 above). The histories of other sciences, in both Britain and the United States, show scientists' wives filling a number of the more senior women's positions in science. For example, Cecil Powell's wife Isobel led the scanning girls in Powell's laboratory, and Janet Landis Alvarez, wife of Luis Alvarez, trained the women bubble-chamber scanners at Berkeley. Among the computers at NACA were a number of engineers' wives. At the Los Alamos Scientific Laboratory, John Von Neumann's second wife, Klara Dan Von Neumann, became a programmer and helped to program and code some of the largest programs of the 1950s. Also at Los Alamos were Kay Manley, wife of John Manley, and Mici Teller, wife of Edward Teller, who performed mathematical calculations for the design of the bomb. For further discussion of couples in the sciences, see Helena M. Pycior, Nancy G. Slack, and Pnina G. Abir-Am, eds., *Creative Couples in the Sciences* (New Brunswick, N.J., 1996). According to Fritz, at least four computers married engineers at the Moore School after 1946. Frances Bilas married Homer Spence, Kathleen McNulty became Mauchly's second wife, and Elizabeth Snyder married John W. Holberton. According to Goldstine, Betty Jean Jennings (Bartik) married a Moore School engineer. Also at the Moore School were Eckert's first wife, a draftsman for the ENIAC project; Alice Burks, whose husband Arthur worked with Eckert and Mauchly on the ENIAC design; and Emma Lehmer, wife of Derrick Henry Lehmer, a computer and table compiler.

38. "Thanks for the Memory," presumably written by WACs at the Moore School, ca. 1943–44, Goldstine Papers.

39. In a retrospective analysis, Goldstine framed the computers' job as a prime candi-

date for mechanization due to its low skill: "Computing is thus subhuman in that it calls on very few of man's manifold abilities and yet is fundamental to many of his other activities, as Leibnitz so clearly perceived. This then is basically why computing was chosen as a human task to be mechanized"; Goldstine, *The Computer from Pascal to Von Neumann*, 343.

40. It is unclear exactly when this shift occurred. It was at least as early as February 1945, when George Stibbitz wrote in a report on relay computers for the National Defense Research Committee: "Human agents will be referred to as 'operators' to distinguish them from 'computers' (machines)." Ceruzzi (n. 15 above), 240.

41. Goldstine interview (n. 11 above). Interestingly, Milkman (n. 2 above) has discussed how jobs perceived as feminine in some places were quintessentially masculine in others — often within the same industry. The idiom of sex-typing, while consistent in individual factories, often differed among factories manufacturing the same product. On the Mark II computer at the Navy's Dahlgren Proving Ground, for instance, operators were male. This area deserves further study.

42. The terms *hard* and *soft*, as used to describe gendered tasks, are significant. For the hard and soft sciences, hard mastery and soft mastery are binary distinctions in science and technology implying that the "hard" ways of knowing are men's domain; "soft" ways of knowing are more feminine. Goldstine, when interviewed, reported that he had resisted "there being a distinction" between hardware and software. He observed: "At the beginning, the hardware was the important thing, but as soon as you get beyond the bottleneck of making the computer," programming software became a new bottleneck. "They've automated the bejeezus out of making chips but not software." Ironically, by the time the process of making hardware was automated programming software had become a man's job and acquired higher status than it had had in the 1940s. See, for example, Phillip Kraft, "The Routinization of Computer Programming," *Sociology of Work and Occupations* 6 (1977): 139-55.

43. Jeanne Holm, *Women in the Military: An Unfinished Revolution*, rev. ed. (Novato, Calif., 1992), 73. Social mores, as well as a variety of rules and regulations, meant that women's qualifications had to surpass men's before they could compete for higher-level jobs within academia (including government-sponsored research) and industry. The army had higher selection criteria for female officers and enlisted personnel "than those for men in the same service" (p. 50). P.L. 110, the legislation converting the WAC to full military status, specified that "its commanding officer could never be promoted above the rank of colonel and its other officers above the rank of lieutenant colonel; its officers could never command men unless specifically ordered to do so by Army superiors" (Treadwell [n. 33 above], 220). Additionally, the War Department in 1943 set the ratio of female officers to enlisted women at one to twenty. Comparable figures for men were one to ten. Using the excuse of a surplus of male officers, it capped WAC officers by limiting entrants to the WAC Officer Candidate School but did not impose a similar limitation on male officers. None of the six women ENIAC operators held high status in academia or the military. Men at the Moore School who were not affiliated with the army, such as Harry Huskey or Arthur Burks, had visible academic appointments. See Rossiter, *Women Scientists in America: Be-*

fore Affirmative Action, 1940–1972 (n. 9 above), for more on hierarchies, promotions, and payment in science.

44. Shurkin (n. 24 above), 188.

45. Kraft (n. 42 above), 141.

46. Fritz (n. 2 above), 19–20.

47. A number of historians have disputed de-skilling assumptions. For example, Sharon Hartmann Strom, "'Machines Instead of Clerks'" (n. 28 above), 64, describes in the case of bookkeeping machine operators how "workers continued to apply hidden skills of judgement and to integrate a number of tasks, particularly to jobs in the middle levels of bookeeping, even though these jobs required the use of machines." Fine (n. 14 above), 84, claims that the stenographer-typist's job was more challenging than the copyist's whom she replaced. For a review of literature on gender, mechanization, and de-skilling, see Nina E. Lerman, Arwen Palmer Mohum, and Ruth Oldenziel, "The Shoulders We Stand On/The View from Here," at the end of this volume. See also Kenneth Lipartito, "When Women Were Switches: Technology, Work, and Gender in the Telephone Industry, 1890–1920," *American Historical Review* 99 (1994): 1075–111.

48. Nina E. Lerman, "'Preparing for the Duties and Practical Business of Life': Technological Knowledge and Social Structure in Mid-19th-Century Philadelphia," *Technology and Culture* 38 (1997): 36 (reprinted as chap. 5 in this volume). Judy Wajcman, *Feminism Confronts Technology* (University Park, Penn., 1991), 37, observes: "Definitions of skill can have more to do with ideological and social constructions than with technical competencies which are possessed by men and not by women."

49. Shurkin, 188.

50. Ibid., 189.

51. C. Dianne Martin, "ENIAC: Press Conference That Shook the World," *IEEE Technology and Society Magazine* 14, no. 4 (1995): 3–10. Because the problem was classified, the equations remained concealed.

52. Goldstine, *The Computer from Pascal to Von Neumann* (n. 3 above), 229. For details of the kinds of calculations performed using ENIAC, see Arthur W. Burks and Alice R. Burks, "The ENIAC: First General-Purpose Electronic Computer," *Annals of the History of Computing* 3 (1981): 310–89. The Burks were another significant husband and wife team, publishing their story together; Alice R. Burks and Arthur W. Burks, *The First Electronic Computer: The Atanasoff Story* (Ann Arbor, Mich., 1988).

53. Fritz (n. 2 above), 20–21. Goldstine recalled bringing Douglas Hartree, a physicist who had built a differential analyzer in Britain, to the United States for a visit. "I got Kay McNulty to be his programmer and she was good and intelligent. The girls soon branched off independently and it was during that period that my wife was making ENIAC into a stored program computer"; Goldstine interview (n. 11 above).

54. See, for example, Bruno Latour, *Science in Action* (Cambridge, 1987).

55. U.S. War Department, Bureau of Public Relations, "Ordnance Department Develops All-Electronic Calculating Machines," press release, February 1946, Goldstine Papers.

56. U.S. War Department, Bureau of Public Relations, "History of Development of Computing Devices," press release, 15–16 February 1946, Goldstine Papers.

57. For media characterizations of ENIAC, see C. Dianne Martin, "The Myth of the Awesome Thinking Machine," *Communications of the ACM* 36, no. 4 (1993): 125, 127; see also Martin, "ENIAC" (n. 51 above), 3–10. Like the laundry industry that made its employees invisible by publicizing the tireless machines, the ENIAC was portrayed as doing almost all of the work; Arwen Mohun, "Laundrymen Construct Their World," in this volume.

58. Steven Shapin, "The House of Experiment in Seventeenth-Century England," *Isis* 79 (1988): 395.

59. T. R. Kennedy, "Electronic Computer Flashes Answers, May Speed Engineering," *New York Times,* 15 February 1946.

60. Ibid.

61. The NACA memorandum (n. 15 above) specifically used *she* to describe the computers in its service. Women played salient roles in the demonstration of many domestic and business technologies, from sewing machines to typewriters to IBM office products, making their omission here all the more pointed.

62. See, for example, *Popular Science Monthly,* October 1946, 212.

63. Herman Goldstine to Captain J. J. Power, Office of the Chief of Ordnance, 17 January 1946, Goldstine Papers.

64. ENIAC file appended to Goldstine to Power, 17 January 1946.

65. Horace K. Woodward Jr. to Adele Goldstine, 23 February 1946, Goldstine Papers.

66. While Adele Goldstine did not receive media acknowledgement, she clearly had some status among her colleagues at the Moore School as the only woman working on the machine's hardware. Initially, she oversaw Holberton. As head of the WAC course, despite her civilian status, she had frequent contact with top administrators at both the Moore School and the Aberdeen Proving Ground. In a publicity folder, biographical profiles on approximately a dozen staff members at the Moore School connected with the ENIAC include J. Presper Eckert, John W. Mauchly, Herman H. Goldstine, John G. Brainerd, Arthur Burks, Harry Huskey, Cpl. Irwin Goldstein, and Pfc. Spence. Adele Goldstine is the only woman included.

67. The affidavit is included in a letter from Harry Pugh, at Fish, Richardson, and Neave, to Herman Goldstine, 12 December 1961, Goldstine Papers.

68. Goldstine, *The Computer from Pascal to Von Neumann* (n. 3 above), 200.

69. Ibid., 330.

70. Ibid., 255 n. 4.

71. "Studies at Penn Aided Artillery," undated clipping from unidentified newspaper, ENIAC Publicity Folder, Goldstine Papers.

72. See, for example, Allen Rose, "Lightning Strikes Mathematics," *Popular Science Monthly,* April 1946, 85, photo caption: "T. K. Sharpless, of the Moore School of Engineering, sets a dial on the Eniac's initiating unit, which contains some of the master controls of the huge, complex mechanics. . . . Mr. Sharpless designed some Eniac equipment."

73. Bruno Latour and Steve Woolgar, in *Laboratory Life* (Beverley Hills, Calif., 1979), 219, point out that "a key feature of the hierarchy is the extent to which some people are regarded as replaceable."

74. Rupp (n. 4 above), 161.

75. Ibid., 161–62.

76. U.S. Department of Labor, "The Outlook for Women in Mathematics and Statistics" (n. 8 above), 9–11. See also U.S. Department of Labor, "A Preview as to Women Workers in Transition from War to Peace," *Women's Bureau Special Bulletin,* 1944; Rossiter, *Women Scientists in America: Before Affirmative Action, 1940–1972* (n. 9 above), chap. 2.

77. U.S. Department of Labor, "The Outlook for Women," 11.

78. Army Service Forces Office of the Chief of Ordnance, Washington, D.C., to personnel at BRL, 29 January 1946, Goldstine Papers.

79. Herzenberg and Howes (n. 10 above).

80. U.S. Department of Labor, "Employment Opportunities for Women Mathematicians and Statisticians," *Women's Bureau Bulletin* 262 (1956): vi.

81. For these women's later employment histories, see Fritz (n. 2 above), 17.

82. Margaret Rossiter, "The Matilda Effect in Science," *Social Studies of Science* 23 (1993): 325–41.

83. U.S. War Department, *You're Going to Hire Women,* booklet produced to persuade managers and supervisors to hire women, cited in Chester Gregory, *Women in Defense Work During World War II: An Analysis of the Labor Problem and Women's Rights* (New York, 1974), 12.

84. For example, the *Women's Bureau Bulletin* 262 (1956) features several pictures of women working with computers and mentions women coding and programming.

85. Gerda Lerner, "The Necessity of History," in *Why History Matters: Life and Thought* (New York, 1997), 119.

PART IV

Industrial Junctions:
Technologies of Industrial Genders

How have the categories "home" and "work" been constructed?
What relationships do we find between gender and the work of pro-
ducing and consuming?
What have been the relationships between industrial values—prog-
ress, efficiency, mechanization, centralization—and domestic spaces?
Between domestic values and industrial practices?

In the middle-class gender ideologies of early industrialization, "home" became the province of woman, who maintained a wholesome, quiet, and moral
refuge for family members who ventured daily into the tumult of the competitive
world outside. This construct has been much discussed by women's historians, who
have analyzed the ways in which the ideology belied the experience of women and
masked the work they performed and the economic value of that (unpaid) work.
Building on this analysis, historians of technology have drawn attention to household technologies as part of larger technological systems and to the household as
a site of women's technological activity and technological knowledge.

By the early twentieth century, when the urban industrial geography of homes,
stores and factories was familiar and the authority of science and rational planning
well established, reformers could view industrial society as a system in need of
a few adjustments. Children, who would one day become American adults, were
too often hungry, illiterate, and, from the point of view of the reformers, without
moral influence. Obviously science and rational planning needed to be applied
to industrial households; new understandings of nutrition and disease and new
energy systems gave the emerging field of Home Economics a full program of education and outreach. These women extended the meaning of "home" into what
they called "municipal housekeeping," improving public facilities and cleaning up
local politics. As Home Economics became an established program in colleges

across the United States, graduates positioned themselves as professionals working for business and government, with expertise in consumption and household management, consumer relations and product development, useful in rural as well as urban settings.

The new positions they created relied on private industry, another constituency seeking to influence industrial households. Manufacturers of a steady parade of stoves, refrigerators, washers, and dryers and the utility companies selling natural gas and electricity all took an interest in encouraging mechanization in as many households as possible. Not surprisingly, their marketing tactics drew on a host of gender ideologies to promote a particular vision of progress and to convince consumers to purchase the latest models of energy-using appliances. They hired home economists in hopes of reshaping technological practices to increase sales.

The articles in this section explore consumer choice, manufacturer goals, and the mediating role of home economists and expertise in an era of increasingly mechanized housework. They allow us to revisit changes in work associated with machinery, including the new technological knowledge and new pacing of work that machines made possible. And they illuminate the industrial junctions of production and consumption, home and work, and gender systems over the course of the twentieth century. Has mechanized housework, like sewing, been viewed as a natural ability in women? When have labor-saving devices been desirable, and for whom? Where has control been located in these processes of technological change? To what degree did centralization and division of labor become part of the workings of industrial households in the twentieth century? Have negotiations at these industrial junctions changed gender systems, and how? Have gender ideologies shaped the negotiations and the resulting technological change?

TWELVE

Economics and Homes: Agency

JOY PARR

Despite the efforts of the laundrymen Mohun discussed in Part I, laundry work by the mid-twentieth century was mostly domestic and decentralized, with many small machines in many houses run by many workers who did not specialize in doing laundry. Joy Parr returns us to the subject of laundering, this time from the point of view of the household-based launderer considering styles of washing machine. Manufacturers touted the joys of the new automatic machines, which washed, rinsed, and spun clothes without human intervention. Canadian women, however, persisted in choosing older wringer machines well into the 1950s, leaving manufacturers puzzled over this seemingly illogical rejection of a "better" technology. What the manufacturers failed to realize, Parr argues, was that Canadian consumers assessed the machines in terms of how they fit into the technological system of the home, choosing, for example, controlled water usage over automatic rinses. What ideas about gender, homemaking, and technology were intersecting at this industrial junction? Was this Canadian junction different from its U.S. counterparts? What role does masculinity play at this industrial junction? Femininity? What different factors besides machine specifications come into play when users choose technologies? How might these choices be gendered?

"What Makes Washday Less Blue? Gender, Nation, and Technology Choice in Postwar Canada," *Technology and Culture* 38, no. 1 (1997): 153–86. Reprinted by permission of the Society for the History of Technology.

The way Canadian women did their wash confounded the appliance managers of American branch plants in the late 1950s. In 1959 wringer washers — a technology little altered in twenty years, and by contemporary engineering standards a technology entirely superseded — outsold automatics three to one in Canada. This was exactly the reverse of the pattern in the United States, where automatics that year accounted for 75 percent of sales.[1] "Theoretically there is no market for ordinary washing machines as everyone should be buying the automatic type," a senior official at Canadian General Electric asserted counterfactually. He added in a bemused attempt at explanation, "I suppose, however, that the big market for ordinary washing machines lies in less developed countries." E. P. Zimmerman, who ran the appliance division at Canadian Westinghouse, yearly through the 1950s forecast a breakthrough for automatic machines in Canada, as did his counterparts at Kelvinator and Frigidaire, and yearly found that sales of wringer machines remained strong. "This is strange," he affirmed (implicitly rejecting the underdeveloped countries explanation), "because usually Canada is much closer to U.S. trends than this."[2]

Readers familiar with the literature on domestic technology might share this puzzlement because the fine work published in the early 1980s by Strasser and Cowan on the United States case has served as the template for understanding household technology in the North Atlantic world. United States production of automatics surpassed wringers definitively in 1951. Although Strasser and Cowan are attentive to distinctions between the priorities of makers and users, in the case of washing machine technology they report not conflict but a quick convergence of interest.[3] They find that automatics were accepted into American households as soon as they were made available by U.S. manufacturers.[4]

Cowan's justly famous parable about how the refrigerator got its hum, which has the giant electrical apparatus and automobile manufacturers successfully championing the condensor cooling technology over which they commanded proprietory rights, only makes the prolonged failure of automatic washers in the Canadian market seem more inexplicable.[5] For it was these same American makers, and for the same reasons, who intended to have Canadian women of the 1950s do their washing in automatic machines. In fact, it was not until 1966, fifteen years later than in the United States, that Canadian automatic sales passed those of wringer washers.[6]

A European observer might not be so befuddled by the Canadian pattern. In the early 1950s when automatics were coming to dominate the U.S. market, fewer than one in five British households owned any washing machine. Still, in 1969 only 5 percent owned automatics. Most domestic laundry was done either in a copper boiler, a variation on the wringer washer which heated the water and used the boiling action rather than a central agitator to circulate the clothes, or in the modern technology of choice, a twin-tub which housed the wash tub and spin dryer side by side in a single casing. Both of these technologies were rare in either Canada or the United States. Even in 1981, there were automatics in only 40 percent of British homes.[7]

Considering the British case, Christine Zmroczek in 1992 observed, "If we want to understand women's experiences of technology, it is important to look closely enough to uncover the differences from country to country and culture to culture, even within Western capitalism." For the interwar period in France Robert Frost draws similar conclusions, pointing out how ill-fitted were American-modeled domestic appliances for French domestic settings. Manufacturers in search of international mass markets have sought to erase differences. Yet recent cross-cultural studies of household technology clearly show that differences have persisted between men and women, and between makers and users. These distinctions mark cultural differences which *frame* as surely as they are framed by historical differences between national economies.[8]

Household technology is centrally different from industrial technology. An industrialist commissions a machine from a producer goods manufacturer, as he might commission a suit from a custom tailor. The machine and the suit, having been made to the user's specifications, upon delivery are promptly put to use. In household technology, this smooth transition cannot be assumed.[9] Makers, as Mme Renee Vautelet, a past-president of the Canadian Association of Consumers, noted in 1958, tended to think of consumers as "the buying side" of themselves.[10] In certain conjunctures, the cultural similarities between domestic machine makers and domestic machine users may transcend those differences made by the market economy and gender. Machines offered for sale then may be accepted unproblematically by women seeking tools for their household work. In the United States, for automatic washers, this appears to have been the case, at least for middle-class urban women.[11] But generally the culture of makers and the culture of users are very different. Machinery which is not made to the specifications of the users, as household technology almost always has not been, often does not satisfy.[12] Here gender, but also class and national differences are at play. Of laundry technology in particular, the Croatian journalist Slavenka Drakulic recalls, "It was only when I first washed my clothes in the States, in 1983, in an American washing machine, that I became aware how differences in tradition influence both the industry and my own attitude towards doing laundry." American machines did not heat the water to 95° Celsius and ran for only a third as long. Though the grandmother who had taught her the secrets of washing "had already passed away," Drakulic writes that "I just could not help remembering her, because, strangely enough, I felt as if my clothes were not properly washed at all."[13]

Understanding what constitutes a proper job is integral to understanding what is acceptable as a proper machine. For many technologies this premise is axiomatic to the design process. For household technologies, particularly domestic technologies used primarily by women, it is not. To understand why women have refused apparently excellent new machines, we need to pay attention to users and include "an examination of the *details* of women's lives" as part of the history of technological change.[14] To understand the effects of machines which were put to use, we need consider how "the form of the household, and the sexual division of labour within it" might actively have shaped domestic technology.[15] A discussion of household technological choice must reckon not only how women's technological preferences as users differed from men's technological preferences as makers and sellers, how engineering and commercial priorities came to prevail, but the possibility that *men, as makers and sellers, did not always get their way.* Here I try to set such a broad context for technological choice, considering how traits of the Canadian political economy, and of makers and marketers, worked with and against the internal politics of households and perceptions of well-being and waste to determine which man-made laundry technology women would (agree to) use.

A wringer washing machine consists of a steel tub, either galvanized or porcelain enameled, upon which the wringer, a pair of wooden or rubber rollers, is mounted. An electric or gas motor suspended beneath the tub drives an agitator to move laundry through the wash water, and revolve the rollers, clamped in tension, to express water from the goods being laundered. The machine was not self-acting. The tub was filled by the operator from a hose or bucket with water heated to the required temperature and then soap or its successor, detergent, and the clothing and linens were added. The woman operating the machine filled a separate tub or pair of tubs with rinse water and then manually lifted the items being washed from the soapy water. She fed them individually through the wringer into the rinse tubs in turn, moving the clothing through rinse waters with a stick. At each rinse she again lifted each piece by hand from the water. After the last rinse she guided the completed washing once more through the wringers, this time into a basket to be carried to the line to dry. After being used for several loads, the soapy water was either siphoned from the tub or disgorged into a floor drain. This process was hard on women's hands and their backs, and except for the ten or so minutes when the agitator was running, required the operator to be actively at work. All in all, this does not seem a technology to inspire devotion among its users.

But by comparison with the technology it replaced in most Canadian homes, the wringer washer was a real improvement. You definitely noticed the difference, Lily Hansen recalled. "I wasn't very good at scrubbing clothes on the washboard, and wringing them at all. You know, you're trying to wring sheets." "When my kids complained about the inconvenient malfunctioning wringer," Martha Watson wrote, "I told them they didn't know when they were well off."[16]

But few denied the limitations of the technology. In rural homes, the machine was stored outside and in winter had to be dragged into the kitchen before washday could begin. In city homes the wringer washer was usually in the basement, and because the machine was not self-acting, "You were running up and down stairs all morning doing this washing." In machines without pumps, "getting the water out of these big tubs, it was heavy work." Even in the best of circumstances, with a nearby pair of concrete tubs, separate hoses from hot and cold taps, and a mechanical siphon to empty the tub, the routine — washing white, colored and then heavily soiled work clothes in sequence, and returning each load in turn to the machine tub for rinsing with the aid of the agitator — was physically demanding. Clothing with buttons or zippers and larger linens had to be folded carefully while still soaking wet before they could be passed through the wringer. Metaphorically, many reported, doing laundry "was a pain."[17]

The machines also could cause more literal discomfort. The early wringer rollers were turned manually with a crank, but by the 1950s most rollers were rotated by the gas or electric engine attached to the machine. A woman who had turned on the powered rollers, and was working close to them watching for exposed buttons or trim, might find she had one hand being drawn through the wringer and the other ill-positioned to reach the release switch. Removing rings before starting the wash could reduce potential damage for women, but children's fascination with

TABLE 12.1. Canadian Homes Having Powered Washing Machines
(percent)

	Electric Wringer	Automatic
1960	73.9	12.1
1961	71.7	14.2
1962	70.1	16.5
1963	68.5	18.3
1964	65.9	20.7
1965	63.1	23.0
1966	59.4	25.6
1967	55.2	29.9
1968	51.6	32.0

Source: Dominion Bureau of Statistics, *Household Facilities and Equipment*, 1951, 1953–68.

the machine remained a concern. Several women reported rescuing their young-sters' limbs from the wringers, and giving thanks that "little kids, you know, their bones are soft." Longer term, family members speculated that their mother's ar-thritis was linked to the many hours in cold basements, standing on a wet floor, woman-handling the wash.[18]

Yet for all this, the transition from wringer to automatic technology was not swift in Canada. Allison Smith, a Commerce graduate from the University of Al-berta, spent the fifties in remote northern villages where her husband was posted as a Royal Canadian Mounted Police officer. In winter she melted ice for the wash on a wood stove, brought the water to a boil in her electric kettle, regretting her accountant's habit of counting as her sons' fifty-four diapers went through each stage of the wash and out onto the line. In her Meadow Lake Saskatchewan kitchen in 1955 she tacked up a picture cut out of a magazine of a Bendix duomatic. Four years later she returned to urban life of modest prosperity, but not until 1973 did she acquire an automatic washing machine.[19]

To understand why, right up until the mid-sixties, more Canadian women each year bought wringers than automatics, why more considered the wringer the proper machine for the job of doing the wash (see table 12.1), we need to look be-yond the relative convenience of the machines. We need to consider the broader context in which the consumption decision was made—what Ruth Schwartz Cowan has called the consumption junction.[20] The washer was not a single ma-chine but an integral part of the mechanical system of the house. The buying deci-sion was similarly complex and political. In the home, major household purchases had opportunity costs. They presented opportunities to some household mem-bers and denied them to others. Within the Canadian political economy, wringer and automatic machines had very different locations. Wringer and automatic machines both washed clothes, but each of the technologies was built upon and had built in distinct assumptions about the relationships between machines and other resources, both human and natural, made in response to their succeeding contexts, assumptions to a degree coherent and common among technologies of their time, as we think of people as bearing affinities of a shared generation. These differences were readily apparent to women of the time, although they are more elusive to us now some forty years distant. To understand the choice between wringer and automatic technology we must disengage from the organizing assumption of Sigfried Giedion's then much-read and still much-cited *Mechanization Takes Command* and feature a history of technology where a good deal more than machinery is at work.[21]

Addressing the Canadian Electrical Appliance Manufacturers in the spring of 1960, A. B. Blankenship, executive vice-president of the leading Canadian consumer research firm, reminded his audience that the images consumers held of the household goods were the keys to understanding their market. He characterized these images dichotomously as "both rational and irrational . . . both real and imagined . . . both conscious and unconscious."[22] Marketers promised to bridge these divides by "getting to really know" the woman longing for a better way to do the wash.[23] But marketers had another pressing promise to keep. Manufacturers wanted them to find, if need be to create in the market, that desiring female subject, that imagined woman, whom makers already implicitly had invented as they engineered the machines.[24] If sales were to be made, the woman that makers theorized as using their machines must be made plausible to women actually doing the wash. Somehow what Robert Frost has called the symbolic and the functional or material sides of the machine had to be made to dwell as happy complements in the laundry rooms of the nation.[25]

Fifties advertisements for wringer washers achieved this symbolic and functional resolution relatively readily by emphasizing tradition, for the firms which made wringers were venerable southwestern Ontario and Ottawa Valley manufacturers, begun as foundries and boilerworks in the 1840s and 1850s. The nameplates affixed to their new washers — McClary, Easy, Beatty, Connor, Clare — were familiar emblems from the fronts of woodstoves and the casings of sink-side kitchen pumps. Thus the ads noted that "McClary quality" had been "famous for more than 100 years," that Connor had long been a Canadian favorite. Mother and daughter frequently were featured together, their gazes both fixed upon a gleaming new machine—a sensible depiction, given the operator attention the wringer washer required. Beatty experimented with pastel yellow, blue, and green machines in 1955, making the machines themselves imagined women by pitching the new model as "An Old Friend in New Dress." The same firm struck the combination expected to sell wringers best with their 1958 Copperstyle, a modern wringer clad in the same heat-conducting metal which had sheathed nineteenth-century stovetop laundry boilers. "Mothers of all ages choose Beatty Copperstyle" ads proclaimed, as a daughter wearing borrowed high heels hurried to join her smiling mother and applauding grandmother in an admiring circle around the washer.[26]

With automatics the marketer's task was more vexed. Most potential buyers of automatics already owned a functioning washing machine, so that, as Susan Strasser has argued in the U.S. case, for the first time the merchandiser was attempting to persuade customers "to move up from an old-fashioned appliance to the newest latest kind, replacing machines that worked perfectly well."[27] The wringer washers in most Canadian households in the 1950s were relatively new and highly prized. Many campaigns for automatics thus emphasized style rather than function, and appealed to (or for) a style-conscious consumer. Stanley Randall, President of the Easy Washing Machine Company, and later an influential Ontario cabinet minister, asserted that women bought washers for three reasons: "1. appearance, 2. features, 3. price. Women will pay $40 to $50 more for an appliance if it appeals to the eye; if you don't sell eye appeal, you don't sell." "The Canadian housewife likes gadgets," he added, likening Easy's latest three-dial, twenty-one setting automatic to a pinball machine, "it lights us and signs off."[28] Certainly Randall, who had been a traveling salesman for Easy during the depression of the 1930s, would have known the importance of proper functioning in a washer, would have known that the machine was first a tool rather than an entertainment. He emphasized gadgetry and eye-appeal, and was silent about function, because in glamour lay the automatic's indisputable advantage over the wringer.

But creating a taste for glamour in the laundry room was a hard sell, as R. J. Woxman, president of American Motors' appliance subsidiary Kelvinator, well knew. If only a domestic appliance family washer "was parked in the driveway," he noted wistfully, "it would be replaced more frequently."[29] The manager of the Ontario Appliance Dealers Association fantasized about annual gala evenings to which an audience clad in evening dress would be summoned by engraved invitation, the presentation of each jewel-like new model invoking admiring applause— a fantasy world where women would value washers, as they valued jewelry, for their appearance and symbolic references alone. This was a possibility which, as he stepped down from the podium, he acknowledged was far-fetched.[30]

Marketers met the concerns and fantasies of real Canadian women best with ads which highlighted the self-acting capacities of the automatic machine. Unlike the displays for contemporary wringers (which showed women looking toward the machines), the graphics in advertisements for automatics more frequently showed a woman turned away from the washer, to smile not at the machine but at the child with whom she was playing or the husband with whom she was about to depart.[31] The claim that the automatic "ended washday" by making it more feasible to do "two or three small washes through the week" may have been the Hobson's choice Susan Strasser describes, between "a weekly nightmare" and an "unending task."[32] Yet Ontario Hydro's promise "a few things each day keeps 'washday' away," of a machine which would "do all the hard work" of the wash and promote busy mothers to the position of "supervisor in the laundry department," could not but tempt women in the home.[33]

Visions of automatics must have danced in the heads of most mothers of infants, for pungent pails of diapers could not be held for a single weekly washday. The woman who owned only two dozen diapers would have been washing them most days by hand. Whatever one's reservations about owning an automatic, who would not have been ready and willing to dream the advertiser's fantasy of a magical machine which all on its own made dirty into clean? Thus, many ads for automatics targeted new mothers. To launch their new washer-dryer set in 1953, Westinghouse worked on the nightmares raised by the thought of arriving twins. Once again the machines were imagined as, and introduced to consumers as, people, in this case as baby twins. This "Blessed Event" campaign used ads showing storks delivering new washer-dryer twin sets. Dealers provided birth certificates, tastefully printed in black and gold, to each buyer who took the mechanical twins home. The firm fused the images of the twins they had manufactured and the twins who would create dirty diapers, by offering a free pair of machines to every mother in Canada who bore twins on the launch day of the new model, the 17th of March of that year.[34] There were echoes of the 1930s Stork Derbys and the celebrated Dionne Quintuplettes in the Westinghouse letters to 15,000 doctors, hospitals and nurses' associations, asking for their intervention to discover lucky candidates and authenticate the births. But the campaign captured well the shared current of pleasure and desperation which flowed about mothers in the midst of the Canadian Baby Boom. The campaign would also have appealed to a singular predisposition among contemporary manufacturers, at once to feature users in the image of their machines and to feature the machines they made as human.[35]

Most advertisements for washers addressed a female audience. Men were invoked infrequently in any capacity in ads to sell wringers, but they began to appear now and then in the late 1950s in campaigns for automatics. The man in a checked hunting shirt an Ontario Hydro ad showed loading an automatic washer, "so easy even a *man* can do it," had only a walk-on part, for the accompanying text quickly turned to address a female reader. But Inglis pitched ads to men twice, first in the "Wife-Saver" campaign of 1958, which attempted rakish double entendre, urging husbands to "save" their wives by "trading them" in on new Inglis washers and dryers. The ads the next year—"Is your Bride still waiting for her Inglis?"—proceeded more cautiously, combining copy written for husbands—"We know you are just as anxious as any husband to save work for your bride but honestly . . . hasn't the family wash been a labour of love too long?"—with an illustration of a young bride holding a large laundry basket rather than a bouquet, intended to catch women's attention. The timing is interesting here. Men portrayed as patriarchs and providers were targeted directly as buyers in 1958 and 1959 as a recession deepened in Canada and manufacturers found sales in replacement markets more difficult to make.[36]

The differences between the imagined users featured in advertisements for wringers and automatics mirrored differences between the makers of the two machines. Almost every wringer washer used in Canada was Canadian made. The leaders in the sector, such as Beatty, Connor, and Easy, had begun manufacturing wringers early in the century as an extension of their long-standing specializations in water pumps and boilers. The first automatics sold in Canada were imported from the United States, and the Canadian manufacture of automatics quickly was dominated by the branch plants of American firms. The leaders in this sector had diversified either like Westinghouse and Canadian General Electric, from making electrical apparatus for industry, or like Kelvinator and Frigidaire, from the mass production of automobiles. A few large American subsidiaries dominated the automatic side of the industry. The makers of wringers were smaller, more numerous, and Canadian-owned.[37]

Wringer manufacturers in Canada in the 1950s built washing machines using the same labor-intensive batch production methods they used to make water pumps and boilers. These manufacturing processes yielded machines which were heavy and thus durable, simply assembled and thus simply repaired. There was little outsourcing; as the technology had been relatively static, few parts were covered by proprietary rights. Still, wringers could be made efficiently in plants producing 10,000–25,000 machines per year, so that economists estimated in the late fifties and early sixties that the Canadian market could have supported at least nine, and possibly as many as twenty-two wringer washer manufacturers.[38] Indeed, the wringer—sold with twelve-year guarantees by manufacturers who had made their reputations handling water—became something of a Canadian specialty, and through the mid-sixties Canadian manufacturers reported strong export sales into United States and overseas markets.[39]

In its manufacture the automatic washer was kin, not to the water pump, but to the other white boxes, the stove, the refrigerator, the dryer, which in their succeeding seasons kept assembly lines steadily producing. In contrast to the batch methods by which wringers then were made, mass production used less labor and fewer materials to create lighter, less resilient automatic machines. The automatics' merchandising emphasis on style was linked to this product engineering decision, to build a machine which could be sold more cheaply but would need more frequently to be replaced, to stimulate the mass consumption which would sustain mass production. One analyst using 1960 data estimated the minimum efficient size of a major appliance plant at 500,000 units per year.[40] As Gomez found in her study of the Spanish industry, longer production runs are still seen as key to least-cost production of automatic washers.[41]

But these economies of scale in mass production were not equally accessible to all producers. The whole Canadian market would not have supported a plant making a half million appliances per year. Equally important, the lower input costs for automatics that U.S. manufacturers passed on to their American consumers did not prevail outside the United States. Automatic washer technology had improved rapidly during the 1950s, but much of this knowledge in the fifties and sixties was still proprietary. Without adequate research and development capacity of their own, U.S. subsidiaries in Canada and Canadian independents had access to these refinements only by purchasing licenses to manufacture or by importing finished parts.[42] The long-run effects of licensing arrangements on manufacturing viability are plain in David Sobel and Susan Meurer's recent *Working at Inglis: The Life and Death of a Canadian Factory.*[43] Inglis, Whirlpool's Canadian licensee, ended their thirty-year association with rights to produce only an obsolete machine. Buying components abroad immediately raised prices. Imported finished parts, valued in 1955 at $2.4 million U.S., made parts bills for a Canadian automatic 10 percent higher than those for an American machine in that year, a difference which persisted through the 1960s.[44] The tariff on washer parts was 22.5 percent. Even in the late sixties, when Canadian automatics finally were selling well, they cost 37 percent more than similar machines being sold in the United States.[45] The product engineering and merchandising of automatic washers presumed mass consumption, that washers could be offered relatively cheaply so that both a conversion and a replacement market would rapidly develop. But the costs of carrying the technology across national boundaries, which made automatic washers relatively more expensive in Canada (and other jurisdictions) than in the United States, made automatics implausible and impractical as objects of mass consumption. For makers, the washer was an object for sale, reaching out toward an imagined consumer. For prospective users (who may or may not have featured themselves as consumers), neither the price nor the promise was as alluring as the makers assumed.[46]

For most of the forties, washing machines of any sort were woefully elusive commodities in Canada. Sufficient metals were released for civilian requirements to produce 126,000 new washers between 1942 and 1945, but that many machines (purchased before the depression began) were wearing out each year by the end of the war.[47] In addition there were 800,000 more households in Canada by the end of the decade.[48] Already in late 1944, 83 percent of housewives who recently had set up housekeeping reported they had tried and been unable to procure a washing machine.[49] By the end of 1945, unmet demand would have exceeded seven years' production at peak prewar rates.[50]

But reconversion to peacetime production was slowed in Canada by the exchange crisis that followed the war. Restrictions on imports and access to materials were more prohibitive for firms making consumer than producer goods, and goods for domestic consumption rather than export.[51] Credit controls designed to dampen consumer demand applied to washers until 1953.[52] In the late forties a woman who wanted a washer had to wait her turn on a dealer's lengthy list. Joan Coffey did get a wringer machine in 1947, when her anguished letter struck a cord with an Eaton's department store manager. "They were just starting to make them and you went on waiting lists for years . . . all the neighbours were aghast. They couldn't figure out how I could get a washing machine. And that [machine] was the love of my life."[53] Many other women turned to commercial laundries, at least for large items such as tablecloths and sheets. This was a sensible way to save their own labor, a reality which caused appliance salesmen considerable dismay.[54] But rather than paying to have the family wash laundered, dried, and pressed, to return home from the market tied in neat paper packages, women from both professional and working-class households reported using the "wet wash" service, which returned the laundry damp, to be dried and ironed at home. This suggests that in Canada, as in Britain, the issue was not labor-saving alone. The lingering array of postwar shortages — of housing with decent plumbing, of washing machines, of cash — made them seek some compromise which would balance the pressures these various scarcities placed on the household.[55]

Early automatics often were bought by men as gifts for their wives. "To purchase gadgets that relieve . . . drudgery and thus promote domestic affection," as Marshall McLuhan observed in 1951 in *The Mechanical Bride,* could be seen as a duty, a species of moral choice. The other leading male commentator on technology of the day, George Grant, was generally critical of American influences upon Canada and wary of transnational technology as a threat to liberty. But he made an exception for "the wonderful American machines" he believed let his wife, Sheila, lead a freer life, acknowledging that "the practical worth of modern technology" was demonstrated "every time Sheila washed the clothes in her machine."[56]

The men who presented their wives with the first automatics were often professionals — geologists or engineers worried about the peripatetic lives their careers imposed upon the family, or university professors who encouraged their wives' dedication to pursuits other than housewifery. They had the income to afford the automatic, enough control over the family budget to make the decision alone, the conviction that manufacturers' promises for the machine would be fulfilled. Beverly Newmarch, in 1948 the wife of a newly hired geology Ph.D. in a British Columbia coal mining town, remembers how she came to have an automatic in her company duplex: "Chuck decided that with what I had had to use, I should now have an automatic. He began to look at want-ads! I was horrified, since automatics were so new, I didn't want to start out with one that had experience! He persevered, however, and found himself a new Bendix, still in the wooden crate. The American consul had brought it up to Victoria and for some reason or other they had not been able to obtain permission to install it in their home — something about the plumbing not being adequate." Ann Brook, married to a navy man frequently away from home, returned from work one day to find the automatic she had declined ("Hum, don't need an automatic washing machine, who needs an automatic washing machine?") already installed. Her husband had conferred once more with their customary appliance salesman, Mr. Beeton, and he and Mr. Beeton had agreed, "maybe you should try it and see." Mrs. Brook did not decline the gift.[57]

If for the men who bought them in the forties and early fifties automatic wash-ers were unambiguously desirable objects which bespoke affection and a better life, for children they are recalled as mesmerising entertainments. The rare front-loading automatics somewhat resembled the even rarer televisions about which most Canadian youngsters had only heard tell before 1955. Most women who got automatics in the early postwar years tell stories of lines of small spectators gather-ing to watch the wash.[58] Women as equipment users had a more complex appraisal to make. Some were persuaded early on. Margaret Shortliffe had first seen an auto-matic Bendix at Cornell University in 1939, noted the merits of its alternating drum technology, and refused agitator substitutes, either wringer or automatic, washing by hand until her husband got a Bendix to Kingston, Ontario, in 1946. Winnifred Edwards, like Shortliffe, got along without a machine until the kind of equipment she had seen in hospital laundries was available for sale in 1952. The consequences of investing in a particular durable tool delayed their purchase of any machine, as it delayed many women's purchase of automatics, for deliberating on an investment takes longer than choosing to consume an object of either personal or altruistic desire.[59]

In such deliberations, price is plainly an important factor. Automatic washers cost more than wringer machines. In 1950, the gap was wide. Standard wringers could cost as little as $90, some automatic models as much as $370. In 1956 the Chatelaine Institute reported best-selling wringer prices ranging from $129 to $259, and automatics from $325 to $469. Over time the gap narrowed, but still in 1966 the average price of wringers advertised in Eaton's mail-order catalog was $146; automatics at an average of $234 cost half as much again.[60]

This was not a negligible difference, particularly in the first decade after the war when couples were equipping homes for the first time. Personal incomes in Canada were not high in this period, in 1947–50 about two-thirds of incomes in the United States.[61] Prices for Canadian consumer goods in nominal and real terms exceeded those in the United States. Hard choices had to be made. The consumer credit con-trols that applied to washers until 1953 required down payments of one-third and full payment within a year.[62] Getting everything at once by running into debt was not an option. Deciding what to get first required considerable juggling. For how long should the household get by without a stove, a mechanical refrigerator, or a washer?

Historians of technology sometimes have been surprised by technological choices because they assumed that the choice was between two technologies for performing the same task, rather than among possible mechanical and nonmechanical improvements. Economists have relied upon the simplifying assumption that like goods are only compared with like. For the economist, the only benefits the person making the choice weighs are his or her own. These assumptions about nonsubstitutability do not apply well to households, as feminist economists recently have demonstrated.[63] Buying a wringer rather than an automatic washing machine was a sensible economy. The savings, for example, would have bought a vacuum cleaner or a radio, and the wash would still be done. The washer was the one place in the basic household consumption package where there was a little discretion. Among the 8,611 Toronto women Eaton's interviewed about their purchases of furniture and appliances between January 1949 and August 1952, the amount paid for refrigerators ($343–348) and for stoves ($205–219) varied little. For washers the range was considerable. Women under twenty-four paid on average $152, women over thirty-five on average $188. More older than younger women were buying automatic machines. New equipment for keeping food cold and making it hot took a relatively fixed amount out of every household budget. A younger homemaker, with more household equipment to acquire at once, could more easily make do with a wringer washer than do without a stove or a mechanical refrigerator.[64] The Central Mortgage and Housing Corporation surveyed 6,600 families who had purchased houses between January and May of 1955 in Halifax, Montreal, Toronto, Winnipeg, and Vancouver. The amount spent varied between cities, on stoves by $62, on refrigerators by $45, but on washers (on average the least expensive of the three appliances) by $127. Only on washers could the new home buyers with the least to spend accrue appreciable savings.[65]

Among the much smaller group of women with whom I spoke and corresponded, a similar pattern emerges. A washer was important, often important enough to risk going into debt for, especially important once children began to arrive.[66] But the choice was not posed in the postwar years as between two laundry technologies, between an automatic and a wringer. Rather, women spoke about the other tasks which might be mechanized and the other obligations of the home. The decision about a washing machine was part of these other decisions about equipping the household and the household's relationship to the wider world. Buying the automatic, the more expensive machine, when a cheaper satisfactory alternative existed could easily seem to foreclose more opportunities than it opened, to be less about liberty than constraint.

In rural homes, the needs of the barn and of the house had to be met from the same purse. Investment in labor-saving equipment for the farm took priority, partly because men made these decisions on their own.[67] Perhaps also, in some parts of Canada as in Iowa and the Palouse region of Idaho and Washington, women saw investments in farm equipment as saving domestic labor because they eliminated the need for hired men.[68] The priority of the barn is plain in the detailed study of 352 Ontario farm families Helen Abell conducted in 1959. On the most prosperous farms, where investment in all labor-saving equipment exceeded $13,000, less than 10 percent of this investment was in domestic technology. Paradoxically and surprisingly, the proportion of the farm family's resources invested in household appliances rose among poorer farmers, to 20 percent for the house when all equipment was valued at less than $7,000.[69] Because there were few satisfactory substitutes in domestic technology, the least mechanized of farm families had to allocate the largest share of their equipment budget to the kitchen and laundry basics, at least in Ontario. Choosing a wringer over an automatic reduced these pressures and might, for example, have brought an electric cream separator into the kitchen.[70] In more highly capitalized operations, a farm woman may more readily have been able to justify the purchase of an automatic machine to free her for farm work outside the house. Jellison finds Iowa women used this argument in the 1950s. The same reasoning may account for why automatic washers more quickly became commonplace in the farms of Quebec, where dairy predominated, than in other Canadian markets.[71]

Both Patricia Cliff and Nettie Murphy linked their purchase of automatic washers to their participation in the work force. Cliff had twins in diapers, and was pregnant again, when she went by herself to Eaton's Warehouse in Victoria in 1960 and asked for "the biggest automatic tub you can give me." But she explained the decision by noting that she had worked for eleven years before the twins arrived and thus had a bank account of her own. Murphy bought her first washer many years into her marriage, during a period when she had paid work. "That made a difference. When you work outside of the home and you have an income, you pack a little more clout."[72] Manufacturers and marketers expected that wives in the work force would buy more automatics than women at home full-time. They assumed that two-earner households would have higher incomes, and that wage-earning wives would feel less "guilty" turning a part of the household budget toward "short-cuts" for themselves. But as Joan Sangster notes, in Canada in the 1950s and 1960s most married women in the labor force came from families in straitened circumstances. Buying an automatic, when for $100 to $200 less a wringer would do, seemed foolish to a woman earning just enough to make ends meet. One Canadian study explicitly addresses this question through a comparison of equal numbers of full-time and part-time wage-earning mothers and mothers at home full-time in Guelph, Ontario, in the early 1960s. There Mary Singer and Sue Rogers found that wage-earning homemakers were significantly less likely to own automatic machines than those at home full-time.[73]

Even in good times, the $100–200 gap between the cost of a wringer and an automatic machine made Canadian women hesitate and consider other household needs. In the late fifties, as manufacturers were expanding their production of automatics, a five-year-long recession began. Many a "wife's pay cheque" was "merely replacing that of a laid-off husband." Marketers began to suspect what researchers later would document, that economic uncertainty had an exaggerated effect upon the purchase of major durable goods. In lean times households were "likely to be cautious about replacing any machine which wasn't actually breaking down" and likely to see the best new machine as the one which put least pressure on other aspects of the family budget.[74]

Makers who featured users making the choice between wringer and automatic washers on the basis of the laundry technologies alone made a more elemental misjudgment. In the early 1950s, the fantasies of Canadian young people were inhabited less by shiny white boxes lined up on showroom floors than by plumbing, wiring, and pipes. The year automatic sales first exceeded wringers in the United States, electrification and running water systems were the stuff of which young Canadians' dreams were made.[75] As a leading Canadian home economist noted in 1946 and was still noting in 1954, it seemed "impractical to discuss the dream houses of the future . . . until more of our houses, urban and farm, have running hot and cold water."[76]

The engineering and marketing of tools for household work often proceeded in isolation from consideration of the mechanical systems which would be required for their support, particularly when domestic appliances were launched into international markets.[77] A 1945 survey of the "Housing Plans of Canadians" found that a third of all Canadian families, and two-thirds of those in rural areas, did not have any running water at all. Lever Brothers, anxious to sell large quantities of the laundry detergents it had designed to replace soap in washing machines, must have been discouraged to discover for itself that year that only 20 percent of Canadian farm homemakers had hot running water and could do a wash without hefting copper boilers on and off the stove.[78] The early acceptance of automatics in Que-

TABLE 12.2. Water Supplies of Canadian Homes
(percent)

	Hot Running Water	Piped from Private Sources
1955	65.3	22.4
1956	67.9	22.0
1957	70.6	25.3
1959	75.3	21.5
1960	78.8	17.8
1961	80.2	17.8
1962	83.1	18.2
1963	84.8	18.0
1964	86.1	18.1
1965	87.4	16.9
1966	88.3	17.6
1967	89.7	16.7

Source: Dominion Bureau of Statistics, *Household Facilities and Equipment, 1955–57, 1959–67.*

bec may be linked to the fact that 63 percent of farms there had inside running water by 1951, a year when only 40 percent of those in Ontario, 30 percent of those in the Atlantic region, and 9 percent of those in the prairies were so supplied.[79]

Automatic machines required not only hot running water but a water supply under strong and steady pressure. Even as the proportion of Canadian homes with hot and cold running water increased over the postwar years (table 12.2), the proportion not connected to community pressure systems remained considerable. In fact this proportion appears at times even to have risen as new suburban dwellings were built beyond the reach of municipal mains.[80] No Whiggish inexorable succession of technologies here. To invest in equipment which would only function well when attached to a city water system might not be wise aforethought. Women who raised their families in the resource economies of western Canada remember having to trade in new technology for old, electric for gas-powered wringers, washers for washboards as they moved to islands or remote mining sites, or from city homes to ranches and farms. Less than an hour's drive from Toronto, Kathy Grensewich used a wringer washer until 1975 because this was the machine the household cistern would support. From such perspectives, automatics were a limited technology, more constrained by the plumbing they required to function than conventional machines.[81]

Important as these straightforward economic and infrastructural constraints were, they do not wholly explain the Canadian preference for wringer washers. By the early sixties, many of the economic and infrastructure considerations which favored wringer washers over automatics had faded. Wages and salaries were rising. Women remembered feeling more prosperous and more confident that prosperity could be sustained.[82] The price gap between wringers and automatics had narrowed. Almost every household had electricity and more than four out of five had hot running water. Yet in 1964 wringers still were outselling automatics by a considerable margin. There was still a mass market for wringers even among Canadian higher income groups. Women who owned wringers still were more likely to replace them with wringers than with automatics.[83] While producers and marketers asserted that replacing a wringer washing machine with an automatic was "trading-up," it is not at all clear that women doing the wash saw the matter in the same way.

Cultural values attach to goods offered for sale. Product engineers build cultural assumptions into the machines they design. Marketers set out to find or to forge a constituency to whom these assumptions make sense. But their sales prospects will not necessarily share makers' values, or make their determination on the basis of marketers' assumptions. The purchase of goods is self-implicating. Thus, as David Nye notes, the possession of electrical appliances "engages the owner in a process of self-definition"; in their operation "the self and object are intertwined." But the cultural current flows two ways. The machine may remake its user ("I was born to use an automatic"), but the user may also refeature the machine ("The automatic is a wasteful extravagance"). Once the constraints of price, plumbing, and income had begun to fall away, it was still not for makers and marketers alone to define how Canadian women would do the wash, or what for them constituted an excellent machine.[84]

Machines are located within moral economies. The tools we use embody values. They may also constrain the field within which we can make moral choices. They "expand or restrict" our "actions and thoughts," reveal or conceal the implications of our decisions.[85] Some machines, by their design, seem to operate with resplendent technological autonomy; others, by design, constantly disclose and allow their operator to monitor the demands of the machine upon the provisioning system of which it is a part. Automatic washing machines are of the first sort, wringer washers are of the second. Canadian women making the choice between them in the 1950s plainly distinguished the two kinds of machines in these terms and gave this signification to the distinction.

A woman filling a washer by hauling water or working a hand pump or standing by a running hose knew how much fresh water she was drawing for the task. She saw the character and the quantity of the waste she was disposing into the yard, the septic field, or the sewer mains when the job was done. A homemaker who relied upon a well and septic system knew she must monitor the capacities of these systems and adapt her domestic routines daily and seasonally to accommodate their limits. Any woman who had run a wringer machine had a clearer sense of the relationships between washing, water, and waste than those of us today who have only used automatic machines that fill and drain through discreet piping, leaving volumes drawn and disposed unobservable and unremarked. For rural women in the 1950s, the new automatics that promised to put each load of laundry through several rinses in fresh water presented an immediate hazard to operation of the farm home. But city women as well had experience with which to recognize the automatics as prodigal, of fuel to heat water, and of water itself. In response to a request for ideas for better washers, Mrs. H.G.F. Barr of London, Ontario, wrote thus in 1955: "I have been appalled at the amount of water that seems necessary to do a normal family wash in the new spin-dry type of machine. I believe one brand boasted that it rinsed clothes seven times, and all of them threw the water out after one use. There is hardly a city or town in Canada that does not have some water shortage in summer months. Large sums are being spent on reforestation, conservation and dams. It would appear that this trend towards excessive use of water should be checked now." Through the 1960s such negative consumer commentaries upon automatic washers remained common, homemakers' rhetoric to describe the new machines more evocative of the manic sorcerer's apprentice in the contemporary Disney film *Fantasia,* than of the regulated modern domestic engineering manufacturers and marketers sought to portray.[86]

Manufacturers, in both their design decisions and marketing strategies, treated the washer as an isolated object rather than as one element in a production process called "doing the wash." It was, after all, the washer alone they had to sell. By contrast, women consumers thought of doing laundry as a task rather than a machine. They appraised the process in the way production processes conventionally are construed, considering their own management priorities and skills and all the noncapital inputs required, as well as the traits of the machinery they might put to work.[87]

In these terms, manufacturers' emphasis on the gadgetry raised alarms among consumers. The early automatics were fragile machines, prone to break down and repairable only by specialized technicians who were not always nearby. "In the search to provide more and more automatic features," Mrs. W. R. Walton of the Consumers Association warned, makers were producing washers "so sensitive and complex, it will take an engineering expert" to fix them.[88] By contrast, in 1958 many wringer washers were being sold with long guarantees, and supported by a dense network of local dealers who by preference specialized in wringer sales. In an economy where all household appliances lately had been in short supply, where couples still were aspiring to an adequate rather than affluent standard of living, buying a delicate automatic seemed both shortsighted and frivolous.

The promise that an automatic machine would do the wash all on its own seemed a threat. Even when intervention was required, the self-regulating features of the automatics defied operator intervention. Tubs that filled by a timer ran half empty when water pressure was low. Loads that became unbalanced under lids that locked for the duration of the wash cycle caused the machine to jostle uncontrollably about the room. Women who in the 1950s expressed a preference for simpler machines over which they could exercise a greater measure of control spoke from well-founded technical, managerial, and resource concerns about the operation of the new automatic laundry equipment.[89]

Between the engineering of the wringer and automatic washing machines lay a generational divide. Wringer washers were made in batches, durable and simple to repair. These product characteristics happily complemented a consumer culture habituated to scarcity and schooled to value conservation and thrift. Automatics were mass produced, designed for a consumer culture which would value innovation over durability and be willing to place convenience for the machine operator ahead of household water and fuel costs, and the social costs of creating more waste. For domestic appliances, at the core of this change was a redefinition of what constituted an excellent machine, a narrowing of the purchasing decision to give priority to labor-saving features over other resource concerns. Many Canadian women in the fifties and sixties were unwilling to cross this generational divide. Their loyalty to the wringer washer technology and their skepticism about the new automatics is a sign of this resistance. The choice between wringer and automatic machines implicated Canadian homemakers in forming distinctions between consumer and user, between gratification and prudence, between production and conservation, between built to last and built to replace. In the circumstances in which they then found themselves, and with the knowledge they then had, it is not hard to see why, red hands and aching back and wet floors notwithstanding, so many resisted the chromium promises of the new machine.

In the postwar years household technologies increasingly were characterized as consumer goods. The rapid rise of a culture of mass consumption, and the more central place consumer goods came to hold in the definition of personal identities and civic values, is well documented for the United States. Sometimes popular knowledge about these goods effected cultural changes, even when the goods themselves were not widely owned. This is the case Robert Frost makes for interwar France.[90] But the process can also work the opposite way.

Consumer goods can be, and have been, refused because of the cultural values they embody. The degree to which mass consumption became institutionalized differed between regions and nations and across classes. In the decade following World War II, these differences varied with the pace of postwar recovery, the precedence given to export or domestic markets, and household versus industrial needs. Centrally, the plausibility of mass consumption was tied to perceptions of plenty and to beliefs about how the national wealth should be husbanded and shared. For consumer goods that are also working tools this dialogue was vigorous and many-faceted. In measure, manufacturers and marketers remade the material and symbolic functions of their machines to address the resistance of consumers. But as long as the purchasers of household equipment continued to think of themselves centrally as users appraising tools, they were declining to be defined solely as consumers. Their choices of what goods to buy bespoke deeper concerns about how much was enough, and for whom, framed in the politics of the households and the communities to which they belonged.

NOTES

This article was first presented at the Technological Change Conference at Oxford University in September 1993. The author is grateful to Shirley Tillotson, Marilyn MacDonald, James Williams, Bea Millar, R. Cole Harris, and Anthony Scott for comments on an earlier draft, and to Ingrid Epp and Margaret-Anne Knowles for research assistance.

1. Canadian Westinghouse Hamilton, Employment Forecast Interview Report (EFIR) May 5, 1959, RG 20 767, National Archives of Canada (NAC); "Thor Gathers Speed after US Agreement," *Marketing*, April 24, 1959, p. 8. *Marketing* was the Canadian equivalent of *Printer's Ink*.

2. Canadian General Electric, October 7, 1958, EFIR December 6, 1962, RG 20 765 23-100-C27, NAC; Zimmerman's comments are in Canadian Westinghouse, EFIR May 6, 1959, RG 20 767, NAC; for Kelvinator see RG 20 773 NAC.

3. Judy Wajcman suggests in her study of refrigerators, however, that Cowan reduced housewives to the role of consumers, responsive only to price, and told the story as a rivalry between manufacturing interests in which user preferences did not figure. Judy Wajcman,

Feminism Confronts Technology (University Park, Pa., 1991), p. 102. It is worth considering whether the precedence of price over use values in consumer decision-making may have been more marked in the United States than in other North Atlantic economies in the 1950s.

4. Susan Strasser, *Never Done: A History of American Housework* (New York, 1982), pp. 267–68; Ruth Schwartz Cowan, *More Work for Mother: The Ironies of Household Technology from the Open Hearth to the Microwave* (New York, 1983), p. 94.

5. Cowan, pp. 128–43.

6. *Home Goods Retailing (HGR)*, January 23, 1967, pp. 1, 7, 23. *Home Goods Retailing* addressed in Canada roughly the same market as *Electrical Merchandising* in the United States. Domestic Appliances — Canadian Manufacturing + Imports-Exports for 1965 and 1966, RG 20 1755 8001-404/34 NAC.

7. T.A.B. Corley, *Domestic Electrical Appliances* (London, 1966), pp. 131–36; Penny Sparke, *Electrical Appliances* (London, 1987), chap. 7. The best British article on women and laundry technology is Christine Zmroczek, "Dirty Linen: Women, Class and Washing Machines, 1920s–1960s," *Women's Studies International Forum* 15 (1992): 173–85. The statistics used in this paragraph are drawn from Zmroczek, pp. 180, 182, tables 1 and 2.

8. Zmroczek, p. 182; Robert Frost, "Machine Liberation: Inventing Housewives and Home Appliances in Interwar France," *French Historical Studies* 18 (1993): 128; Cynthia Cockburn and Rúza Fürst-Diliç, eds., introduction to *Bringing Technology Home: Gender and Technology in a Changing Europe* (Buckingham, 1994), p. 17.

9. There is a fine discussion of this issue in Wajcman (n. 3 above), chap. 4.

10. "Are We 'Selling' the Company to the Consumer?" *Industrial Canada,* July 1958, p. 140.

11. Peter J. McClure and John K. Ryans Jr., "Differences between Retailers' and Consumers' Perceptions," *Journal of Marketing Research,* February 1968, pp. 35–40; the discussion of retailers' understandings of consumer valuation of automatic washers is on pp. 36–37.

12. For example, see the discussion of product engineering in a 1980 Spanish washing machine firm, M. Carme Alemany Gomez, "Bodies, Machines and Male Power," in Cockburn and Fürst-Diliç (n. 8 above), pp. 132–33.

13. Slavenka Drakulic, *How We Survived Communism and Even Laughed* (New York, 1993), p. 53. The whole chapter "On Doing Laundry," pp. 43–54, is compelling reading for those interested in these questions.

14. Zmroczek (n. 7 above), p. 183.

15. Wajcman (n. 3 above), p. 102.

16. In 1993 and 1994, through columns in Victoria and Vancouver, British Columbia, newspapers I recruited twenty-three women, married between 1945 and 1955, to interview about the furniture and equipment they used in their homes up until 1968. These columns were picked up by national news services, and I received additional letters from across the country in response. All interviews were tape recorded and then transcribed. The interview transcripts and letters will be deposited in the Simon Fraser University Archives, Burnaby, British Columbia. Lily Hansen (pseud.), interview by author, New Westminster, BC, May 25,

1994; Tina Wall (pseud.), interview by author, Victoria, BC, June 16, 1994; Martha Watson, Alma, Ontario, letter to the author, July 24, 1993; Irene Newlands, Surrey, BC, letter to the author, October 12, 1993.

17. Allison Simpson, interview by author, Delta, BC, May 18, 1994; Joan Coffey, interview by author, Coquitlam, BC, May 17, 1994; Mary Paine (pseud.), interview by author, Delta, BC, May 17, 1994; Patricia Cliff, interview by author, Victoria, BC, June 16, 1994; Marjorie Barlow, interview by Margaret-Anne Knowles, North Vancouver, BC, March 3, 1994; Lynn Stevens, interview by author, Langley, BC, May 26, 1994; Nettie Murphy, interview by author, Mission, BC, November 24, 1994; Gerry Kilby, interview by Margaret-Anne Knowles, Vancouver, BC, February 1, 1994; "Better Care Longer Wear," *Canadian Homes and Gardens,* October 1950, pp. 68–69; Jane Monteith, "Planning a Laundry for Today and Tomorrow," *Chatelaine,* February 1949, p. 37.

18. Paine, interview; Simpson, interview; Pam McKeen, interview by author, Victoria, BC, June 15, 1995; Winnifred Edwards (pseud.), interview by author, New Westminster, BC, June 9, 1994. *Chatelaine,* the major Canadian women's magazine, noted as advantages for automatic over wringer machines in November 1951 that "your hands never touch the water; weight of wet clothes does not have to be lifted up and down; no dripping water to clear up afterward," and emphatically, "no worry about children playing around the automatic machine."

19. Simpson, interview.

20. Ruth Schwartz Cowan, "The Consumption Junction: A Proposal for Research Strategies in the Sociology of Technology," in *The Social Construction of Technological Systems: New Directions in Sociology and the History of Technology,* ed. Wiebe K. Bijker, Trevor Pinch, and Thomas P. Hughes (Cambridge, Mass., 1987), pp. 263, 278.

21. Sigfried Giedion, *Mechanization Takes Command* (New York, 1948).

22. A. B. Blankenship, "The Consumer of the Sixties," in *Texts of Papers Presented at the 3rd Annual Appliance Marketing Seminar* (n.p., Canadian Electrical Manufacturers Association, June 9, 1960), pp. 33–34.

23. F. W. Mansfield, "A Discussion of the Principles of Marketing Research and How They Can Be Used by the Appliance Industry," in *The Next Fifty Years Belong to Marketing: Texts of Papers Presented at the 2nd Annual Appliance Marketing Seminar* (n.p., Canadian Electrical Manufacturers Association, June 3, 1959), pp. 6, 8, 13.

24. Sparke (n. 7 above), p. 6; Cockburn and Fürst-Diliç (n. 8 above), p. 11; Frost (n. 8 above), p. 127 n.44, citing Bruno Latour and Madeleine Akrich; Susan Ormrod, " 'Let's Nuke the Dinner': Discursive Practices of Gender in the Creation of a New Cooking Process," in Cockburn and Fürst-Diliç, pp. 42–58.

25. Frost, p. 129.

26. "The Beautiful McClary Washer," *Chatelaine,* April 1951, inside front cover; the Connor pitch is described in "Hookers Used in Connor Campaign," *Marketing,* May 15, 1954, p. 16; "Housewives Like Color Washers, Beatty Adds Color Sales Curve with Cromatic Line," *Marketing,* November 11, 1955, p. 20; "Mothers of All Ages Choose Beatty Copperstyle," *Chatelaine,* March 1958, p. 53.

27. Strasser (n. 4 above), p. 267.

28. "The Easy Way to Sell—Coax Man on Store Floor to 'Romance' Your Product," *Marketing*, August 30, 1957, pp. 20–22.

29. R. J. Woxman, "The Consumer Needs and Wants—How Can the Appliance and Home Entertainment Industry Meet the Challenge?" in *Proceedings 5th Annual Appliance Marketing Seminar* (n.p., Canadian Electrical Manufacturers Association, May 17, 1962), p. 41.

30. A. L. Vincent, "The Retailers Viewpoint on Marketing Appliances," in *The Next Fifty Years Belong to Marketing* (n. 23 above), pp. 53–54.

31. "Change Washday into Playday, the New Easy," *Canadian Homes and Gardens (CHG)*, November 1956, p. 45; "Now the New Beatty Washer Saves You," *CHG*, June 1952, p. 34; "Bendix Introduces '53 Line," *Marketing*, January 1953, p. 1; 1958 ad for automatic washer showing mother with toddler in high chair, washer behind them and cover line, "Less time for your laundry, more time for your family," Live Better Electrically file 570, Ontario Hydro Archives.

32. Marie Holmes, "Look What's Happening to Washday," *Chatelaine*, May 1953, p. 78; Strasser (n. 4 above), p. 268.

33. Ads for automatic washers 1959 and 1960, Live Better Electrically file 570.1, Ontario Hydro Archives. The advantages Australian women found in daily laundry are described by Kereen Reiger, "At Home with Technology," *Arena* 75 (1986): 115–16, 117–18.

34. "'Twins for Twins' Promotion," *Marketing*, February 21, 1953, p. 1; "Practical and Emotional Appeals Feature This Consumer 'Contest'," *Marketing*, March 21, 1953, p. 2; "Laundry Dealership Told 'Babies Mean Business,'" *Marketing*, September 5, 1958, p. 46.

35. Marshall McLuhan ponders this conflation in *The Mechanical Bride: Folklore of Industrial Man* (New York, 1951). Dianne Newell pointed out to me this aspect of these essays, particularly apparent in McLuhan's choice of illustrations; see similarly, Richard Sennett, *The Fall of Public Man* (New York, 1974), p. 20; Mariana Valverde, "Representing Childhood: The Multiple Fathers of the Dionne Quintuplettes," in *Regulating Women*, ed. Carol Smart (New York, 1992), pp. 119–46; special Dionne issue of *Journal of Canadian Studies*, Winter 1994–95.

36. Automatic washer ad, 1959, Living Better Electrically 570.1, Ontario Hydro Archives; "Buy a Washer, Save a Wife: Promotion Soaps Up Husbands," *Marketing*, July 11, 1958, p. 30; "After the Wedding, a Washer Inglis Ad Aimed at Husbands," *Marketing*, April 3, 1959, p. 6. The latter campaign ran in both *Maclean's* and *La Patrie*. The *Marketing* stories describe the advertising campaign and the advertiser's rationale for its design.

37. In 1954 the four largest makers of automatics commanded 91 percent of the market. That year the four largest firms producing wringers made only 52 percent of the machines offered for sale. See Clarence Barber, *The Canadian Electrical Manufacturing Industry*, Royal Commission on Canada's Economic Prospects (Ottawa, September 1956), p. 43; U.S. Department of Commerce, *Major Household Appliances: Production, Consumption and Trade—Selected Countries* (Washington, D. C., 1960), pp. 9, 14; on Beatty's history in the field, see "The Costs Are the Same but Value Greater Now," *Financial Post*, May 17, 1958,

p. 28; "Major Appliance Industry Plan," Domestic Appliance Plan 1965–8, Electrical and Electronics Branch, RG 20 1755-404/34 NAC; Beatty Brothers EFIR, October 22, 1963, RG 20 763 NAC; Kelvinator Canada EFIR, April 1956, RG 20 773 NAC.

38. Barber, pp. 29, 51–52, 54; H. C. Eastman, "Electrical Appliances Industry," p. 5, in RG 20 1755 8001-404/35 NAC.

39. Barber, p. 10; Beatty Brothers, interview with R. L. Kerr, EFIR, May 20, 1964, RG 20 763 NAC; Canadian Westinghouse, interview with C. H. McBain and K. E. Waugh, EFIR, June 6, 1962, RG 20 767 NAC. J. H. Connor and Son produced wringer machines in two versions, one for rural and northern markets using cast iron, steel, and heavy aluminum for maximum durability at a higher price, the other using more light metals and plastics for urban users. H. E. English, unpublished report on visit to J. H. Connor and Son Ltd., April 12, 1956, Royal Commission on Canada's Economic Prospects, 3-13-8, RG 33 52 NAC.

40. Eastman, pp. 5, 8, 12; comments on Eastman by G. Q. Rahm, Chief, Appliance and Commercial Machine Division, Trade and Commerce, February 4, 1966, and by Ralph Barford, General Steel Wares, February 23, 1966, RG 20 1755 P8001-404/35 NAC. There is a longer discussion about optimal plant size by Barford, of Beatty and General Steel Wares in P. C. Fedenburgh, site visit, January 5, 1967, RG 20 P8001-270/G47 NAC. Plants that used more outsourcing—that is purchased rather than produced parts—would have been efficient at a smaller scale, but here the national boundary loomed as an obstacle, for even firms with Canadian subsidiaries confined parts manufacture to the United States.

41. Gomez (n. 12 above), p. 132.

42. "Laundry Appliance Firms Sign Agreement," *HRG*, May 15, 1959; P. C. Fredenburgh, "Comparison of US and Canadian Major Appliance Plants," March 22, 1968, p. 23, RG 20 1755 8001-404/41 vol. 2 NAC.

43. David Sobel and Susan Meurer, *Working at Inglis: The Life and Death of a Canadian Factory* (Toronto, 1994), pp. 115, 141, 249–53.

44. U.S. Department of Commerce, *Major Household Appliances* (n. 37 above), p. 14; Barber (n. 37 above), p. 62; "Major Appliance Study," chap. 3, input prices, RG 20 vol. 1755 P8001-404/46 NAC.

45. Fredenburgh (n. 42 above), p. 22.

46. For a discussion of the different ways in which manufacturers using mass and batch production methods feature and attend to consumers, see Philip Scranton, "Manufacturing Diversity: Production Systems, Markets, and an American Consumer Society, 1870–1930," *Technology and Culture* 35 (1994): 476–505.

47. The age of existing stocks—27 percent dating from the 1920s and 30 percent from the early thirties—was estimated by the Wartime Prices and Trade Board (WPTB) research and statistics division using a sample of 6,524 interviews in the summer of 1940: Wartime Prices and Trade Board, RG 64 1452 A-10-9-4 NAC.

48. F. H. Leacy, *Historical Statistics of Canada,* 2nd ed. (Ottawa, 1983), p. A248.

49. WPTB, "A Survey on Household Necessities," RG 64 1460 A-10-9-23 NAC, pp. 2, 8; Lily Hansen, interview by author, New Westminster, BC, May 25, 1994. Hansen remembers buying one of the last washers at Woodward's Department store in 1943.

50. "Production of Durable Goods," August 1946, RG 46 WPTB 1465 NAC.

51. Joy Parr and Gunilla Ekberg, "Mrs. Consumer and Mr. Keynes in Postwar Canada and Sweden," *Gender and History* 8 (1996): 212–30.

52. Department of Finance, "Control of Consumer Credit, PC 1249," March 13, 1951, RG 19 E2C vol. 32 NAC.

53. Coffey, interview (n. 17 above); Watson, letter to author (n. 16 above).

54. See the warnings to salesmen in "Charting Course for Selling," *Marketing,* November 30, 1946, p. 2, and "Selling Via Demonstrations," *Marketing,* July 30, 1949, pp. 8, 12.

55. Cliff, interview (n. 17 above); Margaret Shortliffe, Victoria, BC, interview by author, June 16, 1994, reporting about Kingston, Ontario; Edwards, interview (n. 18 above); Murphy, interview (n. 17 above). See Zmroczek (n. 7 above), p. 183, on the British equivalent of wet wash called bag wash. A recent discussion of commercial laundries in the United States is Roger Miller, "Selling Mrs. Consumer: Advertising and the Creation of Suburban Socio-Spatial Relations, 1910–1930," *Antipode* 23 (1991): 278.

56. McLuhan (n. 35 above), pp. 32, 33; William Christian, *George Grant* (Toronto, 1993), pp. 177, 250. Grant's best-known writings are *Lament for a Nation: The Defeat Of Canadian Nationalism* (Princeton, N.J., 1965) and *Technology and Empire: Perspectives on North America* (Toronto, 1969).

57. Shortliffe, interview; Ann Brook, interview by author, Abbotsford, BC, November 24, 1994; Susan Taylor (pseud.), interview by author, Victoria, BC, May 10, 1994; Joan Niblock, interview by author, Langley, BC, May 25, 1994; Bev Newmarch, Calgary, Alberta, letter to author, September 6, 1993.

58. Olive L. Kozicky, Calgary, Alberta, letter to author, September 14, 1993; Newmarch, letter to author; Elizabeth Perry, Calgary, Alberta, letter to author, September 12, 1993.

59. Shortliffe, interview; Edwards, interview; Liz Forbes, Duncan, BC, letter to author, November 1993.

60. "Laundry's No Problem," *CHG,* October 1950, p. 91; "How to Choose Your Next Big Appliance," *Chatelaine,* November 1956, p. 22; Canadian General Electric, EFIR, June 9, 1965, RG 20 765 23-100 C27 NAC; Tanis Day, "Substituting Capital for Labour in the Home: The Diffusion of Household Technology," (Ph.D. diss., Queen's University, 1987), p. 185. The 1966 prices cited from Day are in 1971 dollars.

61. Jean Mann Due, "Consumption Levels in Canada and the United States, 1947–50," *Canadian Journal of Economics and Political Science* 21 (May 1955): 174–81.

62. Department of Finance, "Control of Consumer Credit" (n. 52 above).

63. Nancy Folbre and Heidi Hartmann, "The Rhetoric of Self Interest: Ideology and Gender in Economic Theory," in *The Consequences of Economic Rhetoric,* ed. Arjo Klamer (New York, 1988), pp. 184–203; Paula England, "The Separative Self: Androcentric Bias in Neoclassical Assumptions," in *Beyond Economic Man: Feminist Theory and Economics,* ed. Marianne A. Ferber and Julie A. Nelson (Chicago, 1993), pp. 37–53.

64. The least spent for refrigerators was $342, the most $353; for stoves the range was between $205 and $219, both narrower differences on larger sums than the $35 gap for washers. "Purchasing of Furniture, Household Appliances and Home Furnishings—Toronto—By Age Groups, 1949–50–51 and 32 weeks of 1952," Market Research 1953–60 S69 v. 25, T. Eaton Company, Archives of Ontario.

65. "Purchasing of home furnishings and appliances by new home owners," Controller's Office, September 2, 1958, Series 165, box 2, file 5.1, T. Eaton Company, Archives of Ontario.

66. McKeen, interview (n. 18 above); Niblock, interview (n. 57 above); Coffey, interview (n. 17 above); Gerd Evans, interview by Margaret-Anne Knowles, Burnaby, BC, April 6, 1994; Mrs. A. B. Botham, Ganges, BC, letter to author, October 16, 1993; Hazel Beech, Lake Cowichan, BC, letter to author, July 22, 1993.

67. This seems to have been the case for Ontario in the 1950s. See Nora Cebotarev, "From Domesticity to the Public Sphere: Farm Women, 1945–86," in *A Diversity of Women: Ontario 1945–80,* ed. Joy Parr (Toronto, 1995), pp. 203, 207. In Quebec, aspiration for domestic comfort took greater precedence; Yves Tremblay, "Equiper la maison de ferme ou la ferme, le choix des femmes quebecoises, 1930–1960," *Bulletin de l'histoire de l'electricité* 19–20 (1992): 235–48.

68. Katherine Jellison, *Entitled to Power: Farm Women and Technology, 1913–63* (Chapel Hill, N.C., 1993), p. 109; Corlann Gee Bush, "'He Isn't Half So Cranky As He Used To Be': Agricultural Mechanisation, Comparable Worth, and the Changing Farm Family," in *"To Toil the Livelong Day": America's Women at Work, 1780–1980,* ed. Carol Groneman and Mary Beth Norton (Ithaca, N.Y., 1987), p. 228.

69. "Report of Findings Concerning Consumer Information; Crafts and Hobbies; Housing (The Farm Home)," Progress Report #7, "Special Study of Ontario Farm Homes and Homemakers 1959," AOS6076, Helen Abell Papers, University of Guelph Archives, pp. 11, 12, 28.

70. J. K. Edmonds, "Keep a Sales Eye on the Farmer's Wife," *Marketing,* May 24, 1957, p. 28.

71. Ibid. Proportion of automatic washing machines among electric washing machines in Quebec homes (%): 1960 (13); 1961 (15.6); 1962 (18.4); 1963 (21.6); 1964 (25.4); 1965 (28.3); 1966 (32.7); 1967 (38.8); Dominion Bureau of Statistics, *Household Facilities and Equipment* (Ottawa, 1960–67). Jellison, pp. 180, 185.

72. The interviews included questions about how buying decisions were made. The response usually was that decisions were arrived at jointly. It was usually difficult to discern whether or how patriarchal privileges or status as breadwinners influences buying priorities. In the analysis of the transcripts and correspondence, I have taken women at their word. Cliff, interview (n. 17 above); Murphy, interview (n. 17 above).

73. J. K. Edmonds, "An Expanding Durable Goods Market: Aim Ad Pitch to Working Wife," *Marketing,* October 28, 1960, p. 42; "The Working Wife—Appliances Target," *Marketing,* December 26, 1958, p. 8; Vance Packard, *The Hidden Persuaders* (New York, 1957), p. 62; Maxine Margolis, *Mothers and Such: Views of American Women and Why They Change* (Berkeley, 1984), p. 167; Joan Sangster bases her finding on detailed studies by the Women's Bureau of the Federal Department of Labour. See her "Doing Two Jobs: The Wage-Earning Mother, 1945–70," in Parr, *A Diversity of Women:* (n. 67 above), pp. 100, 120; Mary E. Singer and Sue Rogers, "Survey of Child Care and Housekeeping Arrangements Made by Homemakers Who Are Employed Outside the Home," June 1966, RE1 MAC AO135, University of Guelph Archives, pp. 22, 25. Newlands, letter to author (n. 16 above); Joyce Cunningham, Victoria, BC, letter to author, October 30, 1993.

74. "The Working Wife—Appliances Target," *Marketing,* December 24, 1958, p. 8; Edmonds, "Aim Ad Pitch to Working Wife" (n. 73 above); Lee Maguire, "Canadian Consumer Buying Intentions: A Study of Provincial and Socio-economic Differences," (master's thesis, University of Windsor, 1967), p. 34. A copy of this thesis is in the University of Guelph Library.

75. Helen Abell and Frank Uhlir, "Rural Young People and Their Future Plans, Opinion and Attitudes of Selected Rural Young People Concerning Farming and Rural Life in Alberta, Ontario and Quebec 1951-52," Canada, Department of Agriculture, 1953, in Helen Abell Collection, University of Guelph Archives.

76. Margaret McCready, "Science in the Home" (typescript, February 1946) and "Whither Home Economics" (typescript, November 1954), in Margaret McCready Collection, AO13518 and 13519, University of Guelph Archives.

77. Anne-Jorunn Berg, "A Gendered Socio-technical Construction: The Smart House," pp. 173-74, and Andjelka Milic, "Women, Technology and Societal Failure in Former Yugoslavia," p. 156, in Cockburn and Fürst-Diliç (n. 8 above).

78. McLean-Hunter, *Housing Plans of Canadians* (Toronto, August 1945), p. 7; Lever Brothers, *Canadian Homes, a Survey of Urban and Farm Housing* (Toronto, 1945), p. 7. The regional variations were considerable. See D. R. White, "Rural Canada in Transition," in *Rural Canada in Transition,* ed. M. A. Trembley and W. J. Anderson (Ottawa, 1966), pp. 39-41. In 1961 two-thirds of prairie farm homes still did not have inside running water.

79. White, pp. 39-41.

80. The association between use of wringer technology and the absence of municipal water systems is apparent in the United States as well. See James A. Carman and J. D. Kaczor, "Ownership Patterns for Major Household Appliances," in *Studies in the Demand for Consumer Household Equipment,* ed. James A. Carman (Berkeley, 1965), p. 118.

81. Evans, interview (n. 66 above); Emma Boyd, interview by Margaret-Anne Knowles, Vancouver, BC, January 27, 1994; Kilby, interview (n. 17 above); Simpson, interview (n. 17 above); Hansen, interview (n. 49 above); Kathy Grensewich, Kitchener Ontario, letter to author, August 2, 1993.

82. Leacy (n. 48 above), p. E49. Women who remembered the fifties as a time when they struggled to get by retrospectively often dated their own postwar prosperity from 1962.

83. In 1964 three-quarters of wringer sales were to households with incomes greater than $5,000; see G. D. Quirin, R. M. Sultan, and T. A. Wilson, "The Canadian Appliance Industry," working paper, University of Toronto, Institute for Quantitative Research in Social and Economic Policy, 1970, pp. 36, 77. Also in 1964, sales of wringer washers amounted to 203,000, while sales of automatics came to 162,900. Of wringer sales, 75.3 percent were to replace wringers; 59.6 percent of automatic sales were conversions from wringers. See "Major Appliance Study, 1964 Study: 1964 sales by types of transactions," Major Appliance Study, RG 20 vol. 1755 P8001-4404/46 NAC.

84. Sparke (n. 7 above), p. 6; Rosemary Pringle, "Women and Consumer Capitalism," in *Women, Social Welfare and the State in Australia,* ed. Cora Baldock and Bettina Cass (Sydney, 1983), p. 100; Susan Strasser, *Satisfaction Guaranteed: The Making of the American Mass Market* (New York, 1989), p. 15; David Nye, *Electrifying America: Social Meanings*

of a New Technology, 1880–1940 (Cambridge, Mass., 1990), p. 281; John Fiske, *Reading the Popular* (Boston, 1989), p. 2.

85. Nye, p. 281; Mihaly Csikszentminhalyi and Eugene Rochberg-Halton, *The Meaning of Things: Domestic Symbols and the Self* (Cambridge, 1981), p. 53.

86. The comment from Mrs. Barr and another similar by Mrs. G. F. Grady of Peterborough are in "Housewives' Ideas for Better Washers," *CHG,* June 1955, p. 66; the filling system of the seven-rinse Inglis automatic is described in Mrs. R. G. Morningstar, "Survey of Time and Motion Studies for Household Equipment," Report 23, 1952, Canadian Association of Consumers, MG 28 I 200 vol. 1 NAC; "Look What's Happening to Washday," *Chatelaine,* May 1953, pp. 79, 81; "Today's Household Equipment," *Chatelaine,* November 1951, p. 90; "Buying Public Loves Laundry 'Automation,'" *HGR,* March 25, 1963, p. 18; "Laundry Market Accelerating Fast in Canada," *HGR,* March 7, 1966; Nye, p. 303; Mrs. Cindy Bolger, Ariss, Ontario, letter to author, July 25, 1993; Forbes, letter to author (n. 59 above).

87. Suzette Worden, "Powerful Women: Electricity in the Home, 1919–40," in *A View from the Interior: Feminism, Women and Design,* ed. Judy Attfield and Pat Kirkham (London, 1989), p. 140; Pringle, pp. 100–101.

88. Corley (n. 7 above), p. 136; Giedion (n. 21 above), p. 570; Mrs. W. R. Walton, paper presented before CEMA, 1962, in *Proceedings 5th Annual Appliance Marketing Seminar* (n.p., Canadian Electrical Manufacturers Association, May 17, 1962), p. 16; Greta Nelson, interview by Margaret-Anne Knowles, Burnaby, BC March 5, 1994; Botham, letter to author (n. 66 above); Newmarch, letter to author (n. 57 above).

89. Perry, letter to author (n. 58 above); for reactions to Canadian women's "doubts and prejudices" about automatics, see Margaret Meadows, "What To Look for When Buying an Automatic Washer," *Chatelaine,* May 1951, p. 84; "Working Up Sales Lather—Market Was Made Sure It Was, Westinghouse," *Marketing,* February 22, 1957, p. 7; "Guarantees Washer for Twelve Years," *Marketing,* March 28, 1958, p. 1; "Consumer Attitude Survey—Fuels and Household Appliances," May 1961, British Columbia Electric Marketing Division, British Columbia Hydro Archives, pp. 5, 39, 41; Bea Millar, interview by author, notes only, Vancouver, BC, 8 May 1996. Millar, head of BC Electric Home Services, noted that the first automatics needed to be bolted to a good cement foundation because the spinning tub caused the washer cabinet to shift.

90. Frost (n. 8 above).

THIRTEEN

Home Economics: Mediators

CAROLYN M. GOLDSTEIN

Utility companies and appliance manufacturers in the United States concluded early on that they needed to reach and enroll homemaking women if they wanted to sell gas and electricity, stoves and refrigerators. Home economists, still developing their field, were also seeking to reach women, to bring science, nutrition, and public health to as many households as possible. Carolyn Goldstein focuses on the home economists hired by these companies, as they manipulated gendered assumptions about technological knowledge and expertise to create a role as mediators between the utility and appliance companies and the women consumers. They believed they could represent the interests of women to the companies and could teach women the new methods of homemaking that gas and electricity made possible.

What gender and technology issues intersect at this industrial junction? Were home economists workers like the meatpackers or the programmers we met earlier? Did it matter that they were women dealing with women? Whose agency played the most important role in shaping uses of the new technologies? Were home economists merely agents of modernization, or did they carve out a space of their own in the negotiated industrialization of the household?

"From Service to Sales: Home Economics in Light and Power, 1920–1940," *Technology and Culture* 38, no. 1 (1997): 121–52. Reprinted by permission of the Society for the History of Technology.

In 1917, the Public Service Electric and Gas Company of New Jersey hired Ada Bessie Swann, a home economist, to develop a program educating homemakers about the benefits of gas and electricity. In less than a decade, Swann transformed a set of simple cooking demonstrations into a "home service department" offering instruction in all aspects of "modern housekeeping." Aimed primarily at adult homemakers who did not have the opportunity to learn home economics in school, Swann's program ranged from laundry, lighting, and cleaning, to food preservation, preparation, and serving. Formats included lecture demonstrations in department stores and before organized groups such as women's and girls' clubs, home management and cookery classes at the utility company facilities, and calls at customers' homes. In addition, through radio programs and individual correspondence, Swann established the New Jersey utility company as a constant presence in the community, available and accessible to all customers and any questions. The large cadre of home economics–trained women she assembled on her staff helped ensure regular contact with homemakers throughout an extensive metropolitan area.[1] Swann's activities typify an emerging group of professional women who found a place in gas and electric utility companies teaching consumers about the new power sources and businesses about consumers.[2]

This case study of home economists' role in the distribution of gas and electric power in the 1920s and 1930s opens a window on the complex gender-technology interactions at what Ruth Schwartz Cowan has called "the consumption junction."[3] It argues the importance of considering production and consumption as part of a single process, one in which home economists occupied an important mediating location. It also refines historical understanding of the relationship between technological change and the notion of "separate spheres." Generations of scholars associated the realms of production and consumption with men and women, respectively. They interpreted men as part of the public sphere and women as tied exclusively to the private. Yet in recent years, historians of women have found so much evidence of men and women crossing these supposedly rigid boundaries that they have called into question the entire "separate spheres" concept.[4] As a group of female technical professionals operating between the realms of production and consumption, home economists used their expertise about con-

sumption and new domestic technologies in ways that similarly challenge interpretations of producers and consumers, men and women, as distinct communities
shaped by rigid, hierarchical power relations. Rather than forcing a dismissal of
these categories, however, the paradoxical and ironic nature of home economists'
experience leads to an exploration of the complex ways in which gender shaped
the very categories of production and consumption themselves. By pinpointing
women as actors in the social process of domesticating new technologies, we learn
that the two realms were not fixed, and that the story of the production, marketing, and consumption of new domestic technologies is a story of constant interplay
between masculinity and femininity, factory and home, public and private. Utility
home economists operated inside a technological system in which the functions of
engineering, sales, and education were gendered for both women and men. Understanding home economists' mediating role requires taking gender into account as
a factor in technological change.

Traditionally, historians of gas and electric systems have focused on the nuts-
and-bolts aspects of urban networks and the political and organizational constraints of their construction, leaving the cultural and art historians to ponder the
symbolic meanings of advertisements and artifacts. The literature on "systems" has
failed to include domestic consumers in the loop.[5] As a result, our understandings
of the technical aspects of urban power systems and of the social significance of
their creation remain disconnected. Historians of housework and domestic technology opened new avenues in this area about ten years ago when they demonstrated the ways new appliances reinforced gender stereotypes even as they made
some specific tasks easier to perform.[6] Recent work in the history of electrification
has begun to paint a more complex picture by examining relationships between
the consumers, social and cultural values, and the builders and managers of new
urban systems. For example, Harold Platt, David Sicilia, and Mark Rose have provided useful, detailed case studies of the growth and marketing of electricity in
Chicago, Boston, and Kansas City and Denver, respectively.[7] More generally, David
Nye's *Electrifying America* demonstrated the ways American culture shaped the
course of electrification in the United States. By situating the development of electrical systems in a broad social context, he raised questions about the process of
negotiation among the consumers, managers, engineers, and salespeople of the
new power sources. Furthermore, Nye identified home economists as influential in
establishing a place for electricity in early-twentieth-century American domestic
life, and he established an important set of contexts for analyzing home econo-

mists' presence at a central nexus in the buying and selling of appliances, a nexus where technical concerns and cultural ideas influenced one another.[8]

In the years following World War I — a time of expansion and competition for utilities — gas and electric company managers in most North American cities enlisted home economists to help them address the "social problem" of carving out a domestic market for their services.[9] Like manufacturers and retailers of packaged food, clothing, and household equipment, utility companies brought home economists into their operations to help demonstrate new technologies and encourage women consumers to buy them. But for utility companies, promoting their systems among women involved a set of goals and tactics that were unique to the power industry and its place in the public eye in the interwar period. Managers of electric power and light systems understood as early as the 1890s that balancing energy load through domestic consumption was critical to the success of central generating systems.[10] Although they did not face the same load problem of their counterparts in electricity, gas company managers confronted analogous pressures to align consumption with massive investments. In the 1920s, as the utilities sought to enlarge the market for power and light to include a growing middle class, they hired home economists to help them establish and reinforce the idea that gas and electricity were fundamental to a whole range of household tasks. In addition, utilities relied on home economists for assistance with a massive program of self-regulation and public relations.

Together, home economists, utility company managers, and industry leaders invented a new institution: the utility company home service department. Its form and programmatic character were shaped by a convergence of the needs of utility companies and home economists' professional agenda. By specifically identifying home service departments as parts of the utility company that were primarily service-oriented and free of sales, they created a system in which the work of women and men, service and sales were complementary. By the late 1920s, the scale and scope of utility home service operations organized along these lines reached impressive proportions. Between 1925 and 1930, the number of home service departments in utility companies rose from 50 to 400. Many of these departments claimed to serve as many as hundreds of thousands of customers each year. Home economists' central presence in utility companies established them as key players within the complex world of power and appliance retail merchandising.[11] Although the balance between service and sales changed in the 1930s, home economists remained important actors in the marketplace through World War II and beyond.

This article focuses on home economists' work in electric utilities. It begins with an exploration of why managers of electrical companies created an institutionalized place for home economists in their operations and how technical and economic considerations, combined with assumptions about gender, informed the ways male leaders in the power and light industry defined the scope of home economics work in the interwar years. It identifies the opportunities such work offered home economists — individually and as a group — and how their professional interests coincided to some extent with the corporate imperatives of utility companies. Based on common assumptions about men's and women's work and the role of the corporation in American life, home economists and utility company managers shared the work of making electrical appliances defining elements of a modern American home. The electric lighting case shows how home economists smoothed the way for the popular acceptance of new products. Finally, it analyzes the circumstances under which home economists' work in electrical utilities changed in the 1930s.

Prior to 1920, the diffusion of electrical appliances was limited to only the wealthiest members of urban communities served by relatively small power grids. Potential customers lived in urban areas, where lighting was first introduced in public buildings and private homes.[12] The expense and novelty of the new power sources placed the appliances they fueled in a luxury market. Electricity tended to be perceived as expensive and dangerous and confined to use by elite, scientifically minded persons. Early showrooms such as Samuel Insull's elegant Electric Shop in Chicago reinforced the idea that electrical appliances were entertaining gadgets for members of the upper classes. Advertisements and displays at world's fairs fostered a popular understanding of electricity as what David Nye calls "a quasi-magical force with utopian potential."[13]

As power companies expanded their distribution networks after 1900, they sought to establish the new power sources as accessible to more Americans by supplementing printed newspaper advertising with sizeable sales forces. Following the example of sewing machine manufacturers who hired women to demonstrate their products as early as the 1850s, electric companies demonstrated home appliances.[14] An increased emphasis on face-to-face relations with consumers opened up opportunities for women trained in domestic science to work as "home service agents," complementing salesmen's efforts to boost appliances. These women were among what Mark Rose calls "agents of diffusion" utility companies hired to promote an increase in the use of electricity through a series of house-to-house demonstrations, cooperative sales incentives, and cooking schools.[15]

After World War I, home economists became part of a bolder, more deliberate corporate strategy to sell power and light to middle-class American homes. As the electrical industry continued to expand and joined the ranks of the nation's most capital intensive big businesses, corporate giants General Electric and Westinghouse channeled their resources into extensive research and marketing organizations which enabled them to maintain monopolistic control of the electrical industry.[16] Electrical merchandise was available in a diverse array of retail settings, but utility companies were the major retailers of appliances in the early 1920s. Besides, the utilities had a double incentive to sell appliances. Although by the late 1930s department stores sold the majority of electrical merchandise in large urban areas, utilities continued to use home service as a way of supporting retailers and cooperating with dealers toward the greater goal of selling power.[17]

Utility investment in home economics expertise reflected a consensus within the industry that fitting electricity to the home required fitting women to the electrical business. Utility executives, local store owners, and appliance manufacturers articulated this assumption in the pages of *Electrical Merchandising,* a trade journal devoted to marketing electricity and electrical appliances. These discussions illuminate the variety of ways men in the industry sought to create demand for electricity and appliances while fostering public relations for the industry at large. These discussions demonstrate how concepts of gender informed the ways men and women in the industry defined the problem and solution of distribution of appliances in the interwar period. Although this trade journal reveals more about what electrical dealers and utility company managers thought they were doing than what they actually accomplished, it nonetheless contains clues about their assumptions concerning consumers, their efforts to construct an ideal clientele, and their ideas about how women and home economists could fit into the larger retail scene.[18]

As electrical dealers shifted their marketing focus away from novelty and luxury appeal and toward more practical benefits, instructing consumers about how to use and maintain electrical appliances became even more of a priority after World War I.[19] Selling electric ranges became a major focus, and company managers tended to view instruction in methods of "electric cookery" as women's work.[20] To persuade consumers to switch from wood or coal to electricity, these demonstrators sometimes had to provide advice about budgets and household management as well as explain the technical details of the range's workings. "Electric range cooking is slowly but surely winning its way in the favor of housewives, but it is undeniably true that much still remains to be done in teaching them this new method of cookery."[21] Although one writer recommended that the appliance salesman keep up-to-date with methods of housework by reading women's magazines, most agreed that only women held the necessary clues with which to create narratives that placed new products in familiar contexts.[22]

Electrical merchandisers grasped after the "woman's viewpoint" in particular because of a perceived market opening; commonplace wisdom held that women purchased about 85 percent of American household products overall. But at least one 1920 observer estimated that women, shopping on their own, bought only 10 percent of electrical goods.[23] Uncertain about how to increase the proportion of women customers, they debated whether a major appliance purchase usually required the approval of the woman's husband and whether homemakers appreciated door-to-door salesman. Identifying middle-class women as the gateway to a larger place for electricity in the home, electrical specialty shop managers sought to add a "feminine touch" to their otherwise utilitarian supply outlets in order to make their shops attractive to women buyers.[24] Many stores, outfitted primarily for repair and technical service, displayed sockets, wiring, and plugs in a state of disarray and discouraged customers from browsing. For assistance in domesticating their sales environments, many electrical specialty shop managers turned to their wives or local women's organizations.[25] F. M. Mosley attempted to make his store resemble home by locating it in a Montgomery, Alabama, residential district and naming it "Mosley's Electrabode." Mosley intended his store to be "a neighborhood affair where women would drop in as casually as they do in a drugstore." The overall effect in many shops was to foster a welcoming ambiance to make middle-class women customers feel comfortable relaxing and socializing, much like that of the lounges in department stores where women could rest during a day of shopping.[26] Utility companies' use of the model Home Electric was another example of the industry's efforts to domesticate electrical appliances.

At the same time, utilities found women useful for public relations. Popular criticism of economic concentration in the electrical industry, as well as charges of false advertising, pressured leaders to go out of their way to promote positive images of themselves and the goods and services they sold.[27] While local independent retailers found women could be useful as "business developers," industry leaders likewise learned that women — and especially home economists — could be assets in their national public relations campaigns designed to foster an image of goodwill in the industry.

In short, by hiring home economists on a systematic basis, utility company managers sought to achieve three interrelated goals: to demonstrate the practical value of a new array of appliances; to increase demand for these products among women consumers in particular; and to improve public relations in the industry. Although men in the electrical industry turned to women of all kinds for advice on meeting these goals, home economists seemed uniquely suited to meeting them.

These goals, and the gendered terms with which male industry leaders understood and expressed them in the early 1920s, dovetailed neatly with the development of the home economics profession. In the decade following the formation of the American Home Economics Association in 1908, increased federal and state government funds enabled land-grant colleges to expand their home-economics programs aimed at training women homemakers, teachers, and extension agents in support of a growing agricultural establishment. The years after 1920 saw a similar increase in home economics graduates at private institutions such as the University of Chicago and Teachers College Columbia University. Diverse in nature, these programs generally stressed the practical application of the natural, physical, and social sciences in the domestic sphere. Women graduated from them with knowledge of the basic chemical and physical principles of such consumer products as textiles, food, and by the late 1920s, electrical and gas appliances.[28] In addition, home economics graduates learned a specific definition of the "ideal home" which designated themselves as critical scientific experts and moral reformers in American domestic life. Although some women reformers at the turn of the century had called for cooperative housekeeping, home economists advocated efficient household management and rational consumption within the context of single-family living.[29] While most home economics graduates spread this vision as homemakers or teachers, increasing numbers of women found practical reasons to welcome work in utility companies and other business enterprises.[30]

For Ada Bessie Swann and other leaders in this growing women's profession, employment in utility companies represented an opportunity to advance their careers, share their knowledge about nutrition and consumer products, and bolster their collective identity as experts about consumers and consumption. Most of the home economists who took jobs as home service agents in the power industry had attended four-year college home economics programs, primarily at land-grant institutions. Many became members of the Home Economics in Business Section of the American Home Economics Association. Beginning in the early 1920s, this group established a set of professional ideals with which to understand and justify their participation in corporations to both their employers and fellow home economists in education and extension work. Their professional ideology revolved around four basic assumptions: first, that their work was educational; second, that as women they could represent the "woman's viewpoint"; third, that their work was scientific and objective; and fourth, that they were "interpreters" and "diplomats" in the consumer marketplace. All these elements of home economists' professional identity rested on and reinforced a belief in progress through corporate capitalism and an optimism regarding the potential of feminine professional participation in that system.[31]

As women and as scientists, home economists were well positioned to help utility companies interpret electricity to consumers while projecting a friendly corporate image. In a 1928 speech to the Home Service Bureau of the National Electric Light Association, Katherine Fisher, director of the Good Housekeeping Institute, the product testing lab allied to the Hearst *Good Housekeeping* magazine, explained how home economists' intuitive access to the "woman's viewpoint" and their scientific training combined to make them strategic links between the manufacturers of equipment and the women who used it. Referring specifically to the utility industry, Fisher posited that if engineers were "statesmen," then women in the appliance field were "super statesmen." She characterized the women who ran utility home service departments as "partner[s]" and "balance wheel[s]" to industrial engineers and brought to their "attention the need for certain refinements which will mean added convenience to the user." While an engineer developed a device, home economists instructed women about the proper way to "use it to advantage."[32]

The utilities' rhetoric of service appealed to home economists' Progressive sensibilities and promised to enhance their vision of a rational consumer society. Beyond the mere selling of appliances, they hoped to influence the development of products and services of a growing consumer economy. Home economists were attracted to the utilities not simply because electricity was modern but also because the new power sources promised, with the home economists' help, to enhance the control and efficiency of household consumption.[33]

Based on a common set of assumptions, then, utility company managers and home economists forged a marketing and public relations strategy that reinforced home economists' professional identity, utility company imperatives, and a gendered construction of both. The name "home service department," rather than "home economics department," reflected the intentions of the "public service corporations" to accentuate their commitment to community needs over profit-seeking motives. This nomenclature also embodied home economists' identity as educators. As Ada Bessie Swann explained in a speech to new home service representatives and utility company managers, she promoted home service departments on the basis of both sales and service: "It is seldom that ideals and commercialism are seen as harmonious teammates in the performance of any job, but home service in a utility, rightly administered, effects this happy fusion. The returns which follow the activities of a well-administered department are both moral and material."[34]

As cornerstones of this "happy fusion," home service departments facilitated a division of labor in which men and women were in charge of the "hard" and the "soft" sell, respectively. For example, a general educational briefing by a home economist often preceded more direct sales pitches from door-to-door salesmen. As one woman in the industry explained: "Home service takes hold before advertising and merchandising begin . . . and it's right there to continue functioning after they stop. Home Service tills the field for appliance sales because it reaches the older women who don't read advertising and the younger women who don't believe all they read. Home Service is a service . . . [I]t goes the salesman one better because it is not primarily interested in individual sales, and so increases customer confidence."[35] Home service classes or "parties" created a comforting cushion around male sales campaigns by introducing women to appliances in a pressure-free context. These demonstrations paved the way for visits from door-to-door salesmen and often helped close a sale afterward.[36] Home service representatives typically accompanied salesmen to a home when a new appliance such as a range, a refrigerator, or a washer was installed to give an initial demonstration of its maximum possible efficiency.

Home service agents also paid "courtesy calls" to new residents and called on consumers in the event of a complaint. Sales and service men found customers reluctant to talk about problems they experienced with new equipment. Around 1925, at the Southern Public Utilities Company and the North Carolina Public Service Company, sales and service men reported that women users of electric ranges "simply would not give them the underlying reasons why they were dissatisfied with the equipment." Charlotte Mobley and her staff of nine young women had more success because consumers felt more comfortable talking frankly with them. Mobley tried to "approach every case as a human problem and to forget that there is a ledger number connected with it." By initiating more general discussions about household problems, Mobley found clues to the underlying difficulties.[37] In the process, home economists often were able to turn these kinds of problems into opportunities for instruction about budgets and household management. Their capacity for "trouble shooting" and for "soothing over distraught feelings" of dissatisfied customers led one commentator to call the home economist "the Dorothy Dix of the appliance business," likening her to the twentieth-century advice columnist.[38]

Home service couched instruction in a warm, welcoming context. As one utility executive explained, "The work of a home service department is 'mass' education to create a greater demand for a product, so subtly done that it creates 'mass' friendship along with the demand."[39] Like Mobley and her staff on home visits, many home economists served as confidential advisers to individual homemakers and also performed as hostesses at public events.[40] Utility home economists gave lectures and demonstrations at the invitation of a church group or a women's club meeting and also to high school and college classes and extension meetings. In effect, many home service departments formed clubs around their operations. This practice formalized and institutionalized the more casual alliances dealers had formed with local women's clubs in earlier decades. For example, the Dallas Power and Light Company offered free membership in the Dallas Electrical Arts Club to all of its customers. Regular classes on weekday afternoons and a weekly radio program taught subjects ranging from scientific cooking, budgeting, child feeding, interior decoration, and party ideas. Various groups could reserve the demonstration halls, equipped with an all-electric kitchen, free of cost for bridge parties, dinners, teas, and club meetings. The Tampa Electric Company also encouraged club women to use its Leisure House, an electric model kitchen and party room, for meetings and luncheons. "It is headquarters for the modern women of the Company's territory," an *Electrical Merchandising* article declared.[41] Through these practices, home service programs extended the presence of the utility company into the community, in some cases going so far as to foster communities of women around itself.

Finessing consumer relations in utilities demanded the emotional, interpersonal labor of surrounding new products and services with "an atmosphere of friendliness."[42] Home economists in utilities pioneered an early form of "personality work" or "emotional labor" that historians and sociologists associate with late twentieth-century service industries. Like the jobs later assumed by airline stewardesses, home service work in utilities involved interacting with consumers in a way that required channeling emotional energy to make the company appear human and responsive to their needs.[43] The language of "hostesses" and "parties" used by utility home economists presaged the central place these terms would have in post–World War II marketing strategies for such products as Tupperware.[44] Business home economists, we have seen, developed a professional identity compatible with the new culture of personality, a culture that Warren Susman argues emerged in the early twentieth century. As corporate communication with a presumed mass of consumers replaced face-to-face relations between individuals, the culture of personality put a premium on a definition of the self that stressed performance and presentation.[45] The importance of personality came up repeatedly in discussions among members of the male sales force in both the gas and electrical industries, but home economists in utilities carried the burden of projecting a feminized version of this ideal.

Home service personality work helped utilities to creatively blur lines between company and community. In 1922, when the Peoples Gas Light and Coke Company of Chicago hired Anna Peterson to direct one of the nation's first utility home service departments, the house organ explained company president Samuel Insull's intentions to promote public relations by offering the women of the city "a service that will make housework easier, show them how to feed their families for less money and tickle 'hubby's' palate so that he will never be late for dinner. Yes, all this without a cent of charge." Peterson was an experienced mother, homemaker, and commercial home economist already well known in the city. "She's just the sort of woman you'd like to have as a neighbor, and she's here to be a good neighbor to all the women of Chicago," company literature proclaimed.[46] Peterson's identity as a friendly maternal figure helped the utility company place itself at the center of an urban community of power consumers.

As in most home service departments, Peterson's operation concentrated on cooking. Peterson and her growing team gave lessons at the company's downtown headquarters and, within a very short time, also on the radio, in churches, at women's club meetings, and at a series of branch locations.[47] Between 1923 and 1926, the utility built nine branches of its Peoples Gas Stores and designed them specifically with home service in mind. For example, the Irving Park store featured a ground floor showroom and a grand staircase leading to a large, high-ceilinged Home Service auditorium on the second floor.[48] In specially equipped vehicles — "Kitchens on Wheels" — staff members also transported their act to more remote ends of the metropolitan area. By 1926, Bernard J. Mullaney, the company's vice president of public and industrial relations, estimated that home service had enabled Peoples Gas to have contact with a total of more than 200,000 customers in the previous year by phone, mail, or demonstrations.[49]

The relationship between gender and technology becomes especially apparent through an examination of home economists' interactions with consumers. For home economists, the cooking lessons offered by utility companies were more than just entertaining sales events. These women took satisfaction in the notion that "scientific cookery is not only possible but necessary with electric ranges."[50] By developing recipes that specified exact temperatures, home economists taught homemakers how to take advantage of new control mechanisms on both gas and electric stoves.[51] Utility cooking lessons made the latest ideas about nutrition integral to this new system and reinforced the larger message about modern, healthy living that home economists in schools and extension services promoted. For example, they pointed out that the relative slowness of electric stoves in boiling water encouraged the more healthful practice of "waterless cookery," in which a minimum of water was used to steam vegetables in order to preserve vitamins.[52] By showing consumers the proper type of pans to enhance range efficiency, and also helping manufacturers to develop such pans, home economists tried to reduce the prohibitive cost of the new ranges and the complete system of cooking they entailed.[53] Whether attracted to home economists' messages or simply out of an interest in learning more about food preparation, middle-class housewives were drawn by the thousands to the cooking demonstrations.[54] One industry member estimated that 12 percent of the entire population had attended an electrical cooking school in 1924.[55] The popularity of utility cooking demonstrations suggests that, at least in matters of food preparation, many American women greeted home economists' instruction. Their selective adoption of the new methods, however, remains unclear.

Encouraged by the popularity of cooking lessons, many home economists sought to use home service as a platform for launching a more general effort to convert the public into rational, tasteful consumers. Thus, Good Housekeeping Institute director Katherine Fisher urged home service workers to move beyond "cookery problems." The focus on cookery, she argued in a 1929 speech, limited the scope of home service departments and detracted from their broader potential: "I am of the opinion that very few of us, even now, adequately appreciate the possible future developments of this service."[56] Utility company managers had their own reasons for broadening the offerings of home service departments. Over time, they found that cooking classes reached a regular, but limited, number of customers; many managers sought to expand their contact through programs about an array of household matters. Peoples Gas's home service department answered consumers' questions on a range of household matters, but the patterns were more pronounced in the electrical industry.[57]

By 1930, as electrification promised to change methods and standards of home decoration, sanitation, health, and lighting, home service helped the utilities communicate the message of "all-electric living." A 1929 editorial in *Electrical Merchandising* noted the growth and expansion of home service in utilities: "Home service is no longer an experiment; it is recognized and valuable central station activity. . . . Women have come to regard the home service woman as the epitome of all housekeeping knowledge. Demands made on the company woman have grown so insistently that home service now includes everything from the handling of the family laundry to bringing up the baby. . . . Gradually, the home service woman has made of her department an educational institution for the women of the community, a housekeeping school and an advisory bureau to which the housewife can turn for guidance on any household problem."[58]

The expanding array of home service activities embraced, for example, instruction in "correct and scientific laundry methods." Lessons in the use of washing machines taught homemakers proper water temperatures, soaps, and techniques for stain removal which suited various fabrics. They stressed the importance of weighing load volumes and identified criteria for sorting separate loads, determining different temperatures, and selecting soaps.[59] Home service agents found it difficult, however, to interest women in laundry and "to get them to attend laundering demonstrations in anywhere near the numbers turning out for food demonstrations."[60] The lack of interest may have been due to urban middle-class women's reliance on domestic servants or commercial agencies to wash their clothes. On the other hand, new and more elaborate methods of food preparation drew large audiences. A 1931 survey of 360 housewives attending home service classes in Brooklyn, New York, showed cooking as the most popular subject, with cleaning and laundering ranking second and third. Three hundred and twenty women expressed interest in "electrical cooking," 227 in "electrical house cleaning," and only 180 in "washing problems."[61] To overcome this relative lack of interest, many home service departments included a short presentation of laundry equipment during cooking demonstrations.[62]

As Ruth Schwartz Cowan has argued, home economists' efforts to introduce science into housekeeping, raise standards of nutrition and cleanliness, and promote modern cooking methods had the cumulative effect of creating "more work for mother" in the twentieth century. Despite the widespread adoption of household technologies, women continued to spend at least the same amount of time cooking and cleaning as they had before the new technologies were available.[63] Clearly, utility home economists delivered messages about housework that encouraged this trend. Yet homemakers' frequent attendance at these demonstrations suggests that women themselves may have contributed to making housework more difficult and time-consuming.

Of course, home economists' prescribed standards of living had the added effect of helping utility companies construct "the female consumer." As electrical merchandisers worried about her characteristics, they wondered how technically inclined she was. Sometimes home service programs contributed to obscuring the technical aspects of electrical appliances while at other times they aimed to make the technology more accessible to women. Some in the field felt reducing the amount of mechanical jargon was the way to keep women interested. In explaining "Why Women Buy Electrical Goods," homemaker Dorothy Blake claimed women were attracted to their "charm and dignity": "Without an electric toaster, [the homemaker] must leave the table, provided she has no maid, watch the toast in the kitchen range, and return to the table hot and flushed. Her mind is on the toast and not the guests. She can't possibly appear, as she would like to appear, calm and interested. But place her behind an electric toaster and the conversation goes smoothly on, while her charm as a hostess is increased a hundred times!"[64] Perspectives such as Blake's warned retailers not to overwhelm consumers by exposing too much of any black box. Ethel Rose Peyser, a former director of the Good Housekeeping Institute and an expert on appliances, distinguished between technical information that women needed and expected to know and information that they preferred to ignore.[65]

Others disagreed. One editorial pointed out: "There actually are some women today who have a fair knowledge of mechanics and whose choice of an electric washer is based on something more than the color of the enameled finish.... It's up to the salesmen to see that that kind of woman is supplied with the facts."[66] As they did so, home economists had to be careful not to become a "too-expert expert" who spoke more about "the miracles of the research laboratory" than those of the kitchen. "The housewife knows what a skillet is and what it is for, but a technical explanation of a calroid unit leaves her dazed — and unsold."[67] Expert demonstrators could apparently make customers feel "self-conscious." For this reason, one store owner recommended letting the prospect try out electric irons on her own.[68]

In electric companies, home service could be more technical in nature as well, directed at homemakers who were, in the words of one woman, "interested in details of mechanism and like to know, like their husbands and sons, 'how it works.'"[69] Convinced that a new generation of progressive women demanded higher standards of quality and specifics about the features of new products, home economists eagerly pitched their instruction to such a critical clientele. For example, as the home service director at the Milwaukee Electric Railway and Light Company, Vera B. Ellwood offered a course on the care and use of electrical equipment. Determined to help women think "concretely" about electricity and "to remove some of the mystery from it," Ellwood taught such topics as how to read a meter, repair an iron, and put a plug on a cord. Homework assignments required students to determine the numbers of amperes, volts, and watts on the appliances they owned as well as to draw diagrams of the circuits and fuses in their homes. A lesson on the electric iron, Ellwood reported, had "one of the women go home and take apart three discarded irons which she had put into the attic, put them together again and find that all three were perfectly good irons with only minor adjustments necessary."[70] While Ellwood's course may have been unusual, it reflected a level of interest in basic technical understanding on the part of home economists and their audiences.[71]

The example of electric lighting demonstrates how home economists helped negotiate the conflicting meanings of a particular product. From the beginning, the electrical industry emphasized both the technical merits and aesthetic qualities of the newer light source and, into the twentieth century, continued to oscillate between these two marketing perspectives. In the decades before World War I, General Electric used advertising to downplay the technical aspects of Mazda lamps by portraying them as luxurious and magical.[72] By 1920, industry promotions stressed the labor-saving aspects of electric lights.[73] But selling lighting on the basis of work also required selling lighting fixtures, which had been understood as decorative items long before the era of electricity.[74] Many electrical dealers of the early 1920s found this meant discussing matters of taste and quality with which they were either unfamiliar or uncomfortable. Assuming women consumers possessed an inherent and exclusive ability to discern "graceful forms," "good workmanship," and suitable fabrics, salesmen felt particularly at a loss in the area of lampshades.

According to one salesman: "[I]t's when she comes to consider the shades for her lamps and fixtures that the lighting salesman stands helpless."[75]

Home service representatives such as Helen Ames Smith at the Rochester Gas and Electric Corporation, rescued salesmen from this kind of predicament by providing advice about the decorative as well as the technical aspects of lighting. Among the most technically trained home service directors in the country and the first woman to receive a B.S. in Electrical Engineering from the University of Michigan, Smith joined the company in 1920 soon after graduation. Her earliest title there was "illuminating engineer," and, for reasons that are unclear, she became home service director by 1927. Most likely she found, as many scientific women of her day, that heading a home service department was the most effective strategy for career advancement.[76] Smith's technical background was helpful, if not entirely necessary, to her job. She wrote: "Just the mere fact that I have an engineering degree counts, I think, as much with the men I meet in business as does the work that I do—at least it is a good introduction." Smith's department instructed the utility's customers about wiring, fixtures, and lamps as part of an overall effort to convince them, in Smith's words, "that there is something about lighting other than buying the fixtures and turning on the current."[77]

Under Smith's guidance, the department also took advantage of popular interest in paper and parchment lampshades and offered classes in lampshade making.[78] Style, taste, pleasure, and economy motivated lamp customers variously, forcing utility home economists to be sensitive to the diverse ways Americans experienced the new "culture of consumption." For even as they promoted a consumer culture centered around new domestic technologies, home economists also played into the aspects of that culture that valued the hand-crafted and homemade.[79]

In home service departments like Smith's at Rochester Gas and Electric, home economists also helped launch a campaign based on the technical merits of electricity. Where the lampshade classes were a product of the assumption that "interest in lighting must be created through the appeal of beauty,"[80] an initiative in the early 1930s attempted to avoid style discussions altogether. After all, even the lampshade, which became the principal means of distinguishing domestic lighting from that in public spaces, had a distinct utilitarian function: to deflect the brightness of electric light.[81] The Better Light-Better Sight campaign promoted "The New Science of Seeing" as established in 1930 by Matthew Luckiesh and Frank Moss at General Electric's Lighting Research Laboratory. In connecting lighting with vision, Luckiesh and Moss sought to make seeing easier by reducing glare and reflection but most importantly by increasing the intensity of illumination. Their studies established "artificial sunlight" as the standard for proper and healthy vision.[82] After engineers developed the "sightmeter," a device that made Luckiesh's concepts measurable and demonstrable, the industry kicked off the Better Light-Better Sight campaign.[83]

In the campaign to sell "seeing instead of fixtures," teams of "home lighting specialists" interpreted this new definition of light for American homemakers.[84] Equipped with a sightmeter and an array of lamps, shades, reflectors, and extension cords, the lighting advisor compared the amount of light used for seeing in daylight with that usually used in the consumer's artificial light. Focusing on improving the lighting conditions of living rooms and children's study areas, she then recommended changes in lamp wattage, arrangement of portable lamps, and the addition of new fixtures to help the homemaker "protect the eyesight of her family."[85]

Like the home economists' demonstrations of electric cookery, the Better Light-Better Sight campaigns equated motherhood with science even as home lighting advisers blended the new lamps with interior decorating.[86] Likewise, in their interpretations of other electrical appliances such as washing machines, vacuum cleaners, and refrigerators, home economists mixed product technicality with middle-class values of domesticity. In the process they sometimes brought consumers closer to technology; other times they helped disguise the fact that electricity was technology at all. Their efforts signaled the fluid boundaries between home and technology. In the interstices emerged a greater role for experts of all kinds, as electrical technology became more domesticated and as home became more closely tied to vast centralized light and power systems.

But just as the place for home economists in utilities became clearly defined in the 1930s, structural change in the electrical industry led utility company managers to reconfigure the parameters of their work. Domestic appliances, as opposed to heavy industrial goods, constituted a growing proportion of overall sales in the electrical industry, and price competition for these goods intensified. In response, industry leaders pursued more aggressive sales and marketing strategies including streamlined industrial design, planned obsolescence, and dramatic world's fairs displays. For example, in the refrigerator market, whereas old-line producers such as General Electric had marketed these large appliances as specialty goods, new manufacturers such as Sears and Norge used mass-market techniques to gain market share in the early 1930s.[87] In the wake of these transformations, conflicts brewed

over the extent to which home economists should be involved with the "commercial" part of the business. Within the National Electric Light Association, a "public
relations" contingent argued that goodwill was the primary purpose of home service departments; "commercial" people, on the other hand, maintained that home
service was "a sales activity, no matter how its functions be disguised." While one
woman who favored public relations pointed out that her staff was "purposely
uninformed" about prices so as to guarantee their educational credibility, a "commercial" woman argued that she could serve customers best by helping with sales
transactions.[88] Economic depression put pressure on utilities to cut back on the
number of women employed and to demand that those they kept on pay close
attention to the "dollars and cents value" of home service departments.[89] The message "Home service pays!" replaced an earlier notion of service which had created
a rhetorical space in which utilities and home economists could bring differing
understandings to their mutual endeavors.[90]

Many home economists resisted the shift to a narrower sales function. Despite
the exigencies of the 1930s, some held out against becoming mere saleswomen.
For example, home service leader Ada Bessie Swann argued that home service
increased consumption of gas and electricity, but that these results could not be
measured.[91] Criticizing those like Swann who were reluctant to embrace the profit-
making aspects of home service, Valentine Thorson, director of home service for
the Northern States Power Company, urged her colleagues in 1936 to "come off the
pedestal" and do more to sell small appliances.[92] These disputes suggest that the
newer definition of home service work took form slowly. Rather than molding
themselves automatically to corporate goals, home economists continued to contend with differing aims almost two decades after their initial entry into the business world.

But by 1940, the circumstances and content of utility home service work had
indeed changed. The public relations aspect of home service remained important,
but the sales dimension dominated.[93] Beatrice C. Jackson expressed the significance of this change in a 1941 letter to Flora Rose, director of the College of Home
Economics at Cornell University. Having worked at Brooklyn Edison Company
since graduation in 1929, Jackson wrote seeking advice from the college's placement bureau about making a career change. "We had a change of organization
some four years ago and as time goes on our work becomes less and less of Home
Economics and more and more of selling and so I feel I would do better elsewhere."
She reported that Florence Freer, Brooklyn Edison's former home service direc-

tor, had left the utility to manage an inn in Connecticut. "I mention Florence so that you can see that I am not the only one, who had found conditions, for a home economics person, so disagreeable."[94]

In the early 1920s, home economists' expectations were realized because initially, home service departments had so little to do with direct sales and so much to do with public relations. As home service departments proliferated and expanded in the 1920s, home economists became part of a gendered division of labor in which women gave instruction about appliances while men sold them. By no means natural, this division of labor reflected late-nineteenth-century middle-class notions of men's and women's separate spheres and it represented utility company managers' attempts to replicate boundaries between public and private, and market and home, within electrical merchandising operations. Teaching had long been well established as an acceptable activity for women, and the "educational selling" focus of home economists' work in utilities made sense in this light. For home economists, not dirtying their hands or corrupting themselves through over-the-counter exchanges of money might have been one way to preserve an understanding of their utility work as part of a reform mission. As home service representatives who instructed and advised but refrained from closing a sale, they reinforced stereotypical images of women as clean, moral, and respectable. They further buttressed a male culture of salesmanship which had been in the making since the late nineteenth century.[95] By drawing the line at the actual point of sale, managers reinforced the cash nexus as a male domain, preserved sales as a male activity of competition and aggressiveness, and justified paying home economists lower wages than salesmen. In the industrial work force, male managers segregated men's and women's work based upon the sexual stereotyping of jobs to buttress male autonomy and authority. Likewise, in utility companies home economists worked alongside salesmen in a separate and unequal, yet critically important, basis which mirrored the division of labor that married middle-class men and women characteristically established at home.[96]

Whether as saleswomen or as educators, home economists functioned as critical mediators between utility companies and their customers in the interwar period, and thus as active agents in the process of technological change. In the transforming marketplace for electrical service and appliances in the interwar period, buying and selling appliances was not a straightforward matter of a simple cash transaction. Advertising and design innovations alone could not sell an overwhelming array of new products. Home economists occupied a critical nexus in the relationship between producers and consumers and played a dynamic translating role between groups who had new reasons to attempt to understand the other. By connecting private homes with the managers of centralized power systems, they bridged and often obscured distinctions between company and community, technology and culture, and the public and private spheres. In this manner utility home economists enabled firms to humanize and animate their products while also legitimizing them in the guise of science and efficiency. Home economists' success in facilitating communication between producers and consumers involved a careful balancing act between their social ability to interact with middle-class women and their technical expertise. At the same time, they promoted their professional agenda by constructing and disseminating new definitions of nutrition, cleanliness, and proper lighting that linked the practical and the aesthetic dimensions of electrical goods. For engineers, electrification was not complete without domestic consumption; increasingly, home economists defined the domestic sphere in such a way as to make it incomplete without electrification. From the point of view of the consumer, this meant increased and often unprecedented contact with experts—and the values they espoused—before, during, and after the actual point of purchase.

Home economists' mediation of technology was a gendered process. By providing a construction of the female consumer to which utility company managers could subscribe, home economists strengthened the construction of consumption as a feminine activity. They facilitated communication in ways that reinforced norms of proper feminine behavior and popular assumptions that women's relationship to domestic technology was merely a matter of consumption. This construction carried contradictory implications. On the one hand, by proclaiming consumption into a legitimate activity for women, home economists made consuming and consumption a respectable enterprise. On the other hand, by feminizing consumption, they also marginalized consumption and, by extension, their own expertise. Most importantly for historians of technology, this work helped to obscure the more complex realities of home economists' own activities as both consumers and producers.

By virtue of their placement between production and consumption, home economists embodied the reality and the contradictions of the two realms' interdependence, despite the increased tendency of the emerging economic and social structures to define them as separate. Home economists' movement between utility managers and homemakers suggests the interconnectedness of the worlds in which processes and products were made and the worlds in which they were used. Their professional identity as translators provides insights into the ways knowledge about the home and knowledge about products became exchanged to make way for the introduction of new domestic technologies. By examining home economists as mediators we can begin not only to explore the complexity of their own goals as reformers and professionals but also to construct a more comprehensive and nuanced picture of how American consumers adapted to new technological systems and how utility companies reached out to consumers. Home economists' mediating activities reveal the many ways in which the acceptance of new household technologies was contingent upon communication between retailers and consumers. In their daily work patterns, home economists explicitly connected these parties to one another. As a result, they experienced issues of production and consumption as connected and inseparable. Their activities at the intersection of these spheres force us to consider "production" and "consumption" as elements in a historically changing relationship, and they point to the explanatory power of using gender as a category of analysis for understanding technological development in terms of both elements.

NOTES

Carolyn Goldstein thanks Anne Boylan, Joan Jacobs Brumberg, David Hounshell, Jan Jennings, Ronald Kline, Robin Leidner, Steven Lubar, Harold Platt, Susan Strasser, and Sarah Stage for comments on previous versions of this essay. For assistance in refining and reshaping her argument, she is especially grateful to Mark Rose, the *Technology and Culture* referees, and the editors of the special issue on gender and technology. Much of the writing was supported by a Fellowship in the History of Home Economics and Nutrition from the College of Human Ecology, Cornell University, Ithaca, New York.

1. By 1931, Swann had a staff of 1,200 home economists and a radio audience of 18,000 women. See Eloise Davison, "Home Economics Invades Business," *Independent Woman* 10 (March 1931): 127. On her career at the New Jersey utility, see Ada Bessie Swann, "Home Economics Work with Electric and Gas Utility Companies," *Home Economist and the American Food Journal* 6 (November 1928): 317–18; "How New Jersey Electric Company Uses the 'Home Electrical' Section of 'Electrical Merchandising' to Reach the Women's Clubs," *Electrical Merchandising* (hereafter *EM*) 29 (February 1923): 3094; and Swann, "What of the Future of Home Economics in Utilities," 1927 speech, folder: "Speeches, Annual Meetings, Etc.," Home Economists in Business, box 1, American Home Economics Association Archives, Alexandria, Virginia (hereafter AHEA Archives). Active within the home economics

profession, Swann also was a founding member of the Electrical Women's Round Table, an organization of women employed in the electrical industry which was established in 1923. In 1935 Swann left the Public Service Electric and Gas Company to head up the Home Service Center at *Woman's Home Companion*. See *EM* 53 (June 1935): 36–37; "Helping 3,000,000 Women to Choose and to Use Electrical Equipment," *EM* 57 (May 1937): 13; and "Companion Home Service Center," *Woman's Home Companion* 62 (November 1935): 98–99.

2. Carolyn M. Goldstein, "Part of the Package: Home Economists in the Consumer Products Industries, 1920–1940," in *Rethinking Home Economics: Women and the History of a Profession,* ed. Sarah Stage and Virginia Vincenti (Ithaca, N.Y., 1997); and "Mediating Consumption: Home Economics and American Consumers, 1900–1940" (Ph.D. diss., University of Delaware, 1994). See also Margaret Rossiter, *Women Scientists in America: Struggles and Strategies to 1940* (Baltimore, 1982); Susan Strasser, *Never Done: A History of American Housework* (New York, 1982), chap. 11; and Harvey Levenstein, *Revolution at the Table: The Transformation of the American Diet* (New York, 1988), pp. 156–58, 198.

3. Ruth Schwartz Cowan, "The Consumption Junction: A Proposal of Research Strategies in the Sociology of Technology," in *The Social Construction of Technological Systems: New Directions in the Sociology and History of Technology,* ed. Wiebe Bijker, Thomas Hughes, and Trevor Pinch (Cambridge, Mass., 1987), pp. 261–80.

4. Dorothy O. Helly and Susan M. Reverby, eds., *Gendered Domains: Rethinking Public and Private in Women's History* (Ithaca, N.Y., 1992); Linda K. Kerber, "Separate Spheres, Female Worlds, Woman's Place: The Rhetoric of Women's History," *Journal of American History* 75 (June 1988): 9–39; Jeanne Boydston, *Home and Work: Housework, Wages, and the Ideology of Labor in the Early Republic* (New York, 1990). For a discussion of the blending of "public" and "private" in the context of the history of technology, see Judith A. McGaw, "No Passive Victims, No Separate Spheres: A Feminist Perspective on Technology's History," in *In Context: History and the History of Technology: Essays in Honor of Melvin Kranzberg,* ed. Stephen H. Cutcliffe and Robert C. Post (Bethlehem, Pa., 1989), pp. 172–91.

5. For examples of "internalist" accounts of urban technological networks, see Carl Condit, *The Port of New York: A History of the Rail and Terminal System from the Beginnings to Pennsylvania Station* (Chicago, 1980), and Michael Massouh, "Innovations in Street Railways before Electric Traction: Tom L. Johnson's Contributions," *Technology and Culture* 18 (1977): 202–17. For examples of studies that focus on the politics of building urban systems, see Charles Cheape, *Moving the Masses: Urban Public Transit in New York, Boston, and Philadelphia, 1880–1912* (Cambridge, Mass., 1980), and Nelson Blake, *Water for the Cities: A History of the Urban Water Supply Problem in the United States* (Syracuse, 1956). For examples of the scholarship on design, see Adrian Forty, *Objects of Desire: Design and Society from Wedgwood to IBM* (New York, 1986); Jeffrey Meikle, *Twentieth Century Limited: Industrial Design in America, 1925–1939* (Philadelphia, 1979); Ellen Lupton and J. Abbott Miller, *The Bathroom, the Kitchen, and the Aesthetics of Elimination* (Cambridge, Mass., 1992); Ellen Lupton, *Mechanical Brides: Women and Machines from Home to Office* (Princeton, N.J., 1993); and Richard Guy Wilson, Dianne Pilgrim, and Dickran Tashjian, *The Machine Age in America, 1918–1941* (New York, 1986). On the role of advertising in promoting electricity and electrical appliances, see Pamela W. Lurito, "The Message Was Electric," *IEEE Spectrum,* Septem-

ber 1984, pp. 84–95; and Roland Marchand, *Advertising the American Dream: Making Way for Modernity, 1920–1940* (Berkeley, Calif., 1985). On the "systems" approach, see Thomas Hughes, *Networks of Power: Electrification in Western Society, 1880–1930* (Baltimore, 1983).

6. Major works include Ruth Schwartz Cowan, *More Work for Mother: The Ironies of Household Technology from the Open Hearth to the Microwave* (New York, 1983), and Strasser (n. 2 above).

7. Harold Platt, *The Electric City: Energy and the Growth of the Chicago Area, 1880–1930* (Chicago, 1991); David Sicilia, "Selling Power: Marketing and Monopoly at Boston Edison, 1886–1929" (Ph.D. diss., Brandeis University, 1991); Mark H. Rose, *Cities of Light and Heat: Domesticating Gas and Electricity in Urban America* (University Park, Pa., 1995).

8. David Nye, *Electrifying America: The Social Meanings of a New Technology* (Cambridge, Mass., 1991). In chap. 6, Nye makes a strong argument for the general influence of the home economics movement on electrification, but he devotes little attention to what home economists actually did to exert such influence. Nor does he explore their work in utilities. See pp. 250–53. For discussions of utility home service work, see Rossiter (n. 2 above), pp. 58–259; Ronald Kline, "Agents of Modernity: Home Economics and Rural Electrification, 1925–1950," and Lisa Mae Robinson, "Safeguarded by Your Refrigerator: Mary Engle Pennington's Struggle with the National Association of Ice Industries," in Stage and Vincenti (n. 2 above); Sicilia, pp. 509–17; and Jane Busch, "Cooking Competition: Technology on the Domestic Market in the 1930s," *Technology and Culture* 24 (1983): 242–44.

9. C. E. Greenwood, "Merchandising Electrical Appliances through Utilities," in *Merchandising Electrical Appliances,* ed. Kenneth Dameron (New York, 1933), p. 92.

10. On the imperatives of efficiency as seen through the eyes of engineers, see Hughes, *Networks of Power* (n. 5 above), pp. 214–26; Forty (n. 5 above), pp. 183–93. On the social implications of electricity in the home, see Nye, chaps. 6–8.

11. Ada Bessie Swann, "Five Years of Home Service," *EM* 46 (July 1931): 27, and "The Home Service Program of the NELA," *NELA Bulletin* 18 (August 1931): 535; National Electric Light Association, *Home Service: Helpful Suggestions for Organizing and Operating Home Service Departments: A Report of the Home Service Subcommittee, Public Relations National Section* (New York, 1930). Because most rural areas remained outside of the electric grid until the late 1930s, home service departments addressed a primarily urban audience. Only 10 percent of farms in the United States had electricity in 1930; 35 percent had it by 1941. See Strasser (n. 2 above), pp. 81–82.

12. Nye; Rose (n. 7 above).

13. Hughes, *Networks of Power,* pp. 223–24; and "The Chicago 'Electric Shop," *Electrical World and Engineer* 53 (1909): 684–89; Nye (n. 8 above), p. 157.

14. On sewing machine demonstrations, see David Hounshell, *From the American System to Mass Production, 1800–1932: The Development of Manufacturing Technology in the United States* (Baltimore, 1984), pp. 84–85, 88. On the importance of load-building, see Hughes, *Networks of Power,* pp. 214–26, and Platt (n. 7 above).

15. Rose, chap. 3. Rose provides an insightful account of Henry Doherty's development and training of a salesforce at the Denver Gas and Electric Company.

16. Nye, pp. 168–81; Alfred D. Chandler Jr., *The Visible Hand: The Managerial Revolution in American Business* (Cambridge, Mass., 1977).

17. Exchanges took place in small specialty shops, department stores, utility company central stations, and a variety of other outlets such as hardware stores, furniture stores, and drugstores. Not one of them controlled the entire market, and each type of outlet's share varied over time and place throughout the interwar period. My account of the structure of the retail trade is based on my reading of *Electrical Merchandising* from 1917 to 1940. Estimates vary about market shares. A 1926 estimate showed that utilities sold 42.5 percent of the appliances and dealers and contractor-dealers sold 26.9 percent. Department stores sold slightly more than 15 percent, as did "all others," a category that included hardware stores and druggists. See "Who Sells Electrical Appliances in 1926?," *EM* 35 (January 1926): 6052. Dameron (n. 9 above), pp. 74–75, cites a 1929 study that reported electrical specialty dealers (including retail operations run by manufacturers) handling 37 percent of sales volume, central stations 29 percent, department stores 13 percent, and "other" 20 percent. On department stores, see Lawrence Wray, "But There's Another New York," *EM* 59 (March 1938): 6–7, 64–65; "The New Electric Shop in One of Pittsburgh's Big Department Stores," *EM* 19 (January 1918): 7; "The Department Store Man's Relation to the Contractor-Dealers' Association," *EM* 19 (January 1918): 14–16; Frank B. Rae Jr., "Department Store Promotes Sales by Other Dealers," *EM* 19 (March 1918): 126–29.

18. Frank B. Rae Jr. launched the journal's predecessor in 1905 while doing publicity work for the American Electrical Heater Company. Soon *Selling Juice,* the house organ Rae edited, attracted the attention of utility company managers. In 1907 Rae split from his employer and renamed his independent journal *Selling Electricity.* He sold the journal to the McGraw Publishing Company in 1916 and the title changed to *Electrical Merchandising.* O. H. Caldwell became the editor, but Rae remained on the staff as a regular writer of articles and editorials. See "How Times Have Changed," *EM* 51 (May 1934): 36–37, 62.

19. On General Electric's shift away from a "disregard" for education in the early 1920s, see David Nye, *Image Worlds: Corporate Identities at General Electric, 1880–1930* (Cambridge, Mass., 1985), pp. 122–24. This shift often emphasized the "labor-saving" qualities of electrical appliances. See also Strasser (n. 2 above), pp. 76–78; and "To Cash in on the 'Kitchen Movies,'" *EM* 17 (February 1917): 59. In 1921 Frank Rae, editor of the trade journal *Electrical Merchandising,* lamented the destruction that could result when the operation and care of appliances was not explained: "[T]oo many clothes are put into a washing machine and too many fed into the wringer. Flatirons are set into a pan of water to cool because the salesman didn't do his job right; washing machine connector cords are allowed to be scuffed about on concrete cellar floors or to lie in the wet; half a dozen appliances are tapped on a single circuit with consequences ruinous to fuses, and women expect dishwashers not only to clean off, but to consume and wholly obliterate all table scraps, including meat bones." Frank B. Rae Jr., "Why Shouldn't Service Pay?," *EM* 26 (December 1921): 287–89. See also "Store Demonstrations When Sale Is Made Save Servicing Charges Afterward," *EM* 31 (April 1924): 4232–33.

20. On home economists' role in promoting both electric and gas ranges, see Busch (n. 8 above), pp. 242–44. On cooking lessons offered by utilities, see also Sicilia (n. 7 above), pp. 509–17.

21. "A 'Cooking Bee' Always Brings the Women Out," *EM* 30 (July 1923): 3497. See also "Making It Easy for You to Put on an Electric Range Demonstration," *EM* 30 (August 1923):

3565. This strategy took off in the 1920s but had its roots at least in the previous decade when the Minneapolis General Electric Company sent recipes to electric range prospects and users. See "Interesting the Ladies in Electric Cooking," *EM* 20 (October 1918): 174. By the 1930s, the orientation toward teaching women how to operate ranges most economically led one woman to write "if you are going to have it run right, and if you are going to sell them, then you have got to have the Home Service Department have a woman go and show that woman of the house how best to use the range." Quoted in Preston S. Arkwright, "We Cannot WISH for LOAD . . . and Get It," *EM* 48 (July 1932): 26.

22. "Should the Salesman be a Housework Expert?," *EM* 28 (November 1922): 108.

23. Advertising and marketing executives started repeating the 85 percent figure as early as the 1890s. See Charles McGovern, "Sold American: Inventing the Consumer" (Ph.D. diss, Harvard University, 1993). For a discussion of it in the interwar period, see Marchand (n. 5 above), pp. 66–69. For a contemporary expression of the "woman's viewpoint" and its importance to utility companies, see Christine Frederick, *Selling Mrs. Consumer* (New York, 1929), p. 281. On General Electric's efforts to target women consumers, see Nye, *Image Worlds*, pp. 128–31. Lidda Kay, "Is Your Store Popular with Women?," *EM* 24 (October 1920): 201–2. Kay claimed "Not more than 10 per cent of all electrical appliances sold are bought by women alone. The other 90 per cent is bought either by husbands and wives together or by men alone."

24. Elizabeth Durkee Kohlwey, "The 'Feminine Touch' and a Woman's Ideas on Store Arrangement," *EM* 23 (June 1920): 312.

25. For example, when Jack Carrigan organized a temporary model electrical home exhibit for Westinghouse in Boston in 1923, he enlisted club women to help him domesticate it. Explaining the effectiveness of this practice, he wrote, "before the women took hold, we merely had an electrical house. It was they who converted it into a home." See "Get Local Women on the Job," *EM* 30 (October 1923): 3662.

26. "'Electrabode' Aims to Please the Ladies, Says Mosley," *EM* 30 (July 1923): 3494. On efforts of department stores to resemble home and to promote and ideal of customer service, see Susan Porter Benson, *Counter Cultures: Saleswomen, Managers, and Customers in American Department Stores, 1890–1940* (Urbana, Ill., 1986), pp. 83–101; William Leach, "Transformations in a Culture of Consumption: Women and Department Stores, 1870–1920," *Journal of American History* 71 (1984): 83–101; William Leach, *Land of Desire: Merchants, Power at the Rise of a New American Culture* (New York, 1993); Jeanne Catherine Lawrence, "Geographical Space, Social Space, and the Realm of the Department Store," *Urban History* 19 (April 1992): 64–83. On similar attempts in French department stores, see Rosalind H. Williams, *Dream Worlds: Mass Consumption in Late Nineteenth-Century France* (Berkeley, Calif., 1982), and Michael B. Miller, *The Bon Marché: Bourgeois Culture and the Department Store, 1869–1920* (Princeton, N.J., 1981).

As this kind of store interior became more common, increasing numbers of women managed and owned their own shops and many retailers hired women to assist in window displays, store arrangements, and sales techniques. See "This Woman Dealer Sets a Fast Pace Selling Lighting Fixtures to Women," *EM* 25 (January 1921): 8–9. An *Electrical Merchandising* editorial noted, "Once upon a time the electrical field was looked upon as man's peculiar province." Since then, "a few women did find a loophole that could be widened into an unlimited circle. From a few pioneers the number of women now engaged in the electrical industry has grown unbelievably and their importance to the industry at large and to their communities in particular should be recognized and encouraged." See "Women in the Industry," *EM* 35 (February 1926): 6118.

27. On the electrical industry's ambitious educational programs in support of public relations, see James David Wilkes, "Power and Pedagogy: The National Electric Light Association and Public Education, 1919–1928" (Ph.D. diss., University of Tennessee, 1973). See also, Tom J. Casey, "Common Sense Merchandising," *EM* 26 (July 1921): 30–33. On the contribution of friendliness and courtesy toward goodwill, see Clotilde Grunsky, "Always with a Smile!," *EM* 31 (March 1924): 4148–49, 4172; and the editorial "Public Relations for the Contractor-Dealer," *EM* 35 (April 1926): 6235.

28. The development of the home economics curriculum merits further study. For a general discussion, see Linda Fritschner, "The Rise and Fall of Home Economics: A Study with Implications for Women, Education, and Change" (Ph.D. diss., University of California, Davis, 1973), pp. 83–89.

29. See Dolores Hayden, *The Grand Domestic Revolution: A History of Feminist Designs for American Homes, Neighborhoods, and Cities* (Cambridge, Mass., 1981); and David Nye, *Electrifying America* (n. 8 above). On rational consumption, see Goldstein, "Mediating Consumption" (n. 2 above).

30. As Margaret Rossiter has shown, home economics–related jobs presented some of the very few employment opportunities for women educated in the sciences. Women such as Mary Pennington (trained in bacteriology), Lillian Gilbreth (trained in psychology and experienced in scientific management), and Helen Ames Smith (trained in electrical engineering) took on identities as consulting "home economists" when jobs in their specific fields tended to be open only to men; Rossiter (n. 2 above). On Pennington, see Robinson (n. 8 above). On Gilbreth's consultation with utility companies about kitchen planning, see Laurel Graham, "Lillian Moller Gilbreth: Extension of Scientific Management into the Women's Work, 1924–1935" (Ph.D. diss., University of Illinois, 1992). Swann and others studied home economics and intended to have conventional teaching careers in the classroom but many found opportunities applying their training in the business world. Most women who worked in utility home service did so for only a few years before marrying and becoming homemakers. However, a significant number followed the path of Ada Bessie Swann and built long-term, lifelong careers in the industry demonstrating appliances, writing instructional manuals, delivering radio programs, corresponding with customers, and overseeing the growth and management of home service departments.

31. For an analysis of business home economists' professional ideals, see Goldstein, "Mediating Consumption." See also Rossiter; Strasser (n. 2 above), chap. 11; and Levenstein (n. 2 above), pp. 156–58, 198. Home economists shared the optimism with other aspiring women professionals whom Nancy Cott describes in *The Grounding of Modern Feminism* (New Haven, Conn., 1987), pp. 225–26. This rhetoric is also quite similar to that employed by advertisers in the same era; see Marchand (n. 5 above).

32. "The Woman in the Industry," *EM* 39 (June 1928): 90–92.

33. For example, under the sponsorship of the declining ice industry, home economists at the U.S. Department of Agriculture's Bureau of Home Economics collaborated with engineer and bacteriologist Mary Pennington on research and publicity projects for ice refrigeration. On Pennington, see Robinson. See also Goldstein, "Mediating Consumption," pp. 147–77.

34. Swann, "Home Economics Work with Electric and Gas Utility Companies" (n. 1 above): 318. See also "Home Service Directors to Meet in Chicago," *EM* 43 (March 1930): 88; Florence R. Clauss, "Home Service—A Major Activity," *EM* 43 (April 1930): 44–46; National Electric Light Association, *Home Service* (n. 11 above); "Five Years of Home Service" (n. 11 above): 27; and "Home Service of the Future is a Community Service," *EM* 43 (June 1930): 90–92.

35. Clara H. Zillessen, "Home Service . . . The Dorothy Dix of the Appliance Business," *EM* 45 (June 1931): 55. Also quoted in Swann, "The Home Service Program of the NELA" (n. 11 above): 535.

36. Florence Freer, "Home Service . . . Opens the Door," *EM* 47 (June 1932): 44–45.

37. S. T. Henry, "Saving the Sale," *EM* 46 (September 1931): 52–53.

38. Zillessen, pp. 54–55, 92, 100. On "soothing distraught feelings," see Swann, quoted in "Making the Consumer Tie-Up," Chronological File: 1926, Home Economists in Business Archives, Westerville, Ohio. Harnett T. Kane, *Dear Dorothy Dix: The Story of a Compassionate Woman* (Garden City, N.Y., 1952).

39. Quoted in Swann, "What of the Future of Home Economics in Utilities" (n. 1 above).

40. Swann, "Home Service of the Future is a Community Service" (n. 34 above): 91; Swann, "The Home Service Program of the NELA" (n. 11 above): 535. At the Philadelphia Electric Company, Clara Zillessen also relied on assistance from "married women of refinement" to serve as "'hostesses'—we never called them demonstrators." See "Philadelphia's First 'Home Electric,'" *EM* 30 (December 1923): 3797.

41. "They Call It Leisure House," *EM* 61 (May 1939): 26. See also "Encouraging Social Contacts with Women's Clubs," *EM* 33 (March 1925): 5162; "Meet Mrs. Mitchell of Our Courtesy Department," *EM* 38 (July 1927): 102–3; "7,000 Joiners," *EM* 52 (September 1934): 58–59; "The Electrical Arts Club of Dallas," *EM* 44 (August 1930): 47.

42. Swann, *Home Service* (n. 11 above).

43. On "emotional labor" of airline stewardesses, see Arlie Russell Hochschild, *The Managed Heart: The Commercialization of Human Feeling* (Berkeley, Calif., 1983). Because home economists actually discussed "personality" as an important requirement for their jobs in business, I prefer Suzanne Kolm's term, "personality work." See "Labor Aloft: A Cultural History of Airline Flight Attendants in the United States, 1930–1978" (Ph.D. diss., Brown University, 1995).

44. See Alison J. Clarke, "Tupperware: Suburbia, Sociability and Mass Consumption," in *Visions of Suburbia,* ed. Roger Silverstone (London and New York, 1996); Nicole Woolsey Biggart, *Charismatic Capitalism: Direct Selling Organizations in America* (Chicago, 1989); W. Hampton Sides, *Stomping Grounds: A Pilgrim's Progress through Eight American Subcultures* (New York, 1992), pp. 216–43.

45. Warren I. Susman, "'Personality' and the Making of 20th-century Culture," in *Cul-*

ture as History: The Transformation of American Society in the Twentieth Century (New York, 1984), pp. 277–78, 280. Within the culture of "personality" itself, there was confusion over gender just as there had been in the culture of "character." The key words Susman identifies in self-improvement manuals around 1900, for example, include *fascinating, stunning, attractive, magnetic,* as well as *dominant, forceful, energetic,* and *efficient.* Out of context, these suggest at once a male and a female construction of personality. This was not the only contradiction the new culture contained; in it, the individual faced the ironic challenge of having both to stand out in a crowd and appeal to it at the same time, to work hard on "being oneself." None of this was considered "natural" to anyone; personality was a matter of performance and it had to be cultivated. Still, some argued that "personality" came more naturally to women than to men. Charles Gray Shaw, professor of philosophy at New York University, claimed that "women have more personality than men; they possess more individuality and this is reflected in their work." Contrasting women's interest in whimsical fashion and dress to express themselves with mens' "regularity and conventionality" of appearance, Shaw praised women's adaptability and faulted men for being "martyr[s] to tradition." Because of their superior personality, Shaw characterized women as "truly the standard-bearers of the race," in a way that echoed nineteenth-century associations of women with character and morality. See Ethel J. Bein, "Men, Women, and Personality: An Interview with Charles Gray Shaw," *Independent Woman* 11 (August 1932): 277, 297.

46. *Peoples Gas Club News* (hereafter *PGCN*), March 15, 1922, pp. 65, 74; "Mrs. Anna J. Peterson," n.d., Peoples Gas, Light and Coke Company Archives, Chicago, Illinois (hereafter PGA). Although many utility companies had hired "home service agents" before World War I, Winifred Stuart Gibbs estimated that Peoples Gas was "one of the first, if not the first public utility that gave the specialist in economics a real opportunity"; see "The Sales and Advertising Value of Home Economics Woman," *Printer's Ink Monthly* 13 (September 26, 1926): 144. On early home service departments, see "Positions for Women at the New York Edison Company," March 31, 1917, Series II, file 101, Records of the Bureau of Vocational Information, Schlesinger Library, Radcliffe College, Cambridge, Massachusetts (hereafter BVI). For other examples, see Florence Freer, "Home Service—Its Functions and Accomplishments," *NELA Bulletin* 19 (December 1932): 733; "Neighborly," *EM* 41 (February 1929): 62–63.

47. On the use of radio by home economists, see Susan Smulyan, "Radio Advertising to Women in Twenties America: 'A latchkey to every home,'" *Historical Journal of Film, Radio, and Television* 13 (1993): 299–314.

48. The branches were at 846 West 63rd Street (near Halsted Street), 1520 Milwaukee Avenue, 11031 Michigan Avenue in Roseland, 75th and Cottage in Cottage Grove, Broadway at Wilson, Larrabee, Commercial Avenue, and Irving Park. See "Opening Dates, Branch Home Service Auditoriums," undated manuscript, PGA; "Our Home Service Department," *PGCN,* November 15, 1923, p. 355; "New Irving Park Store Opens," *PGCN,* October 1, 1926, p. 3.

49. "Our Home Service Department," *PGCN,* June 1, 1923, p. 167; Gibbs, p. 144.

50. Florence Clauss, "How Can We Sell More Ranges?," *EM* 40 (October 1928): 60.

51. Busch (n. 8 above), pp. 224–25.

52. Clauss, p. 62.

53. For example, Margaret Mitchell developed special pans for this purpose for the Wear-Ever Aluminum Company in the early 1930s. See Margaret Mitchell, "Home Economists in the Equipment Field," 1937 speech, box 1, folder: 1930s minutes, Home Economists In Business Section Archives, Westerville, Ohio. At the Corning Glass Works, Lucy Maltby performed similar work starting in 1927. See Regina Lee Blaszczyk, "Imagining Consumers: Manufacturers and Markets in Ceramics and Glass, 1865–1965" (Ph.D. diss., University of Delaware, 1995), and "Where Mrs. Homemaker is Never Forgotten: Lucy Maltby and the Home Economics Department at Corning Glass Works, 1929–1965," in Stage and Vincenti (n. 2 above).

54. Isabell Davie, "Help From Home Service," *EM* 53 (March 1935): 13–15, 68–69. "362 Women Asked of Home Service," *EM* 47 (April 1932): 56–57; Florence Freer, "Organize Laundry Demonstrations," *EM* 47 (February 1932): 54.

55. "How to Conduct a Cooking School," *EM* 34 (August 1925): 5461; "Butte, Mont., Holds Successful Electric-Range Cooking School," *EM* 33 (January 1925): 5053.

56. Katherine A. Fisher, "Home Service Activity Is Selling Higher Standards of Living," *EM* 42 (November 1929): 95.

57. "Some Questions We Answer," *PGCN,* July 1, 1922, p. 176; "Tuning in with Home Service," *PGCN,* February 1, 1924, p. 3. On the electrical industry, see George Leibman, "Home Economics as a Phase of Customer Relations," *NELA Bulletin* 15 (August 1928): 493–501; Florence Clauss, "Making Friends," *EM* 41 (June 1929): 76–77; 113; and Elsie Hinkley, "Report of Subcommittee on Trends in Home Service," in *Home Service: Report of the 1930 Committee,* American Gas Association (New York, 1930), p. 6.

58. "The Growing Importance of Home Service," *EM* 42 (September 1929): 110–11.

59. Florence R. Clauss, "When Is a Washer Well Sold?" The Western Electric Company established a "laboratory-laundry" in the Pershing Building across the street from Grand Central Station in New York City. Here home economists Mildred Nichols and Doris Scott offered information to dealers, salesmen and saleswomen, and to a lesser extent the general public. See "New York Has New Demonstration Kitchen-Laundry—'at Your Service,'" *EM* 31 (February 1924): 4130.

60. Freer, "Organize Laundry Demonstrations," (n. 54 above).

61. Davie, "362 Women Asked of Home Service," (n. 54 above).

62. Davie, "Help From Home Service," (n. 54 above).

63. Cowan, *More Work for Mother* (n. 6 above).

64. Dorothy Blake, "Why Women Buy Electrical Goods," *EM* 23 (June 1920): 292–93.

65. Ethel Rose Peyser, "What She Wants to Know," *EM* 26 (October 1921): 189–90. Rose, who later covered household equipment for *House and Garden,* also wrote *Cheating the Junk-Pile: The Purchase and Maintenance of Household Equipments* (New York, 1922. Rev. ed. 1930).

66. "Not All Women Are Sound-Proof to Technical Talk," *EM* 24 (December 1920): 318. Another dealer expressed more succinctly that "the buyer is not dumb." See Gerald Stedman, "The Buyer Is Not Dumb," *EM* 56 (September 1936): 2–3, 47.

67. "The Too-Expert Expert," *EM* 56 (July 1936): 24.

68. "No Expert Demonstrators," *EM* 50 (December 1933): 28.

69. Camilla Ryneers, "Keep Up with Your Customers on Housekeeping Methods," *EM* 39 (March 1928): 92–93. On electrical work as more technical in nature, see Helen Smith to Beatrice Doerschuk, January 30, 1924, Series II, file 117, BVI.

70. Clauss, "Making Friends" (n. 57 above): 113.

71. The Electrical Arts Club of Dallas offered demonstrations in its "electrical laboratory" where customers could bring their small appliances in for cleaning or testing. "The Electrical Arts Club of Dallas," *EM* 44 (August 1930): 47. See also Wendell Holmes, "Motors, Motors, Everywhere," *Better Homes and Gardens,* January 1939, pp. 12–13.

72. Thomas A. Edison, "The Success of the Electric Light," *The North American Review* 131, no. 4 (October 1880): 295–300; Nye, *Image Worlds* (n. 19 above), pp. 116–23.

73. "Good Light Lightens Labor," *EM* 23 (June 1920): 11.

74. On early fixtures for Edison's incandescent lamps marketed by Bergmann and Company beginning in 1880, see Henry Bartholomew Cox, "Plain and Fancy: Incandescence Becomes a Household Word!," *Nineteenth Century* (Autumn 1980): 49–51.

75. "The Stuff That Lamps Are Made Of," *EM* 29 (January 1923): 3036–37.

76. On Smith, see "Women in Engineering," *News-Bulletin of the Bureau of Vocational Information* (March 1, 1924): 39; Helen Smith to Myra Robinson, May 29, 1928, New York State College of Home Economics Records, Collection Number 23/2/749, Division of Rare and Manuscript Collections, Cornell University, Ithaca, New York (hereafter NYSCHER); newspaper clippings and *Midwest Engineer* (August 1952), University of Michigan Alumni Office. For a general account of home economics as a path scientific women used for advancement, see Rossiter (n. 2 above).

77. Helen A. Smith to Beatrice Doerschuk, January 17, 1924, and Helen A. Smith to Beatrice Doerschuk, January 30, 1924, BVI, Series II, file 117.

78. In 1927, home service representative Jessie Cary Grange conducted four paper lampshade classes a week. First offered as a free service, these sessions grew so rapidly that the utility, in an effort to make them "self-supporting," charged students for instruction and materials. This cost amounted to an average of 55 cents per shade, comparable to the cost of a hand-painted imported silk shade readily available at Sears. Students also had to buy their own frames at about 10 cents a piece. "Lamp Shade Classes Promote Better Lighting," *EM* 38 (October 1927): 88–89; *Sears, Roebuck and Company Spring-Fall Catalog* (Chicago, 1927), p. 327. Two years later, the utility added a Saturday morning class for children. See *NELA Bulletin* 16 (March 1929): 176; Florence R. Clauss, "Home Lighting Progress," *EM* 41 (April 1929): 70. For an early example of a response to this popular interest in lamps and lampshade making, see L. C. Spake, "Buying Hunches from the Fixture Market," *EM* 23 (March 1920): 122–23, and "Teaches Women to Make Own Lamp Shades," *EM* 19 (April 1918): 207. On changing lampshade styles, see Nadja Maril, *American Lighting: 1840–1940* (West Chester, Pa., 1989), pp. 134–39.

79. Richard Fox and T. J. Jackson Lears, eds., *The Culture of Consumption: Critical Essays in American History, 1875–1940* (New York, 1983). Nancy Bercaw argues for the importance of handmade goods to the consumer culture of the late-nineteenth century in "Solid Objects/Mutable Meanings: Fancywork and the Construction of Bourgeois Culture, 1840–1880," *Winterthur Portfolio* 26 (Winter 1991): 231–47.

80. "Lamp Shade Classes Promote Better Lighting," *EM* 38 (October 1927): 89.

81. On this dual function of lampshades, see Alistair Laing, *Lighting* (London, 1982), pp. 67–68.

82. M. Luckiesh and F. K. Moss, "The New Science of Seeing," *Transactions of the Illuminating Engineering Society* 25 (1930): 15–49; Matthew Luckiesh, *Artificial Sunlight* (New York, 1930); and Matthew Luckiesh and Frank K. Moss, *Lighting for Seeing* (Cleveland, 1931). For more on this campaign and Luckiesh's work, see Nye, *Electrifying America* (n. 8 above), pp. 363–66, and Sicilia (n. 7 above), pp. 493–509.

83. One sales manager announced, "artificial lighting is now in competition with daylight, it is no longer in competition with darkness"; Frank B. Rae Jr., "Does Sight Saving Menace Merchandising?," *EM* 53 (February 1935): 25, 30–31. By 1933, Luckiesh and Moss developed their own instruments for measuring brightness. See Matthew Luckiesh and Frank Moss, *The Science of Seeing* (New York, 1937), pp. 322–27.

84. "Pleasure Versus Appliances," *EM* 54 (August 1935): 32.

85. In Utica, New York, a campaign of this type added an average of 220 watts and $8.00 of merchandise to each home visited. "Demonstrating Better Light — Better Sight," *EM* 50 (September 1933): 48.

86. "Lighting Trends Discussed at Forum of 200 Lighting Advisers," *EM* 59 (March 1938): 64–65.

87. As David Nye explains, the electrical industry "was virtually immune to the Depression." See Nye, *Electrifying America* (n. 8 above), p. 348; for a discussion of the structural changes in the 1930s, see chap. 8. Marchand (n. 5 above), p. 300, argues that Depression-era advertising was distinctly "loud, cluttered, undignified, and direct." On the shift in refrigerator production and marketing, see Shelley K. Nickles, "Object Lessons: Designers, Consumers, and the Household Appliance Industry, 1920–1965" (Ph.D. diss., University of Virginia, 1997). On Sears, see Richard Tedlow, *New and Improved: The Story of Mass Marketing in America* (New York, 1990).

88. "Is Home Service a Commercial Activity?," *EM* 40 (July 1928): 71.

89. Laurence Wray, "Cold Figures on Home Service," *EM* 53 (April 1935): 11, 34. See also Clauss, "Home Service — A Major Activity" (n. 34 above): 44; Zillessen, "Home Service" (n. 35 above): 54–55, 92, 100; Elizabeth Stone Macdonald, "Home Service Must Show Tangible Results," *NELA Bulletin* 19 (May 1932): 315.

90. Isabell Davie, "Home Service Sells Refrigeration," *EM* 53 (May 1935): 11–15, 51.

91. Swann, "The Home Service Program of the NELA" (n. 11 above).

92. "Come Off the Pedestal, Home Economists," *EM* 56 (December 1936): 10. On Valentine Thorson, see also "Good Old Roast Beef Closes Range Sales," *EM* 52 (December 1934): 20–21. Likewise, the home service director of a New York gas company criticized university home economics professors for failing to prepare students for the reality that "promotion is necessary education of the general public in the use of available civilized equipment" and that "the fact that it sells is not a sin." See anonymous interview, Home Economics Study, Institute for Women's Professional Relations, September 1935, folder 4279, box 361, Series 200S, Rockefeller Foundation Papers — RG 1.1, Rockefeller Archive Center, North Tarrytown, New York. This document is one of the few surviving interviews conducted as part of the extensive study by Chase Going Woodhouse, *Business Opportunities for the Home Economist* (New York, 1938).

93. "The Dollar and Cent Value of Home Service," *EM* 64 (August 1940): 50, 92. Home economists, then, reinforced utility company aims in patterns that appear similar to the way other experts bolstered corporate power structures. See Loren Baritz, *Servants of Power: A History of the Use of Social Science in American History* (Middletown, Conn., 1960); David F. Noble, *America by Design: Science, Technology, and the Rise of Corporate Capitalism* (New York, 1977).

94. Beatrice C. Jackson to Flora Rose, March 20, 1941, NYSCHER. Jackson remained at Brooklyn Edison, later Consolidated Edison, until 1948, by which point the utility was in the process of phasing out home service altogether. She described her experience as follows: "For sixteen years I was occupied with lecture-demonstrations to women who were home makers. These demos consisted of preparing and cooking meals using electrical equipment and bringing out the labor saving features of the various appliances which we used. We also went into the schools of New York City and gave a lecture on the generation and distribution of electricity using a special demonstration board to illustrate our lectures to the faculty, children, and parent-teacher associations. During the war years we held demonstrations on canning and conservation of food and also gave the Red Cross nutrition classes twice a week to women. For the past two years I worked as a clerk in the office at 1650 Pitkin Avenue where I did filing, answered a telephone order board and kept records of the outside men's calls on manufacturers and the public in general." When she was transferred to clerical work, she was informed by Mr. Holmberg, the manager of Home Service then, that "due to a change of organization of our department it was no longer necessary to have as many women in Home Service." See Beatrice C. Jackson to Esther Stocks, September 20, 1948.

95. On this male culture of salesmanship, see Susan Strasser, "'The Smile that Pays': The Culture of Travelling Salesmen, 1880–1920," in *The Mythmaking Frame of Mind: Social Imagination and American Culture*, ed. James Gilbert et al. (Belmont, Calif., 1993), pp. 155–77; Timothy B. Spears, "'All Things to All Men': The Commercial Traveler and the Rise of Modern Salesmanship," *American Quarterly* 45 (December 1993): 524–57. See also Angel Kwolek-Folland, *Engendering Business: Men and Women in the Corporate Office, 1870–1930* (Baltimore, 1994), chap. 3. On representations of the quintessential man as a businessman, see Marchand (n. 5 above), pp. 188–91.

96. On the sexual stereotyping of occupations, see Alice Kessler-Harris, *Out to Work: A History of Wage-Earning Women in the United States* (New York, 1982), pp. 128, 139–42, 157–58, 203–4, 230–34.

FOURTEEN

Home Ideologies: Progress?

RONALD R. KLINE

Throughout the twentieth century, household appliances were touted as labor saving devices eliminating drudgery, saving time, and increasing leisure. By the 1970s, however, feminist revisions of many standard stories inspired sociologist Joanne Vanek and historian Ruth Schwartz Cowan to revisit the available evidence. Analyzing home economists' research from earlier in the century, they demonstrated that women spent the same amount of time in the 1970s doing housework as their grandmothers had in the 1920s, before the widespread use of most "labor-saving" appliances.

Ron Kline revisits the home economists' careful studies of farm women, their uses of technical devices, and the time they spent on various tasks. He finds that the data in these studies made plain that no time was saved, no leisure gained—but that the analysis at the time drew no such conclusion. Kline argues that home economists, like the promoters of electricity and even the farm women themselves, were too much invested in the progress ideology of technology and in their professional mission to uplift overburdened farm women to think through the data to their logical conclusion. How does Kline explain changing views of the same data in the 1920s and 1970s? How did specific gendered notions about women's role on the farm shape home economists' views of technology, farm women, and home appliances? Where did ideological constructs and technological practices differ? Where did they reinforce each other? When do technological innovations offer possible new roles for women, and when do they reinforce existing social practices? Conversely, which social practices and ideologies promote continuity and which encourage technological change?

"Ideology and Social Surveys: Reinterpreting the Effects of 'Laborsaving' Technology on American Farm Women," *Technology and Culture* 38, no. 2 (1997): 355–85. Reprinted by permission of the Society for the History of Technology.

Among the many strands of technological optimism running through the twentieth century, the idea that "laborsaving" technology would revolutionize the factory, office, field, and home pervaded the discourse of advertisers, engineers, reformers, and social commentators in the United States. While critics might point to the dehumanizing affects of the Machine Age in the 1920s or to widespread technological unemployment in the 1930s, a positive ideology of technological progress seems to have prevailed before the late 1960s.[1] In regard to the household, a wide range of social groups thought that "modernizing" homes with electric lights, running water, washing machines, electric ranges, and vacuum cleaners would eliminate drudgery, save labor time, and increase leisure. Rural reformers, including home economics researchers, hoped these technologies would create an electrical utopia on the farm to match city comforts. Reformers classified many nonelectrical household features, such as hardwood floors and window screens, as "laborsaving," but their belief in a progress ideology of electricity, shared by a wide political spectrum, helped to inscribe "modern" domestic appliances with a largely unquestioned power to transform urban and rural cultures.[2]

Scholars did not seriously challenge this interpretation until the women's movement of the 1970s spawned the pathbreaking and influential work of sociologist Joann Vanek and historian Ruth Schwartz Cowan. Vanek analyzed forty years of time-use surveys to argue that electrical appliances and other "modern" household technologies might have reduced the energy required to perform specific tasks, but ownership of these appliances did not correlate with less time spent on housework by full-time home workers. In fact, time spent by these workers remained remarkably constant — at about fifty-two to fifty-four hours per week — from the 1920s to the 1960s, a period of significant change in household technology.[3] In surveying two centuries of this technology in the United States, Cowan argued that the "industrialization" of the home often resulted in "more work for mother" because the use of such artifacts as coal stoves, water pumps, and vacuum cleaners tended to reduce the workload of married women's helpers (husbands, sons, daughters, and servants) and to promote a "higher" standard of housework. The full-time home worker's patterns also shifted from production to consumption, which included household management, child care, and the post–World War II phenomenon of being "mom's taxi."[4] The irony of laborsaving household technology is now a respected thesis in the history of technology, the sociology of technology, and American social history; it gained wider currency in 1993 through the well-publicized exhibit "The Mechanical Bride," at the Cooper-Hewitt Museum in New York City.[5]

Not as well known is that researchers had identified and analyzed this irony, especially for farm women, several decades before the work of Vanek and Cowan. In fact, these researchers — home economists working in agricultural experiment stations at state colleges throughout the country — designed and conducted the detailed time-use surveys so profitably employed by Vanek and Cowan. In revisiting these studies — of which nearly twenty were published from the mid-1920s to the late 1960s — and in analyzing their reception in the fields of home economics, rural electrification, and rural sociology, I will discuss how home economics researchers and their audiences in the rural social sciences came to grips with data that seemed to contradict a widely held optimistic view of technology. The data from the time-use studies, most of which is from the farm home, show that farm women worked the same number of hours, on average, before and after installing running water and electricity, that modern household technologies often saved the work of their helpers, and that farm women with more technology often worked more hours at "homemaking." Yet neither the researchers, other home economists, nor rural sociologists who knew about these studies used them to challenge the technological optimism associated with electricity and "modern" household technology.

In this article, I ask why these home economics researchers made these optimistic interpretations, privileging certain types of data over others, and in many cases modifying their interpretations to accord with their philosophy of technology. I will analyze their interpretations in terms of a progress ideology of technology (especially that of electrical appliances), an urban domestic ideal that divided household work along gender lines, the increasingly "objectivist" ideal of social science researchers, and the ameliorative goals of the Country Life movement (which commanded the loyalty of most home economics researchers). I will argue that this complex system of beliefs (whose components complemented and contradicted each other) significantly influenced the structure, conduct, and interpretation of the time-use studies.[6]

Although my approach addresses the actions of experts much more than the local practices of rural culture, at the end of the article I will address the question of how we can use the surveys to investigate the work habits of American farm women in the twentieth century. Examining the various ways in which researchers and participants constructed the time-use studies should help us decide how to use them as sources for the social history of rural (and urban) technology.[7]

Country Life Reformers and the "Problem" of the Overworked Farm Woman

The Country Life movement was the main vehicle for melding together the system of beliefs held by the home economists. The researchers who began the time-use surveys in the 1920s worked on a social problem — the overworked farm woman — that had been a staple of Progressive Era reform since the 1870s, when the topic rose to the level of mainstream public discourse.[8] President Theodore Roosevelt's Country Life Commission, called in 1908 to recommend means of improving rural life, gave the issue new urgency: "Realizing that the success of countrylife depends in very large degree on the woman's part, the Commission has made special effort to ascertain the condition of women on the farm." The Commission concluded that because the farm woman's work was not as seasonal as the farm man's, "whatever general hardships, such as poverty, isolation, lack of labor saving devices, may exist on any given farm, the burden of these hardships falls more heavily on the farmer's wife than on the farmer himself. In general her life is more monotonous and the more isolated, no matter what the wealth or the poverty of the family may be."[9]

As the rural arm of the Progressive movement, Country Life reformers attacked this "problem" in traditional Progressive fashion: collecting data and promoting legislation to fund experts to make "objective" scientific surveys. Farm woman Mattie Corson, whose mother had supposedly died from overwork, conducted one of the earliest surveys. With the help of her woman's club, Corson solicited letters from daughters of farm women on whether they would marry a farm man. In 1909 the *Ladies Home Journal* published excerpts from over 900 letters received by Corson, many of which indicted farm men for buying laborsaving technologies for their work in the field and barn before they would buy such things for woman's work in the house. The *Journal,* however, ignored the fact that most farm women had charge of the garden and poultry pen, and that many worked in the dairy and "helped out" in the field on a seasonal basis.[10] A series of five articles in *Harper's Bazaar* by Country Life reformers Martha Bensley Bruère and Robert Bruère, entitled "The Revolt of the Farmer's Wife," took up the "crusade" in late

1912. Addressing the issues of overwork, waylaid education, the "social significance of a bumper crop" in purchasing improved transportation and communication technologies, unhealthy sanitary conditions, and the debilitating effects of growing old on the farm, the Bruères helped to raise the problem of the farm woman to the level of such turn-of-the-century muckraking concerns as urban poverty.[11]

These exposés and pressures from many quarters of the Country Life movement helped to persuade Congress to pass the Smith-Lever Act in 1914, which established the present agricultural extension service in the United States Department of Agriculture (USDA). The act funded two sets of county agents, whose gendered division of labor reflected the USDA's adherence to the urban domestic ideal. Male agricultural agents taught farm men scientific farming, and female home demonstration agents taught farm women the methods of home economics. The USDA also established an Office of Home Economics in 1915 to provide research support for the extension work of the home agents.[12]

Government funding led to more extensive studies of farm women. In 1915 the USDA published a four-part report on replies to a questionnaire sent to the wives of its fifty-five thousand volunteer crop correspondents asking how the agency could help farm women. Most of the two thousand women who replied seemed dissatisfied with farm life. Extracts of letters describing the loneliness, isolation, lack of "modern" household technologies, and drudgery of farm life documented in great detail the complaints raised by earlier surveys. A study of the long, toilsome workday of fourteen hundred farm women by the New York State College of Agriculture at Cornell prompted the *Literary Digest* to title its 1919 article on the subject "Some Solid Reasons for a Strike of Farm-Wives." The original story in Cornell's *Extension Service News* reported that the monotonous character of the closely confined work of farm women "is believed to offer an explanation of the prevalence of insanity among rural women." (This common view, which dates to the 1870s, only began to be overturned with a survey of rural mental health in the 1920 Census and the publication of related studies in the 1920s.)[13]

In 1920, Florence Ward, head of extension work with women in the USDA, published the results of a survey of ten thousand farm women by home demonstration agents in thirty-three northern and western states (including the 1919 New York survey). Although Ward implicitly denied the insanity charge by saying that the average farm woman enjoyed better living conditions than her city sister, the survey found "five outstanding problems" of the farm woman: a long work day (over eleven hours), regular performance of heavy manual labor, low standards of beauty and comfort in the house, perilous health of the mother and child, and few income-producing home industries. To overcome these problems, Ward recommended that farm women adopt measures that had long been advocated by home economists: "improved home equipment," "more efficient methods of household management," and education in nutrition and child care. Ward also thought that farm women should have an awareness that cultivating the home was "perhaps the only means of stopping the drift of young people to the city." The lack of labor-saving technology uncovered by the survey underlined Ward's conclusions.[14]

What was behind the great interest in the farm woman? Historian Katherine Hempstead has argued that it was part of broader concerns about both the quantity and quality of rural-to-urban migration in this period. Although Ward and others feared that depopulation would weaken rural institutions, most agrarian reformers (including rural leaders) thought some outmigration was economically necessary and socially desirable because of the increased productivity caused by farm mechanization (beginning with the horse-drawn reaper in the Midwest). But many Country Life reformers (especially urban elites, probably a majority of the movement) worried that the farm men and women left behind would be of inferior quality (e.g., less enterprising and intelligent) and would, thus, create an inferior rural life and send inferior, surplus youth to the city (a fear compounded by the aforementioned reports of insanity among farm women). Many reformers also worried that immigrants from Southern and Southeastern Europe would migrate to the farm in the place of those who had left, thus creating a peasant (agri)culture in the United States. Thus, "to many, the Country Life problem was in no small part an eugenics problem, and the need for the 'best' people [i.e., native-born whites from Northern European stock] to stay on the farm and reproduce freely served more than strictly agricultural ends."[15]

Although Hempstead notes that only a few authors, such as Assistant Agricultural Secretary William Hayes in 1911, referred explicitly to the science of eugenics, many reformers shared Hayes's view that, under normal conditions, "folks are the best crop of the farm [and] the farm is the best place to raise folks." Farm journal editors, rural sociologists, farm women, and other commentators on rural society expressed this ideology in the early part of the century, using similar metaphors about "child crops" that drew on the long-standing agrarian myth of the primacy of rural life.[16] In *The Woman on the Farm* (1924), reformer Mary Meek Atkeson argued that urbanites should support the Country Life movement because the countryside continued to provide most of the nation's leaders and suitable (i.e.,

Figure 14.1. Louise Stanley, chief of the USDA's Bureau of
Home Economics, which helped coordinate the studies of
time spent on homework by rural and urban women.
(*The Home Economist,* cover, February 1928.)

Americanized) workers for its factories: "In payment for this interest shown by
the city people the farm will return not only its corn and hogs and cattle, but also
a steady stream of bright-eyed young people to carry the best American traditions
into every city in the land. As the farmer will tell you jokingly, his young folks are
indeed the 'best crop' of his farm." Leading home economists adopted a similar
rhetoric. Speaking at an agricultural conference in 1926, Louise Stanley, chief of the
USDA's Bureau of Home Economics (see fig. 14.1), justified her agency's efforts to

improve the farm home by quoting her former boss, the recently deceased Henry Cantwell Wallace, Secretary of Agriculture and son of Henry Wallace, an original member of the Country Life Commission. Stanley quoted from the younger Wallace's *Our Debt and Duty to the Farmer* (1925) that the "surplus of this crop [of children] makes the steady stream of fresh, virile, potent blood which flows into the cities and is such an important contribution to the vital human force of the nation."[17] Because healthy women and healthy rural institutions (like the home) were needed to (re)produce both a healthy surplus youth "crop" and the important institution of rural life, the allegedly overworked farm woman stood at the center of the Country Life problem in this period.[18]

While reformers agreed on the main problems of farm women and succeeded in popularizing the issue,[19] many rural leaders objected to the stark portrayals of farm life. In 1909, Henry Wallace, founder and editor of the midwestern *Wallaces' Farmer,* said the *Ladies Home Journal*'s allegations against farm men were slanderous and blamed farm women for preferring old-fashioned ways of doing housework to buying laborsaving technology. Clarence Poe, editor of the southern *Progressive Farmer,* thought the "descriptions of the farm woman frequently given in the magazines seem to us unrecognizable caricatures," but he also suggested how farm men could address the problems documented by the USDA's 1915 survey.[20]

More commonly, articulate farm women defended their way of life. "Mrs. T. R." wrote *Wallaces' Farmer* in 1909 to complain, "Some of our magazine writers are fairly waxing eloquent over the tragedy of the farmer's wife, portraying the farmer as a moneyed miser whose only ambition is to acquire more land, in the meantime sacrificing his wife's health and happiness to satisfy his ambition." Her response to this common view of rural patriarchy was to retort that "God created man first as head of the house, then he created women to manage him."[21] Adeline Goessling, the home editor of *Farm and Home,* asked her readers in 1920 what they thought about recent newspaper and magazine descriptions of the hardships of farm women. Nearly ten thousand women wrote in to deny the charges. Extracts from these letters, published in the magazine and in *Literary Digest,* extolled the clean air, healthy food, neighborliness, the satisfaction of honest hard work, and the urban-style technologies available in the country.[22] In 1922, farm women on a committee of a national agricultural conference "resented keenly the present fashion of magazines and newspapers to belittle the country woman, in stories representing her as having few home conveniences and, apparently, fewer brains."

The women conceded that many farm women were "overworked and inefficient," but these did not represent "normal or usual conditions" in their view. As late as 1925 one farm woman wrote the *Rural New Yorker* to complain about magazine writers who held up "that old rag-baby scarecrow of the preponderance of insanity among farm women. And this in the face of the fact that reliable statistics have proved it time and again to be utterly false."[23] Responding to this outpouring of protest from farm women, Emily Hoag, an economist for the USDA, interviewed farm women around the country and studied (mostly) upbeat letters solicited by *Farmer's Wife* in 1922. Although many women complained about the lack of laborsaving technology and overwork, Hoag's report de-emphasized the "tragedy" of the farm woman in favor of the virtues of rural life praised by this group of women.[24]

All of these commentators—irrespective of their position on the status of farm women—viewed household technology as a progressive social force, as a solution to the problems of farm women or as an indication that they were better off than the muckrakers suggested. To paraphrase the title of a recent book on technology and colonialism, machines were the measure of farm men and women, tangible evidence of how far they had progressed toward being "civilized."[25] Ironically, the 1920 U.S. Census showed that slightly higher percentages of farm households owned telephones and automobiles than did nonfarm households, but these statistics did not upset the prevalent view of the backwardness of rural life portrayed by Country Life reformers and in many forms of urban popular culture (especially in newspaper cartoons of "hayseeds" and "rubes"). "Modern" household technology, especially electricity, increasingly became a marker that divided the up-to-date city from the old-fashioned countryside.[26]

Although a progress ideology of technology prevailed in this period, a few commentators on rural life recognized some ironies of supposedly laborsaving household technology before it was documented by the time-use studies. In 1910, Ruth Stevens, a home economist and the first woman's-page editor for the *Progressive Farmer*, told her readers, "household appliances have been invented in many instances by men to sell and many are more than useless. . . . Any household appliance should save more time and labor than it takes to adjust it and clean it after use." Thomas Carver, an agricultural economist, wrote in 1914 that "it must be remembered that these laborsaving improvements [in the home] seldom reduce the amount of work. They merely enable people to accomplish more with the same effort." Mary Atkeson echoed this view in her 1924 book on the farm woman when she said that older women had expected that the sewing machine would "liberate them from toil." Yet styles changed accordingly and "women simply elaborated their garments until they were as much work as before. Such a tyrant does even the best labor-saving invention sometimes become!" A New York farm man replied in a 1926 survey of electric home-plants that electricity "only makes the work easier and less tiresome with the exception of the vacuum cleaner. There the work is cut nearly 1/2." [27] But if Cowan's thesis holds in the countryside, we should expect that it was the man's work and that of his sons in removing carpets from the house and beating them on a clothes line that was cut in half, not necessarily that of the farm woman.

The Time-Use Studies

This complex of beliefs — the progress ideology of technology, the urban domestic ideal, and the Country Life goal of "saving" the overworked farm woman — shaped the interpretations of researchers who began studying the time habits of farm women in the mid-1920s. Funded by the Purnell Act, passed by Congress in 1925 to support "economic and sociological investigations" to improve the "rural home and rural life," and performed at state agricultural colleges under the guidance of the USDA's Bureau of Home Economics, the studies were conducted by home economists imbued with these beliefs and the (often conflicting) goal of conducting "objective" social science research. Although home economics had focused on the fields of nutrition and "domestic science" since its founding in the late nineteenth century, only four home economics departments in fifty agricultural college experiment stations surveyed were conducting research when the Purnell Act went into effect. By late 1926, the Act was funding research in food and nutrition, clothing and textiles, and rural home management (including five time-use studies) in home economics departments at thirty-six state colleges. [28]

The work of home economist Ilena Bailey provided much of the background for the time-use surveys that began with the Purnell Act. A member of the USDA's Office of Farm Management, then the Office of Home Economics, Bailey collected a year's worth of records about home management from thirteen farm women in 1912 and investigated the conditions of and labor patterns in ninety-one farm homes in southern Michigan in 1917. Both studies substantiated a long work day for these farm women: an average nine and three-quarter hours to thirteen hours in the first study, and an average of eleven and one-half hours per day for the second one. Florence Ward's survey of ten thousand farm women revealed an average workday of eleven and one-third hours.[29]

Bailey resumed this work when she was transferred in mid-1923 to the new Bureau of Home Economics, a full-fledged research arm of the USDA, and became head of the Bureau's Division of Economic Studies. Time-use studies formed one of the group's four major tasks when Hildegarde Kneeland took over the Division in early 1924. A home economist who had done graduate work in nutrition, as well as in statistics, sociology, and economics at Columbia University and the University of Chicago, Kneeland took up Bailey's preparation of a clock chart of the farm woman's day and then helped coordinate an extensive series of investigations, "Present Use of Time by Homemakers," one of the national projects funded under the Purnell Act.[30] Conducted by home economists at agricultural experiment stations, these surveys sought to discover how much time farm women spent on "homemaking" (generally defined as housekeeping, household management, and child care), farm work (e.g., work in the garden, poultry pen, dairy, and field), leisure, and sleep by asking them to fill out the Bureau's standard forms or time charts, graded in five-minute intervals, at the end of each day for a "typical" week (see fig. 14.2). (In this article, I will use "housework" to mean the range of activities called "homemaking" in these studies.) The researchers contacted farm women through the Farm Bureau (a business-oriented farm organization), home-demonstration groups, and women's clubs. Often, they excluded non-English speaking women in order to obtain data that could be used by practicing home economists to help the "average" farm woman.[31]

This methodology, in addition to the taxing requirements of keeping a detailed daily log for a week, biased the samples toward well-educated, middle-and upper-class, native-born, white women (the female half of the traditional clientele of the USDA Extension Service).[32] The methodology also underscored the gendered division of labor sanctioned by the home economists' urban domestic ideal by dividing women's work into housework and farm work, a division that was probably artificial to most farm women.[33] One of the ways the time-use studies helped home economists promote this ideal on the farm was to create many more, and more finely distinguished, categories for analyzing housework than for analyzing farm work.

The Bureau of Home Economics faced a severe problem in its attempts to construct guidelines for this classification. Extensive correspondence between Knee-

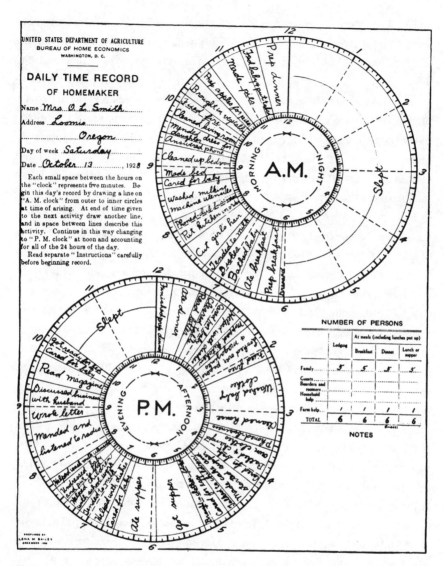

Figure 14.2. Sample Daily Time Record. (From Maud Wilson, "Use of Time by Oregon Farm Homemakers," Oregon State Agricultural College Agricultural Experiment Station, *Bulletin* No. 256, November 1929, p. 41.)

land and one experiment-station researcher, Inez Arnquist in Washington State, reveals that a good deal of negotiation was possible between extension workers and the Bureau in Washington, D.C. Kneeland initially did not require Arnquist to follow the Bureau's classification scheme for her report, probably because the Bureau was preparing its own summary of the records. But Kneeland sent Arnquist a preliminary version of its two-hundred page classification guide, which the Bureau did not complete until the spring of 1928, and Arnquist regularly wrote Kneeland's staff about where to place such troublesome time-chart entries as "scrub[bing] milk house" and taking "message to husband (farmer)" in the field. Kneeland's staff classified the former activity as "Farm Work" (under "Dairy Work"), as one would expect, but placed the latter task under "Homemaking" because being a messenger—even about farming operations—came under the category of "Care of Members of Household." Category decisions like this one simultaneously illustrate the Bureau's adherence to the urban domestic ideal and how artificial the home-farm classification of labor could be for women. Kneeland began to coordinate the studies more closely when she called the Purnell researchers together for a conference in mid-1928 and later informed Arnquist that the Bureau was entitled to some editorial control over her report before it was published. Arnquist acquiesced and told Kneeland the "time study has been difficult and I have never been less sure of myself on anything that I have attempted."[34]

Utilizing these and other methodologies, home economists conducted detailed time-use studies of about two thousand farm women. Researchers used BHE time-charts and guidelines to survey women in Idaho, Washington, Rhode Island, Oregon, South Dakota, and Montana from the mid-1920s to the early 1930s. Most researchers then switched to interview methods and studied the time habits of farm women in New York, Vermont, Wisconsin, and Indiana from the mid-1930s to the mid-1960s (see table 14.1). Town or urban women were surveyed in five of these studies and in four separate nonfarm investigations. A second study of Rhode Island investigated rural and urban women employed outside the home. An agricultural economist performed a time-use survey of over three hundred Nebraska farm homes in 1924 using his own methodology.[35]

In large measure the home economist researchers shared the Country Life goal of ameliorating the lot of farm women and thus conducted surveys that were similar in methodology and purpose to the prescriptive social-science surveys common to urban social work, rural sociology, and agricultural economics in this period (the Purnell Act also funded the latter two areas).[36] Trained in sociology,

TABLE 14.1. *Time-Use Studies by Home Economists of Farm Women*

Publication Date	Researcher	State	Methodology	Number of Records			Average Weekly Hours of Farm Women	
				Farm	Rural Nonfarm	Urban[a]	House-work	Farm Work
1927	Crawford	ID	BHE Charts	49	0	32	62.7	9.7
1929	Arnquist & Roberts	WA	BHE Charts	137	21	39	53.0	9.9
1929	Whittemore & Neil	RI	BHE Charts	102	0	0	54.1	4.5
1929	Wilson	OR	BHE Charts	288	71	154	51.4	11.3
1930	Wasson	SD	BHE Charts	100	0	0	53.8	11.6
1933	Richardson	MT	BHE Charts	91	0	0	53.7	9.2
1940	Warren	NY	Interviews	497	0	0	52.1	6.6
1946	Muse	VT	Interviews	183	0	0	56.5	12.0
1954	Wiegand	NY	Interviews	95	0	155	53.2	7.0
1956	Cowles & Dietz	WI	Time Charts	85	0	0	52.8	8.0
1968	Manning	IN	Interviews	41	17	53	55.4	

[a] Includes "town" women in some studies

Source: Adapted from Joann Vanek, "Keeping Busy: Time Spent on Housework, United States, 1920–1970" (Ph.D. diss., University of Michigan, 1973), Table 3.2, pp. 80–82. I have added the following studies, which Vanek lists in Table 2.1, pp. 53–55, but not in Table 3.2: Elizabeth Wiegand, "Use of Time by Full-time and Part-time Homemakers in Relation to Home Management," Cornell University Agricultural Experiment Station, *Memoir* No. 330, July 1954; and Sarah L. Manning, "Time Use in Household Tasks by Indiana Families," Purdue University Agricultural Experiment Station, *Research Bulletin* No. 837, January 1968.

Kneeland stressed the survey as a method of social reform. She and her colleagues wanted to establish an "objective" empirical research base which they could use to improve the working conditions of farm women by recommending how they could increase their "efficiency," reduce their workday, and spend more time on leisure. Yet the reformist ideal (based on a progressivist ideology of technology) often clashed with the objectivist goal of the social survey. Louise Stanley, head of the BHE, said in 1926 that "we need . . . to know what equipment is going to help the homemaker most. Time studies will show us this."[37] It came as something of a surprise, then, when many of the time studies showed that "laborsaving" household technologies—which home economists had been recommending for some time—did not live up to their billing.

Home economist Maud Wilson conducted one of the most perceptive studies on this score when she surveyed about six hundred farm, rural nonfarm, and town and urban households in Oregon from 1926 to 1927. Her survey showed that women in "well-equipped" houses (those which had both electricity and plumbing, representing 19 percent of the sample) spent an average of three hours and twenty minutes less per week on meal preparation, routine cleaning, and washing clothes than women in "poorly equipped" houses (those which had neither electricity nor plumbing, representing 57 percent of the sample). But the technologically advanced women spent an average of two hours and ten minutes more on ironing, sewing, child care, and care of the grounds. Both groups spent nearly the same amount of time on specific tasks, and the town women, who had a much higher percentage of "laborsaving" devices than farm women, spent almost exactly the same amount of time on housework as their country sisters (51.5 hours per week, versus 51.6).

Why? Wilson, who was working on her Ph.D. in home economics from the University of Chicago, gave four possible explanations: (1) "no time reduction is possible, and the equipment is of value because it makes the job more pleasant or because it reduces [the] energy requirement"; (2) "time habits tend to persist, with the result that the family living standard is raised"; (3) the "homemaker spends more time on the parts of the task which she most enjoys doing"; and (4) "time given by other members of the family or by hired help is reduced rather than her own time." Although Wilson thus articulated what Cowan would later call the "ironies of household technology" and said many people had "an exaggerated idea of the time reduction possible" from using this technology, she promoted it as a means to improve housework. The caption of a photograph showing a collection of household appliances in her report read "Good equipment means better housekeeping at the *same* time cost."[38] The caption expresses an interpretative shift made by many time-use researchers. When confronted by the intractability of apparently paradoxical data, they turned from advocating technology as a means to reduce labor time, to promoting it as a way to increase the quality and productivity of housework.

Before World War II at least four other researchers identified these ironies of household technology and offered somewhat different explanations for it than did Wilson. Inez Arnquist in Washington and Margaret Whittemore in Rhode Island elaborated on Wilson's second point by speculating that women allotted a specific amount of time to a task and thus either slowed down when an appliance allowed them to be ahead of schedule or raised their standard of living by doing more work in the same amount of time. Whittemore also advised the farm woman to "be careful to see that her [washing] machine does not cause her to work harder" as had happened when the sewing machine was first introduced.[39] Jessie Richardson of Montana published data in 1933 showing that slightly more time was spent on washing clothes and preparing food by women with electric or gasoline-powered washing machines, running water, and refrigerators (three minutes, twenty minutes, and forty-five minutes per week, respectively). Rather than criticizing the baneful effects of technology—a common practice in the debates over technological unemployment during the Great Depression—Richardson thought this equipment made household tasks more enjoyable and "thus induce[d] the worker to spend more time than formerly."[40] The data published by Jean Warren in New York in 1940—which she collected using a different methodology, the interview—showed a more complex picture. About five hundred farm women spent less time in individual tasks by using a washing machine, electric iron, and vacuum cleaner, but women with running water and those having both running water and electricity spent more time on housework than those having neither technology (7 percent more time in both cases). Warren explained the "paradoxical" data by noting that technology was not the only factor (others included family size and composition, number and size of rooms, kitchen arrangement, etc.). Like Wilson, she implicitly advocated the use of "modern" technology by arguing that women with running water and electricity did a better job of "homemaking" than those without these "conveniences."[41]

Three other time-use researchers in the 1920s and 1930s used the same methodology and time charts as these home economists but did not comment on a negative correlation between household technology and time spent on housework (even when some data seemed to indicate this phenomenon). A study in South Dakota did not present data to compare ownership of technology with time spent. One in Idaho attributed large amounts of time spent on housework to lack of "conveniences," but its data also showed that both women in the group who used the most amount of time had kitchen sinks. The third study, of married women in Rhode Island who worked outside the house, did not comment on data showing that women without running water tended to spend more time on housework (although this was not a linear relationship).[42]

In 1929 a survey of ten test farms in Illinois before and after they received electricity, conducted by agricultural engineer Emil Lehmann at the University of Illinois, even claimed that electricity reduced the amount of time women on five of these farms spent on laundry, house care, and leisure, but increased the time they spent on farm work. However, Lehmann's research was not comparable to the Bureau of Home Economics studies because he used an extremely small sample and excluded the increased time spent on meal preparation. More to the point, the Illinois State Electric Association, a utility group, financed the study, which Lehmann conducted during a sabbatical leave in which he worked for the Illinois Power and Light Company. Lehmann was so optimistic about the benefits of electricity that he even explained the decreased leisure time of the five farm women by saying that since electricity eliminated the drudgery of household tasks, they did not have to rest so much![43]

The Bureau of Home Economics had an ambivalent view of the time-use research. In a 1928 report on records received from seven hundred farm women in five states, Hildegarde Kneeland lamented the large amount of time still being spent on housework. "Are we, then, to conclude that the farm home has not been touched by the industrial revolution? . . . Such a conclusion would, of course, be unjustified. For though the working hours of the farm woman are still long, they were undoubtedly even longer 50 years ago."[44] But as Wilson, Arnquist, and Whittemore began to prepare reports for publication that explicitly questioned the time-saving capabilities of household technology, Kneeland softened her position. In a 1929 magazine article, she repeated her argument that women had worked much longer in the past, but added other reasons to explain the continued long hours, including the fact that "much of the gain which the Industrial Revolution has so far brought has gone into reducing the work of the household to a one-worker job."[45] When Lehmann wrote the Bureau of Home Economics in 1930, asking for estimates on time saved by electrical appliances for a pamphlet he was preparing for a utility group, Kneeland wrote a memo to her boss that she could not "give Mr. Lehmann the information which he wants, as our [summary] study shows a negative correlation between time expenditure and equipment, due undoubtedly to the fact that the housewife with better equipment raises her standards of housekeeping or slows up in her time expenditure."[46] However, neither this analysis nor that of the ironies of household technology by Wilson and others made its way into the Bureau's summary report of the prewar time studies.[47]

To summarize, by 1940 researchers were split about evenly on the issue of whether ownership of modern household technology correlated with reduced time spent on housework for farm women. Five studies showed that a direct correlation did not exist between technology, especially electrical appliances, and reduced time. The coordinator of the time-use studies at the Bureau of Home Economics was ambivalent, admitting, not in print, but in a private, internal memo that a negative correlation existed. On the other side of the issue, three studies assumed that technology was laborsaving without analyzing the correlation; and one study done with a very small sample and from the point of view of the electrical industry generated data showing that electricity reduced time spent on housework.

It is difficult to gauge what practical effect this research had. Home economists at the Rural Electrification Administration (REA), for example, either did not know about the prewar studies or ignored them. Neither the REA's *Rural Electrification News* nor its annual reports mention the studies.[48] The only applicable archival reference I have found shows that REA administrator Morris Cooke learned of the existence of the time-use surveys in 1936, shortly after the REA was established, when he asked the USDA if it had "any dependable statistics on the hours per week of farm women."[49] At least two electrification experts connected with the REA independently noted the irony of laborsaving devices. Yet they, like Lehmann, interpreted their results in a pro-electrification manner. In 1937, Clara Nale, chief home economist for the REA, wrote that "in using an electric range, the farm wife is greatly reducing the amount of work necessary for her husband in felling trees and sawing and splitting the logs, as well as eliminating work for herself." But Nale maintained that saving the work of farm women's helpers was a selling point for the range, not a detriment to her clients, because she thought the range freed both men and women for other farm work.[50] In 1939, George Kable, editor of a commercial rural electrification magazine, told an REA conference that he had put a pedometer on a Maryland farm woman before and after she had electricity. To his surprise, she took more steps after electrifying the house because she now had more time to work with her chickens, a task she preferred over housework. As a proponent of household electrification, Kable probably dropped what commitment he might have had to the urban domestic ideal in order to interpret the unintended result as showing the benefits of electricity.[51]

Home economists and rural sociologists not connected with electrification received the time-use surveys more warmly. Some agreed with the interpretation of Maud Wilson and her colleagues that laborsaving devices did not necessarily save time. In 1932 the Home Management Committee of President Herbert Hoover's Conference on Home Building and Homeownership included in its report a compilation of the Bureau of Home Economics studies of 1,500 farm and nonfarm households. However, unlike Kneeland, the committee publicly noted "little difference between the time spent by homemakers who have good and those who have poor equipment." Home economist Jean Dorsey, author of this section of the report, cited the work of Wilson, Whittemore, and Arnquist to back up her conclusion that household technology made work easier and enabled a higher standard of living but did not reduce the time spent on housework. Yet she also quoted results from Lehmann's study, in order to suggest that electricity may save time for future "homemakers."[52]

At least two rural sociologists also employed the time-use studies to discuss the irony of laborsaving devices. In *Rural Social Trends* (1933), John Kolb of the University of Wisconsin and Edmund S. de Brunner of Columbia College interpreted the studies to mean that "increased conveniences do not decrease the length of the working day so as to provide more time for reading, for participation in social organizations, or for leisure-time activity. Instead they seem merely to free time for other work or themselves lead to more work, as more laundry, keeping the house cleaner, and so on." Perhaps picking up on Maud Wilson's speculation that time habits tend to persist, Kolb and de Brunner explained the phenomenon on the basis of sociologist William Ogburn's theory of cultural lag: farm women's housekeeping habits lagged behind changes in technology. (The explanation may have been suggested by Ogburn, who was editor of the series in which the book was published and research director for *Recent Social Trends*, the famous study commissioned by Hoover, which published a shorter version of Kolb and de Brunner's monograph.) Kolb and de Brunner also cited a Wisconsin study showing what might be called "more work for father." Farm men "with tractors work longer days on the average than the farmers without tractors," an intriguing topic that needs more research.[53]

Although Kolb and de Brunner repeated their interpretation of the time-use studies in their influential college textbook, a *Study of Rural Sociology* (1935), also published in a series edited by Ogburn, they deleted it by the third edition in 1946. Jean Dorsey also did not refer to the irony of laborsaving devices in *Management in Family Living*, a popular college textbook that she coauthored with home economist Paulena Nickell in 1942. Dorsey analyzed at length the same time-use surveys she had discussed in 1932. But rather than repeating her statement that technology did not necessarily save time, she finessed the topic by stating that the "effect of various kinds of working equipment upon the homemaker's energy expenditures was shown to be very great in the discussion on the energy costs of homemaking activities." Readers would have to be fairly astute to interpret her statement to mean that household technology did not save time.[54] Another home management book, published by Ruth Bonde in 1944, interpreted the time-use data to mean that rural women spent more time on food preparation than urban women because they had fewer "conveniences."[55]

The postwar era, of course, saw a rapid "modernization" of the farm home. The electrification of farms, which had steadily increased under the REA's loan program that established electrical distribution cooperatives, proceeded apace during the prosperity and mass consumerism after the war. The percentage of electrified farms in the United States increased from 10 percent in 1935, to 33 percent in 1940, to 78 percent in 1950.[56] While researchers in the prewar period could easily compare farm houses with and without electricity in order to correlate time spent on housework with the ownership of electrical technology, these distinctions were becoming less noticeable after the war. Even though many prewar researchers had not noticed any time savings with electricity, their postwar colleagues expected that the average time spent on housework after the war would decline because of the larger number of "modern" farm homes. They were surprised, then, when their data showed that farm women still spent about fifty-two to fifty-four hours a week on housework, the same amount of time as before the war.

As in the prewar period, the time-use researchers responded in a variety of ways to data on the irony of laborsaving household technology. Marianne Muse of Vermont was surprised that she found no correlation between time spent on preparing meals and the availability of running water and that her data showed a negative correlation between time spent on sweeping and the ownership of hand sweepers and vacuum cleaners. But rather than use the prewar studies as a guide to investigate a general relationship between technology and time spent on housework, she explained the apparently paradoxical data on the basis of not being able to isolate one "cause" from another.[57] Elizabeth Wiegand, a student of Jean Warren's, followed up her mentor's work by investigating the same area of New York state eighteen years later in 1954. Although her mentor had de-emphasized the irony of household technology, Wiegand stated it clearly: "Modern equipment may make it possible for the homemaker to do better home-making work and save her some energy, but it does not necessarily save her time." She noted that some technologies saved time (running water and electric or gas stoves), while others did not (freezers and electric mixers). Comparing her results with Warren's, she noted that farm women seemed to have shifted their time from food preparation to care of clothes and marketing, a point Wilson had made in 1929.[58] In a 1956 study of Wisconsin farm women, May Cowles thought it "remarkable" that the time spent on housework had not changed since the mid-1920s, despite the many changes in household technology. Like Wiegand, she noticed a shift from the prewar to the postwar period in time spent on food preparation to that spent on purchasing, management, and family care.[59] But Sara Manning's 1968 survey of Indiana farm women only noted that the ownership of dishwashers, washing machines, and ironers did not correlate with less time spent on meal preparation. She did not analyze the phenomenon or note any shifting patterns of time usage.[60]

How were the postwar studies received? Nickell and Dorsey, in their popular textbook on home management, analyzed the studies at length in each new edition. By the third edition in 1963, they observed that time spent on housework had not changed since the 1920s. They also noted changes in the pattern of time use, such as decreased time spent on food preparation and increased time spent on care of clothes. They explained the latter phenomenon by saying that washing machines enabled women to wash more often. In all four editions, Nickell and Dorsey de-emphasized the irony of laborsaving devices, which Dorsey had clearly stated in 1932, by shifting the emphasis from the savings of the homeworker's time to the savings of her energy.[61]

This situation changed after sociologist Joann Vanek published her dissertation, "Keeping Busy," in 1973. Before Vanek, every researcher had held household technology as an independent variable. Although the home economists differed about how they identified and interpreted data which (usually) showed no correlation between this variable and time spent on housework, they thought it worthwhile to include it in their studies. After Vanek showed in an exhaustive manner that time spent on housework had remained remarkably constant from the mid-1920s to the mid-1960s, despite the obvious technological changes that had occurred, at least one major study dropped technology as an independent variable. In 1976, Kathryn Walker, third in a line of time-use researchers at Cornell, noted, "Contrary to the opinion of many, average time used by wives for household work has not been drastically reduced because of technological developments in automatic equipment such as dishwashers, washers, and garbage disposers." Like Maud Wilson in 1929, Walker noted that women did more work with modern technology and shifted their work patterns toward consumption and child care.[62]

Understanding the ideologies of the home economics researchers helps us explain their remarkably similar interpretations of time-use data from the 1920s to the 1960s. The complementary (and long-standing) ideology of technology as a progressive social force, the urban domestic ideal, and the Country Life goal of "saving" the overworked farm woman shaped the manner in which these supposedly "objective" social-science studies were structured, conducted, and analyzed. Home economists, whom I have called "agents of modernity" elsewhere,[63] interpreted their research to reflect the benefits of household technology, which they saw as a positive cultural power. When confronted with paradoxical data indicating that laborsaving devices did not save time, all of the researchers—to varying degrees—shifted the stated purpose of their studies from finding ways to give farm women more leisure, to promoting modern technology as a means to increase the quality and productivity of "homemaking." For many, this meant spending more time in "higher" pursuits like home management and child care. Home economists did not drop their original goal of giving farm women more leisure, though, because researchers after World War II were still surprised to find that the time spent on housework had not changed in the era of electrical home appliances.

What changed, of course, was the climate of opinion about the social implications of technology. With the widespread criticism of technology and the rise of the women's movement in the late 1960s and 1970s, scholars outside of home economics—in sociology and history—developed a critical attitude toward technology. This ideology seems to have encouraged them to emphasize the ironies of household technology and to argue that technology had not created less work for farm (or urban) women—a phenomenon that home economists had documented repeatedly and analyzed (to some extent) since the 1920s. To a large degree, the home economists' data and observations supporting this point were masked by a positive ideology of technology and the modernizing virtues of the technologies they promoted.[64]

The Time-Use Studies and Farm Women

How does this analysis help us use the time studies to investigate the work habits of American farm women? My focus on how the studies were constructed should not blind us to the fact that they contain an immense amount of data pertaining to the social history of technology in rural life. In fact, the researchers' beliefs about technology, domesticity, Country Life, and "objective" social science motivated them to collect reams of data about a major topic of interest to the history of technology: the relationship between technological change and social change. The detailed records of nearly 2,000 women, taken over a period of forty years, provide a wealth of statistical information on technology and time spent on housework, farm work, and leisure. The uncannily close fit between the number of hours worked in the house per week by rural and urban women has illuminated general relationships between technology and housework patterns for the middle-class and upper-class women whom Cowan and others have studied. More research is needed on the suggestion that these data represent a cultural norm of fifty-two to fifty-four hours for the work week of full-time houseworkers for this period, just as forty hours became the cultural norm for full-time office and factory workers by mid-century.[65]

While the researchers' beliefs thus helped produce data of use to the history of technology, the same ideologies have limited the applicability of that data. The time-use studies, after all, only illuminate the stratum of one economic and social class of farm women, most of whom seem to have accepted the urban domestic ideal to a great extent (probably through the agencies of advertising and home economics clubs). The "typical week" of these women represents the ideal desired by home economists. Researchers did not want records showing that farm women spent about as much (or more) time working in the garden, poultry pen, dairy, and field than in the house during some weeks, and thus did not collect them. Although Wilson, Arnquist, and Wasson gave figures for work done during different seasons of the year, they treated as "typical" those weeks in which seasonal farm work did not overwhelm housework. As we have seen, the majority of rural home economists were more interested in studying (and promoting) their view of women's work in the house, not in the barn or field.[66]

In spite of this methodology, the time-use studies contain a great deal of information about the farm work of this class of women. Many researchers break down their data into the number of hours spent in the garden, caring for poultry, milking cows, and tending livestock. Although the number of hours is much less than that devoted to housework, it is far from being insignificant (see table 14.1). Ranging from 4.5 hours per week in Rhode Island to 11.6 hours per week in South Dakota before World War II, the figures did not decline dramatically with the increased consumerism following the war. The constant level of farm work for this class of full-time home workers thus complements historian Katherine Jellison's argument that many *other* postwar farm women (i.e., those who had not adopted the urban domestic ideal to such a large extent) used the time savings possible with some technologies not to do more housework in the same amount of time, but to work off the farm, or to work more in the field, especially driving tractors and running to town for spare parts for farm machinery.[67]

In regard to technological determinism, the time-use studies reinforce the conclusions drawn by recent historians and sociologists that household technologies and other so-called urbanizing inventions did not transform everyday life in a wholesale manner. Instead, urban and rural people wove new technologies like the telephone and the automobile into existing social patterns to a large extent.[68]

In this case, the time-use surveys show that a large number of middle- and upper-class farm women employed some "laborsaving" household technologies to do more work during the culturally defined, standard work week in the house. Indeed, social norms seem to have played as large a role as the availability of "modern" technology in determining what constituted the length of a "typical" work week for these women.

The time-use studies thus have a double valence. A contextual analysis of these reports provides some insight into the complex set of ideologies held by the home economics researchers and how their beliefs informed the construction and interpretations of their data. The importance of ideology is underscored in this story by the fact that an ideology more critical of technology emphasized data showing the ironies of modern household appliances that had been masked by a progressive ideology of technology. On the other side of the coin, understanding the system of beliefs of these researchers enables us to use their studies selectively to investigate the relationship between technology and social change for this class of farm women.

Missing from my story are the ideologies of the farm women themselves. Research into how these women helped construct the time-use studies and how their beliefs and work practices were, in turn, shaped by helping to generate and collect this data would help close the analytical circle. It would allow us to study more fully the mutual construction of ideology, artifacts, and household work patterns in the twentieth century.[69]

NOTES

An early version of this article was presented at the annual meeting of the Society for the History of Technology in October 1993. The author thanks members of the audience at that session, Margaret Rossiter, Carolyn Goldstein, and the *Technology and Culture* referees for their comments, and Carolyn Goldstein for assistance in using the Archives of the Bureau of Home Economics of the Department of Agriculture, as well. Research for this article was supported by National Science Foundation Grant Number SBR-9321180.

1. For recent work on these issues, see Merritt Roe Smith, "Technological Determinism and American Culture," Michael L. Smith, "Recourse of Empire: Landscapes of Progress in Technological America," and Leo Marx, "The Idea of 'Technology' and Postmodern Pessimism," in *Does Technology Drive History? The Dilemma of Technological Determinism,* ed. Merritt Roe Smith and Leo Marx (Cambridge, Mass., 1994); Howard Segal, *Future Imperfect: The Mixed Blessings of Technology in America* (Amherst, Mass., 1994); and John Jordan,

Machine Age Ideology: Social Engineering and American Liberalism, 1911–1939 (Chapel Hill, N.C., 1994). I use "ideology" in the nonpejorative sense of cultural codes as advocated by Clifford Geertz in *The Interpretation of Cultures* (New York, 1973), chap. 8. For a similar usage in the history of technology, see Eric Schatzberg, "Ideology and Technical Choice: The Decline of the Wooden Airplane in the United States, 1920–1945," *Technology and Culture* 35 (1994): 34–69.

2. On the progressive ideology of electricity, see James W. Carey and John J. Quirk, "The Mythos of the Electronic Revolution," *American Scholar* 39 (1970): 395–424; Thomas P. Hughes, "The Industrial Revolution That Never Came," *American Heritage of Invention and Technology* 3 (Winter 1988): 58–64; and Ronald Kline, *Steinmetz: Engineer and Socialist* (Baltimore, 1992), pp. 252–61. For expressions of this ideology in regard to the rural home, see *Rural Electrification News,* published from 1935 onward by the U.S. Rural Electrification Administration.

3. Joann Vanek, "Keeping Busy: Time Spent in Housework, United States, 1920–1970" (Ph.D. diss., University of Michigan, 1973); "Time Spent in Housework," *Scientific American* 231 (November 1974): 116–20; "Household Technology and Social Status: Rising Living Standards and Status and Residence Differences in Housework," *Technology and Culture* 19 (1978): 361–75; and "Work, Leisure, and Family Roles: Farm Households in the United States, 1920–1955," *Journal of Family History* 5 (1980): 422–31.

4. Ruth Schwartz Cowan, "The 'Industrial Revolution' in the Home: Household Technology and Social Change in the 20th Century," *Technology and Culture* 17 (1976): 1–23, and *More Work for Mother: The Ironies of Household Technology from the Open Hearth to the Microwave* (New York, 1983).

5. Judith A. McGaw, "No Passive Victims, No Separate Spheres: A Feminist's Perspective on Technology's History," in *In Context: History and History of Technology: Essays in Honor of Melvin Kranzberg,* ed. Stephen H. Cutcliffe and Robert Post (Bethlehem, Pa., 1989), pp. 172–91; Ellen Lupton, *Mechanical Brides: Women and Machines from Home to Office* (Princeton, N.J., 1993), pp. 15–27. For citations of the thesis, see, for example, Susan Strasser, *Never Done: A History of American Housework* (New York, 1982), p. 251; Christine E. Bose, Philip L. Bereano, and Mary Malloy, "Household Technology and the Social Construction of Housework," *Technology and Culture* 25 (1984): 53–82; Glenna Matthews, *"Just a Housewife": The Rise and Fall of Domesticity in America* (New York, 1987), pp. 111–12, 245; David E. Nye, *Electrifying America: Social Meanings of a New Technology, 1880–1940* (Cambridge, Mass., 1990), p. 258; Judy Wajcman, *Feminism Confronts Technology* (University Park, Pa., 1991), chap. 4; and Katherine Jellison, *Entitled to Power: Farm Women and Technology, 1913–1963* (Chapel Hill, N.C., 1993), pp. xx, 53–54.

6. My analysis draws upon recent work in the sociology of science and technology, which stresses the "interpretative flexibility" of scientific theories and technological artifacts. See Harry Collins, *Changing Order: Replication and Induction in Scientific Practice* (London, 1985); and Trevor J. Pinch and Wiebe E. Bijker, "The Social Construction of Facts and Artifacts: Or How the Sociology of Science and the Sociology of Technology Might Benefit Each Other," in *The Social Construction of Technological Systems: New Directions in the Sociology and History of Technology,* ed. Bijker, Thomas P. Hughes, and Pinch (Cam-

bridge, Mass., 1987), pp. 17–50. In this article I address the issue of how stable interpretations of this data were constructed (the issue of "closure mechanisms," to use the sociologist's terminology) in terms of the ideologies of the researchers.

7. On the large number of social science sources for rural history and general remarks about their interpretation, see Hal S. Baron, "Rural Social Surveys," *Agricultural History* 58 (1984): 113–17, and Harold T. Pinkett, "Government Research Concerning Problems of Rural Society," ibid., pp. 365–72.

8. Alan I. Marcus, *Agricultural Science and the Quest for Legitimacy: Farmers, Agricultural Colleges, and Experiment Stations, 1870–1890* (Ames, Iowa, 1985), pp. 7–12; John Mack Faragher, *Sugar Creek: Life on the Illinois Prairie* (New Haven, 1986), pp. 232–33.

9. *Report of the Commission on Country Life* (1909; reprint, New York, 1917), pp. 103, 104. On the Country Life Commission and movement, see Clayton S. Ellsworth, "Theodore Roosevelt's Country Life Commission," *Agricultural History* 34 (1960): 155–72, and William L. Bowers, *The Country Life Movement in America, 1900–1920* (Port Washington, N.Y., 1974). A later study shows that the Commission largely ignored the data it collected. Only 1 percent of the respondents to the Commission's inquiries were farm women, and the "*majority* of the total sample and of the farmers were *unqualifiedly* satisfied" with the condition of farm homes, sanitary conditions on farms, and communication services. See Olaf F. Larson and Thomas B. Jones, "The Unpublished Data from Roosevelt's Commission on Country Life," *Agricultural History* 50 (1976): 583–99, on pp. 588, 597 (italics theirs).

10. "Is This the Trouble with the Farmer's Wife," *Ladies Home Journal,* February 1909, p. 5. On the work patterns of farm women in the twentieth century, see Deborah Fink, *Agrarian Women: Wives and Mothers in Rural Nebraska, 1880–1940* (Chapel Hill, N.C., 1992); Jellison (n. 5 above); Mary Neth, *Preserving the Family Farm: Women, Community, and the Foundations of Agribusiness in the Midwest, 1900–1940* (Baltimore, 1995); and Jane Adams, *The Transformation of Rural Life, Southern Illinois, 1890–1990* (Chapel Hill, N.C., 1995).

11. Martha Bensley Bruère and Robert Bruère, "The Revolt of the Farmer's Wife!" *Harper's Bazaar,* November 1912, pp. 539, 550, 580; December 1912, pp. 601–2, 621; January 1913, pp. 15–16, 37; February 1913, pp. 67–68, 92; March 1913, pp. 115–16; and "After the Revolt," *Harper's Bazaar,* May 1913, pp. 235, 248. I would like to thank Kathleen Babbitt for drawing my attention to this series.

12. David B. Danbom, *The Resisted Revolution: Urban America and the Industrialization of Agriculture, 1900–1930* (Ames, Iowa, 1979); Katherine Hempstead, "Agricultural Change and the Rural Problem: Farm Women and the Country Life Movement" (Ph.D. diss., University of Pennsylvania, 1992), chaps. 5–6; Jellison, chap. 1.

13. USDA, Office of the Secretary, *Social and Labor Needs of Farm Women, Domestic Needs of Farm Women, Educational Needs of Farm Women, and Economic Needs of Farm Women* (Washington, D. C., 1915, reports nos. 103–106); *Literary Digest,* Dec. 20, 1919, pp. 74, 78; and New York State College of Agriculture, *Extension Service News* 6 (1919): 77–78, on p. 78. On research in rural-urban mental health, see Pitirim Sorokin and Carle Zimmerman, *Principles of Rural-Urban Sociology* (New York, 1929), pp. 264–73; and Sorokin, Zimmerman, and Charles Galpin, eds., *A Systematic Source Book in Rural Sociology* (Minneapolis, 1932), 3: 236–50.

14. Florence Ward, "The Farm Woman's Problems," *Journal of Home Economics* 12 (1920): 437–57, on pp. 449–50. On Ward, see Gladys L. Baker, "Women in the U.S. Department of Agriculture," *Agriculture History* 50 (1976): 190–201.

15. Hempstead, pp. 101–21, on p. 116. On the urban base of the Country Life movement, see Bowers (n. 8 above).

16. Hempstead, p. 201 (quotation). See, for example, Henry Wallace, "The Socialization of Country Life," *Wallaces' Farmer*, January 7, 1910, p. 2; USDA, *Domestic Needs of Farm Women*, p. 66; Hempstead, p. 201 (citing Martha Crow in 1915); Charles Galpin, *Rural Life* (New York, 1918), chaps. 5–6; and "Why Young Women Are Leaving Our Farms," *Literary Digest*, October 2, 1920, pp. 56–57. On the agrarian myth, see Richard Hofstadter, *The Age of Reform* (New York, 1955), chaps. 1–3, and David Danbom, "Romantic Agrarianism in Twentieth-Century America," *Agricultural History* 65 (1991): 1–12.

17. Mary Meek Atkeson, *The Woman on the Farm* (New York, 1924), pp. 21–24, on p. 24; Louise Stanley, "The Development of Better Farm Homes," *Agricultural Engineering* 7 (1926): 129–30, on p. 130.

18. Another concern was that, in the face of a rising urban population, too much out-migration might lead to food shortages and rising food prices. Martha and Robert Bruère asked in 1912, "Shall the nation go hungry because the farmers' wives don't like their jobs? For, after all, a man will not live on the farm without a wife." See "The Revolt of the Farmer's Wife: The War on Drudgery," *Harper's Bazaar*, November 1912, pp. 539, 550, 580, on p. 539.

19. See, for example, *New York Times*, May 30, 1915, V14–15; and *Literary Digest*, October 2, 1920, p. 56.

20. "The Trouble with the Farmer's Wife," *Wallace's Farmer*, February 19, 1909, p. 262; and Clarence Poe, "Is the Farm Woman Getting a Square Deal? *Progressive Farmer*, April 17, 1915, p. 381. Poe's letter to the Secretary of Agriculture had prompted the USDA survey of the crop correspondents.

21. Mrs. T. R., "In Defense of the Farmer's Wife," *Wallace's Farmer*, September 17, 1909, p. 1164. Deborah Fink tends to support this view of patriarchy, while Nancy Osterud, Mary Neth, and Katherine Jellison emphasize mutuality within a flexible patriarchal system. See Fink (n. 10 above); Osterud, *Bonds of Community: The Lives of Farm Women in Nineteenth-Century New York* (Ithaca, N.Y., 1991); Mary Neth, "Gender and the Family Labor System: Defining Work in the Rural Midwest," *Journal of Social History* 27 (1994): 563–77, and *Preserving the Family Farm* (n. 10 above), chap. 1; and Jellison (n. 5 above), pp. 181–86.

22. "Farm Women Who Count Themselves Blest by Fate," *Literary Digest*, November 13, 1920, pp. 52, 55.

23. Atkeson (n. 17 above), p. 297; Frances Gilbert Ingersoll, "A Farm Wife's Protest," *Rural New Yorker*, March 14, 1929, p. 482.

24. Jellison, pp. 27–30. Neth argues that Hoag's portrayal of mutuality is more representative of the actual status of farm women than the crop-correspondent survey. See Neth, *Preserving the Family Farm*, p. 237.

25. Michael Adas, *Machines as the Measure of Men: Science, Technology, and Ideologies of Western Dominance* (Ithaca, N.Y., 1989).

26. See Claude S. Fischer, *America Calling: A Social History of the Telephone to 1940*

(Berkeley, 1992), pp. 93, 102 [census statistics]; James H. Shideler, "Flappers and Philosophers and Farmers: Rural-Urban Tensions of the Twenties," *Agricultural History* 47 (1973): 283–99; Don S. Kirschner, *City and Country: Rural Responses to Urbanization in the 1920s* (Westport, Conn., 1970); and Nye (n. 5 above), chap. 7.

27. See *Progressive Farmer,* February 12, 1910, pp. 130–31; T. N. Carver, "The Organization of a Rural Community," in Department of Agriculture, *Yearbook 1914* (Washington, D.C., 1915), pp. 135–36; Atkeson, p. 304; and Frank Sincebaugh's survey form, in R. F. Buchanan, "An Economic Study of Farm Electrification in New York State," field records for Delco plants, 1927, College of Home Economics Papers, Cornell University Archives, microfilm reel 143.

28. Sybil B. Smith, "Development of Home Economics Research at the Agricultural Experiment Stations under the Purnell Act," Department of Agriculture, *Report on the Agriculture Experiment Stations, 1926* (Washington, D.C., 1927), pp. 89–96; H. C. Knoblauch et al., *State Agricultural Experiment Stations: A History of Research Policy and Procedure,* Department of Agriculture, Miscellaneous Publication No. 904, May 1962, pp. 222–23 (text of Purnell Act); Joel P. Kunze, "The Purnell Act and Agricultural Economics," *Agricultural History* 62 (1988): 131–49. On the history of home economics, see Barbara Ehrenreich and Deirde English, *For Her Own Good: 150 Years of the Experts' Advice to Women* (New York, 1978); Matthews (n. 5 above), chap. 6; Margaret W. Rossiter, *Women Scientists in America: Struggles and Strategies to 1940* (Baltimore, 1982), pp. 200–201, 258–59; and Carolyn Goldstein, "Mediating Consumption: Home Economics and American Consumers, 1900–1940" (Ph.D. diss., University of Delaware, 1994), chap. 1.

29. Ilena Bailey, "A Study of the Management of the Farm Home," *Journal of Home Economics* 7 (1915): 348–53, and "A Survey of the Farm Home," *Journal of Home Economics* 13 (1921): 346–56; and Ward (n. 14 above), p. 440.

30. Hildegarde Kneeland, "Report of the Work of Division of Economic Studies for Year Ending June 30, 1924," attached to memo from Kneeland to Louise Stanley, July 24 (?), 1924, Bureau of Home Economics Archives, National Archives, Washington, D.C. (hereafter cited as BHE Archives), RG 176, entry 2, box 550; Goldstein, chap. 2; "Research in Home Economics at the State Agricultural Experiment Stations," *Journal of Home Economics* 19 (1927): 154–57; Paul Betters, *The Bureau of Home Economics: Its History, Activities and Organization* (Washington, D.C., 1930), pp. 56–64. The BHE also furnished time charts to home economists who were not participating in the Purnell-funded studies. See, for example, Eloise Davison to Ilena Bailey, October 21, 1924, BHE Archives, entry 6, box 611; and Louise Stanley to Martha Van Rensselaer, October 26, 1925, BHE Archives, entry 1, box 8.

31. See, for example, Jean Warren, "Use of Time in Its Relation to Home Management," Cornell University Agricultural Experiment Station, *Bulletin* No. 734, June 1940, p. 6. For an example of home agents working with ethnic women, see Joan M. Jensen, "Crossing Ethnic Barriers in the Southwest: Women's Agricultural Extension Education, 1914–1940," *Agricultural History* 60 (1986): 169–81.

32. Vanek, "Keeping Busy" (n. 3 above), pp. 59–60, briefly addresses this point. One researcher noted that only 33 percent of those contacted through women's clubs "submitted usable reports. The remaining 67 percent comprised not only those who lacked interest or

initiative but also those who felt that their circumstances were such that they could not hope for a 'typical' week for some time to come. It also included many who were working under such physical or mental strain that they shrank from adding to their burdens by contracting to keep the record. Summer reports were especially difficult to obtain." See Maud Wilson, "Use of Time by Oregon Farm Homemakers," Oregon State Agricultural College Agricultural Experiment Station, *Bulletin* No. 256, Nov. 1929, p. 11. On the bias of the USDA toward serving the upper classes, see Neth, *Preserving the Family Farm* (n. 10 above), chap. 4. For an argument that home-demonstration agents served more than the upper two-thirds of farm families, see Gladys Gallup, "The Effectiveness of the Home Demonstration Program of the Cooperative Extension Service of the United States Department of Agriculture in Reaching Rural People and in Meeting Their Needs" (Ph.D. diss., George Washington University, 1943), pp. 29–30.

33. Jellison (n. 5 above), chap. 1; Neth, *Preserving the Family Farm,* chaps. 1 and 8; Cynthia Sturges, "'How're You Gonna Keep 'Em Down on the Farm?': Rural Women and the Urban Model in Utah," *Agricultural History* 60 (1986): 182–99; Dorothy Schwieder, "Education and Change in the Lives of Iowa Farm Women, 1900–1940," *Agricultural History* 60 (1986): 200–215; Kathleen Babbitt, "The Productive Farm Woman and the Extension Home Economist in New York State, 1920–1940," *Agricultural History* 67 (1993): 83–101; Jane Adams, "Resistance to 'Modernity': Southern Illinois Farm Women and the Cult of Domesticity," *American Ethnologist* 20 (1993): 89–113; and, for a later period, Sara Elbert, "Amber Waves of Grain: Women's Work in New York Farm Families," in *"To Toil the Livelong Day": America's Women at Work, 1780–1980,* ed. Carol Groneman and Mary Beth Norton (Ithaca, N.Y., 1987), pp. 250–68.

34. Inez Arnquist to Hildegarde Kneeland, February 23, 1927 (quotation); Arnquist to Laura Brossard, March 31, 1927 (quotation); Brossard to Arnquist, April 9, 1927; Kneeland to Arnquist, May 22, 1928; Kneeland to Arnquist, June 27, 1929; Arnquist to Kneeland, July 5, 1929 (quotation); and other correspondence between Kneeland, Brossard, and Arnquist, from September 25, 1926, to July 13, 1929, BHE Archives, entry 6, box 607. A copy of the classification, "Study of the Use of Time by Homemakers," March 1928, is in BHE Archives, entry 8, box 641.

35. Vanek, "Keeping Busy" (n. 3 above), pp. 53–55; Blanche M. Kuschke, "Allocation of Time by Employed Married Women in Rhode Island," Rhode Island State College Agricultural Experiment Station, *Bulletin* No. 267, July 1938; J. O. Rankin, "The Use of Time in Farm Homes," University of Nebraska College of Agriculture Experiment Station, *Bulletin* No. 230, December 1928. Vanek lists four town and urban studies, and two compilations of the farm studies.

36. See Lowry Nelson, *Rural Sociology: Its Origins and Growth in the United States* (Minneapolis, 1969); Harry C. McDean, "Professionalism in the Rural Social Sciences, 1896–1919," *Agricultural History* 58 (1984): 373–92; Stephen P. Turner and Jonathan H. Turner, *The Impossible Science: An Institutional Analysis of American Sociology* (Newbury Park, Calif., 1990), pp. 24–28, 41–57; Martin Bulmer, Kevin Bates, and Kathyrn Kish Sklar, eds., *The Social Survey in Historical Perspective, 1880–1940* (New York, 1991); Goldstein (n. 28 above), pp. 110–11.

37. Goldstein, pp. 21, 22, 141; Stanley (n. 17 above), p. 130.

38. Wilson (n. 32 above), pp. 37, 39, 46 (italics hers).

39. Inez F. Arnquist and Evelyn H. Roberts, "The Present Use of Work Time of Farm Homemakers," State College of Washington Agricultural Experiment Station, *Bulletin* No. 234, July 1929, pp. 26–27; Margaret Whittemore and Bernice Neil, "Time Factors in the Business of Homemaking in Rural Rhode Island," Rhode Island State College Agricultural Experiment Station, *Bulletin* No. 221, September 1929, pp. 19–20.

40. Jessie E. Richardson, "The Use of Time by Rural Homemakers in Montana," Montana State College Agricultural Experiment Station, *Bulletin* No. 271, Feb. 1933, p. 23.

41. Warren (n. 31 above), pp. 42 (quotation), 49, 74, 77–79.

42. Ina Z. Crawford, "The Use of Time by Farm Women," University of Idaho Agricultural Experiment Station, *Bulletin* No. 146, January 1927; Grace E. Wasson, "The Use of Time by South Dakota Farm Homemakers," South Dakota State College Agricultural Experiment Station, *Bulletin* No. 247, March 1930; Kuschke (n. 35 above).

43. E. W. Lehmann and F. C. Kingsley, "Electric Power for the Farm," University of Illinois, Agricultural Experiment Station *Bulletin* No. 332, June 1929, pp. 375, 401–3. On Lehmann's employment by Illinois utility companies, see Lehmann to Rachel Mason, August 21, 1928, and Max Hoagland to Lehmann, December 11, 1928, E. W. Lehmann Papers, University of Illinois Archives, Urbana, Ill. (hereafter cited as Lehmann Papers), box 5.

44. Hildegarde Kneeland, "Women on Farms Average 63 Hours Work Weekly in Survey of 700 Homes," Department of Agriculture, *Yearbook of Agriculture, 1928* (Washington, D.C., 1928), pp. 620–22, on p. 621. Although Kneeland does not identify the researchers, the areas of the country she mentions and the dates of the work indicates that her data included that gathered by Wilson and Arnquist, with Crawford and Warren providing partial data on their ongoing projects.

45. Hildegarde Kneeland, "Is the Modern Housewife a Lady of Leisure," *Survey Graphic* 62 (1929): 301–2, on p. 302.

46. Hildegarde Kneeland to Louise Stanley, July 17, 1930, BHE Archives, entry 2, box 550. On Lehmann's project, see Eloise Davison to Lehmann, July 8, 1930; Lehmann to Vera Meacham, July 16, 1930; and Lehmann to E. A. White, September 17, 1930, Lehmann Papers, box 5.

47. Department of Agriculture, Bureau of Nutrition and Home Economics, "The Time Costs of Homemakers—A Study of 1,500 Rural and Urban Households," 1944, mimeographed copy in National Agricultural Library, Washington, D.C. Kneeland also ignored the question of time saved by household technology in other articles that she published on the time studies. See Kneeland, "Women's Economic Contribution in the Home," *Annals of the American Academy* 143 (1929): 33–40; Kneeland, "Leisure of Home Makers Studied for Light on Standards of Living," Department of Agriculture, *Yearbook of Agriculture, 1932* (Washington, D.C., 1932), pp. 562–64; and Kneeland, "Home-making in This Modern Age," *Journal of the American Association of University Women* 27 (1934): 75–79.

48. In fact, one home economist with the Colorado Extension Service said in an article for the REA that "Let us suppose that electricity in the home can reduce the labor load to the eight-hour day for five and one-half days a week making a forty-four hour week

which has been generally established in business." See Exine Davenport, "Electricity's Part in America's First Industry," *Rural Electrification News,* July 1940, 13–14, on p. 13.

49. Morris Cooke to Paul Appleby, January 24, 1936; and Appleby to Cooke, January 31, 1936, BHE Archives, entry 3, box 579.

50. Clara Nale, "Electric Aids for Farm Wife Make Her Self-Sufficient, *Rural Electrification News,* December 1937, pp. 9–10, on p. 9. On Nale, see Ronald Kline, "Agents of Modernity: Home Economists and Rural Electrification in the United States, 1925–1950," in *Rethinking Women and Home Economics in the 20th Century,* ed. Sara Stage and Virginia Vincenti (Ithaca, N.Y., 1997).

51. George W. Kable, "Lines and Loads, or Farms and Folks," *REA Annual Administrative Conference* (Washington, D.C., Jan. 9–13, 1939), pp. 623–43, on pp. 632–33, booklet in Morris L. Cooke Papers, Franklin Delano Roosevelt Library, Hyde Park, N.Y., box 139.

52. John M. Gries and James Ford, eds., *Household Management and Kitchens,* vol. 9 of the President's Conference on Home Building and Home Ownership (Washington, D.C., 1932), pp. 26–32, on p. 30. Effie I. Raitt was chair of the Household Management Committee. The figure of 2,500 families in this report (p. 27) is probably a misprint of 1,500, the usual number given in later references to the BHE compilation.

53. John H. Kolb and Edmund de S. Brunner, *Rural Social Trends* (New York, 1933), pp. 65–66, on p. 66. For similar comments regarding farm men, see Rankin (n. 35 above), p. 7; and Neth, *Preserving the Family Farm* (n. 10 above), p. 227. On Ogburn and *Recent Social Trends,* see Jordan (n. 1 above), pp. 179–84.

54. Paulena Nickell and Jean M. Dorsey, *Management in Family Living* (New York, 1942), pp. 56–65, 86–93, on p. 91.

55. Ruth L. Bonde, *Management in Daily Living* (New York, 1944), pp. 81–86, on p. 83.

56. Bureau of the Census, *The Statistical History of the United States from Colonial Times to the Present,* rev. ed. (Stamford, Conn., 1965), 1: 510; Nye (n. 5 above), chap. 7.

57. Marianne Muse, "Time Expenditures on Homemaking Activities in 183 Vermont Farm Homes," University of Vermont Agricultural Experiment Station, *Bulletin* No. 530, June 1946, pp. 43–44, 49, 52–53, 59–60.

58. Elizabeth Wiegand, "Use of Time by Full-time and Part-time Homemakers in Relation to Home Management," Cornell University Agricultural Experiment Station, *Memoir* No. 330, July 1954, pp. 39 (quotation), 42–43.

59. May L. Cowles and Ruth P. Dietz, "Time Spent in Homemaking Activity by a Selected Group of Wisconsin Farm Homemakers," *Journal of Home Economics* 48 (January 1956): 29–35.

60. Sarah L. Manning, "Time Use in Household Tasks by Indiana Families," Purdue University Agricultural Experiment Station, *Research Bulletin* No. 837, January 1968.

61. Nickell and Dorsey (n. 54 above), 2d ed., 1950, pp. 111–24; 3d ed., 1963, pp. 103–12; 4th ed., 1967, pp. 127–32. Another popular book on home management just cites the BHE studies without much discussion; see Lillian M. Gilbreth, Orpha M. Thomas, and Eleanor Clymer, *Management in the Home* (New York, 1954), pp. 28–29; 2d ed., 1959, pp. 28–29.

62. Kathryn E. Walker and Margaret E. Woods, *Time Use: A Measure of Household Production of Family Goods and Services* (Washington, D.C., 1976), p. 32.

63. Kline, "Agents of Modernity" (n. 50 above).

64. This tension between reform and research for research home economists mirrors that between sales and education for home economists working for private companies and the REA. See Goldstein (n. 28 above), and Kline, "Agents of Modernity."

65. Vanek, "Keeping Busy" (n. 3 above), p. 88, briefly discusses this point.

66. At least one study by home economists did investigate more thoroughly the non-household work of women in gardening, taking care of poultry, milking, and farming in the field. See M. Ruth Clark and Greta Gray, "The Routine and Seasonal Work of Nebraska Farm Women," University of Nebraska College of Agriculture Experiment Station, *Bulletin* No. 238, January 1930.

67. Jellison (n. 5 above), chap. 6. See also Elbert (n. 33 above).

68. See, for example, Fischer (n. 26 above); Michele Martin, *"Hello Central?": Gender, Technology, and Culture in the Formation of Telephone Systems* (Montreal, 1991); and Ronald Kline and Trevor Pinch, "Users as Agents of Technological Change: The Social Construction of the Automobile in the Rural United States," *Technology and Culture* 37, no. 4 (1996).

69. I address the issue of the mutual construction of ideology, artifacts, systems, and cultural practices in *Consumers in the Countryside: Technology and Social Change in Rural America* (Baltimore, 2000).

The Shoulders We Stand On/The View from Here: Historiography and Directions for Research

NINA E. LERMAN, ARWEN P. MOHUN, AND RUTH OLDENZIEL

Our approaches to "gender and technology" are unabashedly inter-disciplinary. Scholars studying technology and scholars explicating gender systems have provided us with a versatile set of tools for thinking about interactions of gender and technology in historical context. The articles collected in this book draw on the work of scholars within the community of the history of technology and on that of gender theorists, gender historians, and feminist sociologists of science and technology. In this essay we lay out a brief historiography, a map of the work we rely on.[1] We gather it as an invitation to others to join and extend a complex conversation.

Interdisciplinary exchange is a scholarly strategy more often prescribed than practiced. Its benefits are easy to recognize: fresh directions for research and new techniques for understanding familiar topics. Achieving them can be more difficult, however, requiring struggles with different standards for evidence and different uses of language. We caution that moving from prescription to practice requires tolerance: this discussion stands at the confluence of several streams of scholarship, each of which has strengths and weaknesses. For example, feminist sociologists of technology have been very productive in analyzing the relationship between gender and technology. These studies, most of them ethnographies of contemporary innovations like computers and genetic engineering, tend to pay careful attention to rapid and recent patterns of technological change, but they can be far less sophisticated in understanding slower-paced transformations in gender ideologies over time. Conversely, researchers studying gender in historical context tend to leave technological issues unexamined, neatly relegated to the "black box" that scholars of technology have been at pains to open. Historians of technology, meanwhile, have been sluggish in their attention to masculinity as a crucial cultural dimension of the material they have traditionally studied and reluctant to challenge familiar taxonomies. In addition to making this conversation possible, a tolerant attitude will reduce the amount of energy spent on reinventing wheels: the existing scholarship is rich, if diverse, and fragmented more now by academic divisions than by neglect of subject matter. Because of this diversity, the historiographic discussion below is followed by a topical treatment of more recent work.

Early Work on Women, Gender, and Technology

To discuss early work in women, gender, and technology is to discuss work in the history of technology, in women's history, and in the philosophy of science. The interdisciplinarity of this conversation has a long history, or rather several long histories, which converge at important points.

One of the great strengths of the Society for the History of Technology (SHOT), and of the journal *Technology and Culture,* has been a willingness to make room for a persistent group of explorers charting the integration of both women and gender analyses into a stubbornly masculine subject area. In the late seventies and early eighties, several scholars within the community attempted to introduce historians of technology to the new field of women's studies and feminist critiques of science and technology, scholarship concerned most centrally with the historical invisibility of women and the power attached to domains traditionally inhabited by men. Ruth Schwartz Cowan began this conversation in 1976 with her classic essay on the industrial revolution in the home. Instead of accounting for the supposed absence of women in the technological realm, she showed that women's traditional sphere constituted a significant site of technological activity and that women were important technological actors rather than passive recipients of finished products. Her arguments were to have enduring significance.[2]

Soon thereafter, Martha Moore Trescott edited the first feminist collection of historical research on technology, *Dynamos and Virgins Revisited,* published in 1979. Trescott grouped the essays primarily by locale, in keeping with the "separate spheres" ideology women's historians were examining at the time. The essays in the first section dealt with "women as active participants in technological change" in industry, invention, and science; those in the second section treated "some of the effects of technology on women in the more private sphere of life," primarily the home and reproduction.[3]

Three years later Judy McGaw took the news of these and other research efforts back to the women's studies community in an essay appearing in *Signs: Journal of Women in Culture and Society*.[4] Although she addressed women's studies audiences, her essay served historians of technology equally well. It broached important historiographical questions and offered a detailed discussion of the existing literature with extensive footnotes. Our discussion here will focus on works published since her essay appeared in 1982.

If Cowan's arguments had recast women's traditional activities as a significant force in technological change, McGaw's essay convincingly argued for the industrial workplace, as well as the household, as a site where women's interaction with technology should be studied. This call for research built on a growing body of literature on the sexual division of labor and skill.[5] It was of clear relevance to a history of technology community that was already discovering labor history through the work of David Noble, Merritt Roe Smith, Hugh Aitken, and others.[6] McGaw's essay also explained to historians of women's work that discussions about skill and deskilling were by definition discussions of technology. And she insisted that the then-underexplored field of consumption, in addition to production, would require more careful examination from historians of women and of technology.

The same period produced another milestone when political scientist Joan Rothschild gathered within the pages of *Machina Ex Dea* the work of an interdisciplinary group of women scholars. Largely programmatic statements supported less by empirical studies than by theoretical insights, these rich if sometimes uneven essays introduced themes that would be central to the research agenda for the next ten years. They also issued calls for action and challenged what Rothschild called "a built-in resistance to the subject of women and technology" within the SHOT community.[7] This omission of women as actors in technological change was soon documented in more detail by John Staudenmaier.[8]

A product of their time, many of the essays in *Machina Ex Dea* went beyond historical analysis, offering a critique of the status quo and suggesting that women's values might come to be embodied in the design of dominant technologies.[9] But for historians, Rothschild made clear that it would not do simply to insert a few women into the historical record: "the very approaches to technology studies will have to undergo serious rethinking if work on women and technology is to be fully included and its impact felt in the entire field of study."[10] Autumn Stanley explained what that impact might look like: it would change the understanding of technology from what men do to what people do and would alter our ideas about what constituted significant technology.[11]

In 1985 another important collection, this time edited by sociologists Judy Wajcman and Donald MacKenzie, offered a sampling of work in both the history of technology and sociology of technology.[12] Among other strengths, the volume gave sustained attention to gender in the construction of technology, in its introduction and in the selections presented. For example, the work of sociologist Cynthia Cockburn introduced historians of technology to the ideas of exploring masculinity in traditional technological subjects and of treating access to know-how as a source of social and political power. For sociologists, Ruth Schwartz Cowan's articles on the industrial revolution in the home and the development of the refrigerator demonstrated that the home, too, must be considered a site of technological activity. The work of Cowan and others situated technological change in historical context. Camouflaged as an undergraduate reader, the book provided another opportunity for the convergence of multiple streams of scholarship.

These developments in various fields close to the SHOT community paralleled an enormous avalanche of new developments in women's history in those years, as scholars began studying the full range of women's historical experience. Within the broad community of women's studies, historians made it clear that older essentialist interpretations were inadequate. Women's gender roles were not fixed but always changing; there was no innate, eternal woman's culture; the meanings of masculinity and femininity had been continually renegotiated through time and place and in relation to each other. Most of this early scholarship focused on middle class women in nineteenth-century America, for whom prescription dictated concern with a "domestic sphere" separate from a "public" one. Historians of women soon expanded their analyses to encompass other regions, nations, and time periods. Sustained attention to the lives of women of different social categories further deepened and complicated these understandings.[13]

By the mid-1980s, Joan Scott, a historian of French social and labor history, could draw on this wealth of women's history and new cultural studies approaches. Her writings represented not only a culmination of the many historical studies that retrieved women's stories from the past but also a new era in which scholars pushed the lessons of these studies to a higher analytical level by recasting them using the concept of gender. Scott announced a new analytical beginning. Many historians encountered the concept of gender first through Scott's work, finding her approach particularly useful because she incorporated historical contingency into her theoretical apparatus.[14]

Scott's essay "Gender: A Useful Category of Historical Analysis" made clear that gender theory could be not only a useful but also a central approach to understanding important historical themes. It laid out some of the layered ways in which she understood gender systems to operate. First, gender is part of human identity, the ways people see themselves and their lives as men and women. Second, gender plays structural roles in societies, defining barriers and boundaries, embedded in all manner of institutions, such as education and labor markets. Finally, gender ideologies show up in symbolic forms: images and ideals of manhood and womanhood, gendered metaphors, masculine and feminine connotations of objects and activities. We have found these categories useful in understanding technology's history.

Finally, historians of technology also learned about gender theory from feminist philosophers of science. Their explanations of the nature of scientific knowledge, fragments of which appeared in early statements like Rothschild's *Machina Ex Dea*, have provided insights about gender and the power of knowledge and provoked new questions in approaches to technology. Unfortunately, the strength of their contributions to science studies has in some ways proved a hindrance to explorations of gender and technology in general and historical research in particular: focusing on science, they have tended to privilege it in a way that ironically parallels its privileged stature in modern Western society; exploring primarily recent developments, they have encouraged the presentist conflation of science and technology (sometimes now referred to as "technoscience"). Nonetheless, by reexamining and carefully situating claims to expertise and definitions of knowledge, a range of authors including Evelyn Fox Keller, Sandra Harding, Helen Longino, and Donna Haraway continue to provide tools for rethinking approaches to technological knowledge and the layered meanings of technological activities.[15] Their approaches can be particularly helpful to historians of technology grappling with the relation of engineering knowledge to scientific knowledge.[16]

Current Work and Opportunities for Research

The range of current research in and opportunities for further study of gender and technology are challengingly diverse. In this section we describe some topics that have received scholarly attention and point to aspects ripe for rethinking, borrowing, or continued exploration. This survey could be organized in a variety of ways; to some degree it defies organization. In the spirit of interdisciplinary conversation, we have eschewed organization by discipline, choosing instead a topical approach grounded in the work at hand. This scholarship emphasizes, variously, technological actors, particular technologies, and sites of technological activity. In a loose fashion, we follow suit.

In surveying the literature of the past decade or so, some general trends emerge. In women's and gender studies, theoretical approaches and research strategies reflect the shift from exploring women to exploring gender described above. Earlier work sought to restore women to the historical narrative, focusing on women in male-dominated domains or acknowledging women's undervalued activities. In these studies, the question of whether technological changes had been a boon or a bane to women was often central. More recent work focuses on gender relationships, on both men and women, and asks questions about cultural practices and social systems.[17] In technology studies, scholars have in general continued to develop their emphasis on contextualization and on technological networks and systems. Both historians and sociologists have devoted increasing attention to social factors shaping technologies and to the "mutual shaping" of technology and society. In technology as in gender studies, recent work focuses on questions about cultural and social practices.

Scholarship on gender and technology begins to integrate these insights. So far, this scholarship tends to challenge traditional conceptions either in gender studies or in technology studies, while uncritically accepting familiar traditions in the other. For example, beginning with traditional technological subjects (presumed masculine), some scholars have explored particular tools, machines, or systems and their designers and inventors, seeking to restore women as contributors to technological change. Such explorations have then led to analysis of gender in relation to the subject at hand, whether invention or artifact. Here categories tend to be organized by particular technologies, such as cars or stoves, and incorporate gender by crossing boundaries from production to consumption. The importance of the artifact remains unchallenged. Alternatively, based on their understanding of the separate masculine and feminine spheres of paid workplace and domestic household, other scholars have documented the presence of technology in the household and of women in the industrial workplace. These discussions, too, have moved to gender systems and relationships, for example by exploring gender divisions of labor in a particular locus. The separation of private and public spheres is left intact. The challenge of full integration remains.

Technological Actors

New trends have had an impact even in the most entrenched and traditional areas in the history of technology. Older approaches, combined with popular connotations of "technology," have tended to make the very male world of invention and engineering look "normal," and thus even more exclusively male than is actually the case. Scholars seeking to restore women as contributors to technological change have relied on two primary strategies. Beginning in early efforts, such as those appearing in *Dynamos and Virgins Revisited,* and continuing in the larger projects they spawned, scholars have explored the individual struggles of women inventors, patent holders, and engineers to make a place for themselves in what has been defined as a traditionally male realm.[18] Examining broader social trends, historian of science Margaret Rossiter and sociologist Sally Hacker, among others, have tracked and documented the patterns of struggles and strategies of female engineers and scientists. They argue that women in the technical professions should not be seen as anomalies but should be considered an essential part of the process of (male) professionalization.[19] Other scholars, following the lead of Ruth Schwartz Cowan, reframed the household as a site of meaningful technological activity, and women as technological actors wherever they interact with technology. In combination, these scholars have demonstrated effectively that despite social barriers and stereotypical assumptions women have been absent only from the history of technology as written, not from its history as experienced.

As supposedly "natural" roles have been challenged, men and their interactions with technology have also become subjects of gender analysis. In the early eighties, British sociologist Cynthia Cockburn's pathbreaking work repeatedly demonstrated the importance of exploring gender in the male workplace to understanding technological change. Labor historians looking at gendered patterns of skill and automation followed close behind.[20] Within the SHOT community, Judy McGaw's 1989 essay "No Passive Victims, No Separate Spheres" challenged historians of technology to engage this emerging literature and include masculinity in their work, a challenge taken up by Carroll Pursell at the 1991 SHOT annual meeting in Madison, Wisconsin. In his address to a plenary session at that conference, and in a subsequent article, Pursell sought to encourage historians of technology to use analysis of masculinity as a tool for further understanding the engineers and tech-nologies they knew well.[21] Historians of technology sit on a treasure trove of detail about masculinity, most of it continuing in the age-old tradition of treating the male as norm and "normal," and thus genderless. For example, Robert Post's fine examination of the world of automotive hot rodding never explicitly discusses the pervasive masculinity of this milieu.[22] Similarly, "shop culture" and "engineering culture" can be viewed as varieties of masculinity, as a wide array of scholars have pointed out.[23] Sustained attention to masculinity and its relation to femininity can illuminate the role of technologies in the construction of gender, as well as the modern definition of technology as a male pursuit and the power relations at stake in such a definition.

Broadening the definition of technological actor has made the concept a more flexible one, as not only inventors and designers but also factory workers and users can now be understood as agents in processes of technological change. Some scholars suggest that the concept of "actor" can encompass even nonhuman com-ponents of technological and social networks, broadening the definition still fur-ther.[24] For many historians, granting apparent agency to oceans and microbes, like considering pilot and airplane as a single "organism," runs counter to the exploration of human experience they see themselves undertaking. Nonetheless, the perspectives provided by considering all the people who interact with tech-nology to be "technological actors" make social relations more visible and have transformed approaches to both artifacts and sites of technological activity.

An Emphasis on Artifacts

Most studies of technology have focused on a particular technology: refrig-erators, steam engines, sewing machines. The increasing recognition of questions not only about what men and women do but also about how masculinity and femininity can be used symbolically has begun to inform examinations of mod-ern technologies from automobiles to nuclear weapons.[25] In relation to both cars and airplanes, for example, historians have argued that women connoted safety and comfort even as they also drove racing cars and flew as barnstormers.[26] More importantly perhaps, recent studies have been able to show that cultural practices are never external, but always part and parcel of technological development: if the projected use of the bicycle for women prompted the successful development of the safety bicycle that was practical and safe, the failure of the electric car and the success of the combustion engine can also be traced to their gendered associations.[27] These case studies show that gender matters a great deal in the trajectory of technological change.

In addition, social scientists pursuing the "social shaping" or "social construction" of technology have come to emphasize the connectedness of all phases of technological development as relevant to questions of technological change. Studies focusing exclusively on invention or use, for example, are by definition limited and limiting. Integrating examination of design, manufacture, marketing, purchase, and use, on the other hand, allows a range of social and cultural factors, including gender, to become more apparent. Technological change is not a one-way street flowing from design to use; rather, design and use (and everything in between) mutually shape each other. Cockburn and Ormrod's *Gender and Technology in the Making*, a contemporary ethnographic study, encompasses all these stages. Following microwave ovens from factory to kitchen, the authors explored people's interactions with the artifact all along the way. Their work begs for historically situated companions.[28]

Several categories of artifacts merit further discussion, and further research. Studies of media and imaging technologies—what might be called "technologies of representation," such as photography, film, television, as well as medical diagnostic procedures such as ultrasound—mark a notable trend. These case studies benefit from more sophisticated understandings of symbolic and ideological aspects of gender systems as well as from an attention to the integration of production and consumption.[29] Because images of women have been so pivotal in modern gender systems, this burgeoning field suggests ways in which technologies and gender ideologies are closely intertwined. It promises riches for the historian in the years to come.

Particular reproductive technologies, too—as Ruth Schwartz Cowan pointed out twenty years ago—deserve attention by historians of technology. While scholars in women's studies and women's history have produced a vast literature on reproductive and medical technologies, the pages of *Technology and Culture* have rarely reflected the extent of this research.[30] Early work focused on the question of whether new technologies would liberate women or provide tools of male intrusion and control of women's bodies.[31] Although political issues remain central, current scholarship analyzes the complicated ways in which gender ideologies and reproductive technologies are shaping each other.[32] Recent historical work, too, has moved away from dichotomous boon-or-bane questions, as scholars seek to explore the technologies of childbirth, abortion, and contraception available to and chosen by women of the past, technologies often deliberately camouflaged in the written record.[33]

Tight connections between technologies and gender roles are also evident in a range of what we might call "technologies of identity," such as clothing, cosmetics, and other means of personal adornment. An increasingly sophisticated literature on fashion from corsets to suits and work on the cosmetics industry points to a range of subjects worth further examination with attention to technological change.[34] In addition to this external redecorating, technological reshaping and redefining of the human body is also possible: studies of cosmetic surgery, genetic engineering, and sex change show that technological change can also take place below the surface of the skin, shaping gender identity in ways that have not yet reached the pages of *Technology and Culture*.[35]

As these discussions suggest, scholars have come to regard the body—like the household or the factory—as a site of technological activity. Pacemakers, radiation, and hormone therapy allow machines to be part of bodies, and bodies to be regulated like machines. Often discussed under the rubric "cybernetic organisms" or "cyborgs," this blurring of boundaries between human and machine has profound implications for gender ideologies: for example, because the body has conventionally been a site and symbol of women's nature, these technological manipulations can raise crucial questions about traditional roles.[36] A careful historical perspective, however, is often missing from these discussions; from eighteenth-century automata to mechanistic views of human physiology, the nature of the interface between bodies and machines has a long and complex history. A prehistory of cyborgs, particularly one that takes gender into account, remains to be written.[37]

Separate Spheres, Separate Sites

More traditionally, discussions of sites of technological activity have been based in a conception of gender defined by separate male and female "spheres": "workplace" and "household." Linked as related aspects of "industrialization" by Ruth Schwartz Cowan twenty years ago, workplace and household have more often been considered as separate academic spheres, the former explored by labor historians and the latter by historians of women. This split has been particularly evident among North American scholars. Such distinctions are rooted in historiographical traditions that lie beyond the scope of this essay.

Attention to the experiences of women and to gender has deeply affected the field of labor history. Over the last fifteen years, historians examining women's work and the gendered division of labor have questioned the overriding impor- tance of class as a structural category; have broadened their purview to include workplaces such as department stores and offices; and have argued that the com- munity as well as the shop floor must be studied to fully understand worker's experiences. Studies analyzing patterns of sexual division of labor, gendered con- struction of skill, and mechanization and deskilling have perhaps been of greatest immediate interest to historians of technology.[38] Despite the extent of this litera- ture, however, these approaches rarely focus on the workplace as a technological site where gender and technology interact.[39] Treating workers, as well as managers, as technological actors profoundly influenced by gender ideologies can help ex- plain attitudes toward safety, risk, new machines, and other workers. New research must also recognize that technology is meaningfully present even in workplaces that do not manufacture objects out of metal. We must also explore how gendered ideas about technology get translated between the workplace, the community, and the home.

The factory, of course, is not the only industrial workplace: the household may have industrialized differently but is unquestionably a site of work and of tech- nological change, as noted above.[40] Beginning in the early eighties, scholars en- deavored to document the work and technological activity of the household. They established that domestic tasks from laundry to childcare involve effort, skill, and sophisticated understandings of the material world, and have increasingly moved beyond the initial question of whether household technologies saved or increased labor in the home.[41] Scholars have explored differences between rural and urban women and begun to compare domestic work across national boundaries.[42] Some have turned attention to technologies of childrearing and childhood, and to the ways children's socialization and education replicates or redefines relationships between gender and technology, although these areas remain underexplored.[43] Re- cent work addresses the complex and reciprocal relationships between industrial home, industrial workplace, and industrial community. These studies seek not only to document the presence of technology, of work, or of women, but also to explain their chronic invisibility.[44]

Indeed, as Judy McGaw argued in 1982, running a household in a consumer society is "real work"—and consumption deserves the careful attention we are used to giving production in the history of technology.[45] Since then, historians from a variety of fields have created a substantial literature on consumption, on which historians of technology can draw. The earliest of these studies focused primarily on marketing and advertising—the "production" side of consumption—rather than on the "work" of consumption in the sense that McGaw meant.[46] More recent studies have focused on consumers as active agents and on the role of gender in shaping consumer-producer relationships.[47] Historians have only begun to examine technology, consumption, and gender together. More emphasis is needed on the role of technological change in the reciprocal relationships of patterns of consumption and the nature of what is being consumed, on mediators between consumers and producers, and on consumers as active users and modifiers—what some have called "coproducers"—of technology. For instance, scholars have only begun to look at how technology and culture interact in consumer's decisions to repair broken or malfunctioning technology rather than replace it.[48] By moving away from a production- and business-oriented focus on the artifact alone, these more integrative approaches also move away from the old connotation of technology as inherently masculine.[49] They begin to break down old taxonomies and make room for new understandings of the interplay between gender and technology.

Exploring the vast terrain mapped out here means crossing old boundaries. This means recognizing the importance of examining both sides of familiar dichotomies. Pairs such as masculinity and femininity, production and consumption, home and work—or even technological and social—must now be studied as reciprocal categories, dependent on and defined in relation to each other. In this view, balancing studies of production with separate studies of consumption, or studies of factories with studies of households, may be remedial, but it is not the final outcome of sustained attention to the questions raised in this volume.

It is as impossible to understand gender without technology as to understand technology without gender. As the work gathered here suggests, future research must attend to masculinity as well as femininity, instead of assuming male normality and female exceptionalism. And it must attend to women's technological activity and technological power, recognizing women as important makers and doers of things whether in the kitchen or in the factory. But it is also crucial to recognize that gender is not the only social boundary used to render technological activity invisible, nor is it the only one bolstered by technological associations. If we pay attention to social ideologies and power in our studies of technological change, we can begin to understand technological ideologies as well. Then we can address questions of why some technologies acquire status while others become invisible, and explore the mutual shaping of social categories and technology as systems for the distribution of power — systems which sometimes work in surprising ways. This integration across old boundaries emphasizes the contraction in *Connection* "technology and culture," the importance of context in understanding technology, and the importance of technology in understanding human society.

NOTES

Our views of the material presented here have been shaped in myriad discussions with a substantial thought collective, many of whom appear in the notes that follow. The original version of this essay benefited from readings and comments by Paul Edwards, Gabrielle Hecht, Christian Gelzer, Carolyn Goldstein, Helen Longino, Steven Lubar, Robert Post, Erik Rau, and John Staudenmaier. We dedicate this essay in thanks to the many foreparents cited here, in gratitude for their support, their example, and their scholarship. The present version leaves the text essentially unchanged; the notes have been updated.

1. Most importantly, we recognize the modern Western bias of this discussion, a bias that reflects both the North American focus of the articles that follow and the Western focus of the scholarship treating gender and technology explicitly. Francesca Bray's important *Technology and Gender: Fabrics of Power in Late Imperial China* (Berkeley, 1997) marks a major correction to this exclusive focus. Although they do not explicitly address questions of gender, we are indebted to Michael Adas and Bryan Pfaffenberger for clear arguments about the meanings attached to technology and the ways those meanings construct power relationships. See Michael Adas, *Machines as the Measure of Men: Science, Technology, and Ideologies of Western Dominance* (Ithaca, N.Y., 1989), and Bryan Pfaffenberger, "Technological Dramas," *Science, Technology and Human Values* 17 (1992): 282–312. Further, the literature discussed here explores primarily dominant white gender systems, often in a fashion similar to the normalized male bias it seeks to correct. For an introduction to ideas about race and technology, see Venus Green, *Race on the Line: Gender, Labor, and Technology in the Bell System, 1880–1980* (Durham, N.C., 2001); Bruce Sinclair, ed., *Technology and the African-American Experience* (Cambridge, Mass., forthcoming); and articles by Rebecca Herzig, Roger Horowitz, and Nina Lerman in this volume.

2. Ruth Schwartz Cowan, "The 'Industrial Revolution' in the Home: Household Technology and Social Change in the 20th Century," *Technology and Culture* 17 (1976): 1–24, and "From Virginia Dare to Virginia Slims: Women and Technology in American Life," *Technology and Culture* 20 (1979): 51–63. In fact, Cowan brought technology to women's historians as early as the Berkshire Conference of 1973; see her brief essay "A Case Study of Technological and Social Change: The Washing Machine and the Working Wife," in *Clio's Consciousness Raised*, eds. Mary Hartman and Lois Banner (New York, 1976).

3. Martha Moore Trescott, ed., *Dynamos and Virgins Revisited: Women and Technological Change in History* (Metuchen, N.J., 1979), quotes pp. 2–3. On women's changing roles, see the essays by Judith A. McGaw, Susan J. Kleinberg, and Susan Levine; on women as

active participants, see the essays by Deborah Warner, Margaret Rossiter, and Trescott in the section titled "Women Inventors, Engineers, Scientists, and Entrepreneurs." Cowan's essay "The Industrial Revolution in the Home" was reprinted here as well. Trescott's title is a play on Lynn White, *Dynamo and Virgin Reconsidered: Essays in the Dynamism of Western Culture* (Cambridge, Mass., 1971); the original hardcover edition of White's book appeared as *Machina ex Deo* (Cambridge, Mass., 1968), and the title of that edition similarly inspired the title of Joan Rothschild, ed., *Machina ex Dea: Feminist Perspectives on Technology* (New York, 1983).

4. Judith A. McGaw, "Women and the History of American Technology," *Signs* 7 (1982): 798–828.

5. Beginning with Harry Braverman's seminal *Labor and Monopoly Capital: The Degradation of Work in the Twentieth Century* (New York, 1974), the function of mechanization in relations between employers and workers has been widely discussed. For critiques of Braverman, see Paul Thompson, *The Nature of Work: An Introduction to Debates on the Labour Process* (London, 1983); Heidi Hartmann, "Capitalism, Patriarchy, and Job Segregation by Sex," *Signs* 1 (1976): 137–69, and "The Unhappy Marriage of Marxism and Feminism: Towards a More Progressive Union," *Capital and Class* 8 (1979): 1–33; Anne Phillips and Barbara Taylor, "Sex and Skill: Notes Towards a Feminist Economics," *Feminist Review* 6 (1980): 79–88.

6. David F. Noble, *Forces of Production: A Social History of Industrial Automation* (New York, 1984), and "Social Choice in Machine Design: The Case of Automatically-Controlled Machine Tools," in *Case-Studies on the Labour Process,* ed. Andrew Zimbalist (New York, 1979); Merritt Roe Smith, *Harpers Ferry Armory and the New Technology: The Challenge of Change* (Ithaca, N.Y., 1977); Hugh G. J. Aitken, *Scientific Management in Action: Taylorism at Watertown Arsenal, 1908–1915* (Cambridge, Mass., 1960).

7. Rothschild, *Machina ex Dea,* (n. 3 above), quote p. xvii. See also Joan Rothschild, ed., *Women, Technology and Innovation* (New York, 1981), which appeared as a special issue of *Women's Studies International Quarterly* 4 (1981). These years produced several other feminist tracts; see, for example, Jan Zimmerman, ed., *Technological Woman: Interfacing with Tomorrow* (New York, 1983), and *Once Upon the Future: A Woman's Guide to Tomorrow's Technology* (London, 1986); and Wendy Faulkner and Erik Arnold, eds., *Smothered by Invention: Technology in Women's Lives* (London, 1985).

8. John M. Staudenmaier, "What SHOT Hath Wrought and What SHOT Hath Not: Reflections on Twenty-Five Years of the History of Technology," *Technology and Culture* 25 (1984): 707–30, and *Technology's Storytellers: Reweaving the Human Fabric* (Cambridge, Mass., 1985).

9. Arguments based on the premise that women are fundamentally different and therefore employ different values in choices of technological and scientific development are most often referred to as essentialist feminism and/or ecofeminism. Oft-cited examples include Evelyn Fox Keller, *A Feeling for the Organism: The Life and Work of Barbara McClintock* (San Francisco, 1983), and Carolyn Merchant's influential *The Death of Nature: Women, Ecology, and the Scientific Revolution* (San Francisco, 1980), a summary of which appeared in Rothschild, *Machina ex Dea,* as "Mining the Earth's Womb, Dea Ex Machina." *Machina ex Dea*

prominently featured this vein of scholarship; see, for example, Ynestra King, "Toward an Ecological Feminism and a Feminist Ecology," pp. 118–29. For more recent reiterations of this point of view, see Knut H. Sorensen, "Towards a Feminized Technology? Gendered Values in the Construction of Technology," *Social Studies of Science* 22 (1992): 5–32, and Rosalind Williams, "The Political and Feminist Dimensions of Technological Determinism," in *Does Technology Drive History? The Dilemma of Technological Determinism,* eds. Merritt Roe Smith and Leo Marx (Cambridge, Mass., 1994). See also related ideas in Janine Morgall's *Technology Assessment: A Feminist Perspective* (Philadelphia, 1991), which argues for integrating women's interests in the early design stages of developing technologies, and in Sherry Turkle, *The Second Self: Computers and the Human Spirit* (New York, 1984). For a different variety of ecofeminism, see also Stacy Alaimo, "Cyborg and Ecofeminist Interventions: Challenges for an Environmental Feminism," *Feminist Studies* 20 (1994): 133–52, and the work of Donna Haraway (n. 15 below). Judith McGaw discusses the idea of "feminine technologies" in her essay in this book.

10. Rothschild, *Machina ex Dea,* p. xvii.

11. Autumn Stanley, "Women Hold Up Two-Thirds of the Sky: Notes for a Revised History of Technology," in ibid.

12. Donald A. MacKenzie and Judy Wajcman, eds., *The Social Shaping of Technology* (Philadelphia, 1999 [1985]).   *2ᵈ ed with numerous new articles*

13. Any choice fails to do justice to the sheer quantity and quality of this scholarship. A survey of article-length work can now be found in a massive and exhaustive compendium, Nancy Cott, ed., *History of Women in the United States: Historical Articles on Women's Lives and Activities,* 20 vols. (Munich and New York: K. G. Saur, 1992). Often-cited work includes Barbara Welter, "The Cult of True Womanhood, 1820–1860," *American Quarterly* 18 (1966): 151–74; Carroll Smith-Rosenberg, "The Female World of Love and Ritual: Relations between Women in Nineteenth-Century America," *Signs* 1 (1975): 1–29; Nancy F. Cott, *The Bonds of Womanhood: "Women's Sphere" in New England, 1780–1835* (New Haven, Conn., 1977); Linda Kerber, "Separate Spheres, Female Worlds, Woman's Place: The Rhetoric of Women's History," *Journal of American History* 75, no. 1 (1988): 9–39. For a sampling of attention to the diversity of women's experience in the United States, see Vicki L. Ruiz and Ellen Carol DuBois, eds., *Unequal Sisters: A Multi-Cultural Reader in U.S. Women's History,* 3d ed., (New York, 2000), and earlier editions. On essentialist feminism see n. 9 above.

14. Joan W. Scott, "Gender: A Useful Category of Historical Analysis," *Journal of American History* 91 (1986): 1053–75. An expanded—and better known—version of this essay appeared in her *Gender and the Politics of History* (New York, 1988). Sandra Harding articulated a similar vocabulary in the field of feminist philosophy of science in *The Science Question in Feminism* (Ithaca, N.Y., 1986). Although influential in the late eighties, neither was the first; see Ann Oakley's use of "gender" in *Sex, Gender and Society* (London, 1972).

15. For recent samplings of this literature, see Sally Gregory Kohlstedt, ed., *History of Women in the Sciences: Readings from Isis* (Chicago, 1999); *Osiris 12. Women, Gender, and Science: New directions,* eds. Sally Gregory Kohlstedt and Helen E. Longino (Chicago, 1997); and Evelyn Fox Keller and Helen E. Longino, eds., *Feminism and Science* (New York, 1996). For older work, see Sandra Harding and Jean F. Barr, eds., *Sex and Scientific Inquiry*

(Chicago, 1975); Sandra Harding and Merrill Hintikka, eds., *Discovering Reality: Feminist Perspectives on Epistemology, Metaphysics, Methodology, and Philosophy of Science* (Boston, 1983), which includes Nancy Hartsock's "The Feminist Standpoint: Developing Ground for a Specifically Feminist Historical Materialism," among other essays; Hilary Rose, "Hand, Brain, and Heart: A Feminist Epistemology for the Natural Sciences," *Signs* 9 (1983); Evelyn Fox Keller, *Reflections on Gender and Science* (New Haven, Conn., 1985); Donna Haraway, *Simians, Cyborgs and Women: the Reinvention of Nature* (London, 1988), which includes her now-classic essay, "A Manifesto for Cyborgs: Science, Technology, and Socialist Feminism in the 1980s," originally published in *Socialist Review* 15 (1985): 65–107; Harding, *The Science Question in Feminism;* Nancy Tuana, ed., *Feminism and Science* (Bloomington, Ind., 1989); Helen E. Longino, *Science as Social Knowledge: Values and Objectivity in Scientific Inquiry* (Princeton, N.J., 1990) and her useful overview "Essential Tensions—Phase Two: Feminist, Philosophical, and Social Studies of Science," in *The Social Dimensions of Science,* ed. Ernan McMullin (Notre Dame, 1992); Lorraine Code, *What Can She Know? Feminist Theory and the Construction of Knowledge* (Ithaca, N.Y., 1991); and Hilary Rose, *Love, Power and Knowledge: Towards a Feminist Transformation of the Sciences* (Bloomington, Ind., 1994). Both Judith McGaw and Joan Rothschild have drawn on this literature; see Judith A. McGaw, "No Passive Victims, No Separate Spheres: A Feminist Perspective on Technology's History," and Joan Rothschild, "From Sex to Gender in the History of Technology," in *In Context: History and the History of Technology,* eds. Stephen H. Cutcliffe and Robert C. Post (Bethlehem, Pa., 1989).

16. For an overview of discussions of engineering knowledge, and further citations, see Walter Vincenti, *What Engineers Know and How they Know It: Analytical Studies from Aeronautical History* (Baltimore, 1990). For further discussion of gender and varieties of technological knowledge, see Nina Lerman, "The Uses of Useful Knowledge: Science, Technology, and Social Boundaries in an Industrializing City," *Osiris* 12 (1997): 39–59, and Jennifer Light's article in this reader.

17. A good introduction to the scholarship on gender and technology is provided by Judy Wajcman, *Feminism Confronts Technology* (Cambridge, 1991). The bibliography is exceptionally good and includes literature from outside North America. Some additional bibliographical references may be found in Cynthia Gay Bindocci, *Women and Technology: An Annotated Bibliography* (New York and London, 1993). Patrick D. Hopkins, ed., *Sex/Machine: Readings in Culture, Gender, and Technology* (Bloomington, Ind., 1998) is a useful collection of reprinted articles from various disciplinary perspectives.

18. See Deborah J. Warner, "Women Inventors at the Centennial," in Trescott, *Dynamos and Virgins Revisited* (n. 3 above), and essays by Martha Moore Trescott and Autumn Stanley in the same volume. Trescott continued her work in "Lillian Moller Gilbreth and the Founding of Modern Industrial Engineering," in Rothschild, *Machina ex Dea* (n. 7 above), pp. 23–37, and "Women Engineers in History: Profiles in Holism and Persistence," in *Women in Scientific and Engineering Professions,* eds. Violet B. Haas and Carolyn C. Perrucci (Ann Arbor, Mich., 1984), pp. 181–205. See also Stanley's article "The Patent Office as Conjurer: The Vanishing Lady Trick in a Nineteenth-Century Historical Source," in *Women, Work, and Technology: Transformations,* ed. Barbara Drygulski Wright et al. (Ann Arbor, Mich., 1987),

and her substantial book *Mothers and Daughters of Invention: Notes for a Revised History of Technology* (New Brunswick, N.J., 1995 [1993]). For women inventors, see also Anne L. Macdonald's anecdotal but rich *Feminine Ingenuity: Women and Invention in America* (New York, 1992), and Deborah J. Merritt, "Hypatia in the Patent Office: Women Inventors and the Law, 1865–1900," *American Journal of Legal History* 35 (1991): 235–306. For a critical assessment, see Judith A. McGaw, "Review Essay: Inventors and Other Great Women: Toward a Feminist History of Technological Luminaries," *Technology and Culture* 38, no. 1 (January 1997): 214–31.

19. See Sally L. Hacker, *Pleasure, Power, and Technology: Some Tales of Gender, Engineering, and the Cooperative Workplace* (Boston, 1989), and Margaret W. Rossiter's brilliantly researched tomes, *Women Scientists in America: Struggles and Strategies to 1940* (Baltimore, 1982) and *Women Scientists in America: Before Affirmative Action, 1940–1972* (Baltimore, 1995). See also Amy Sue Bix, "Feminism Where Men Predominate: The History of Women's Science and Engineering Education at MIT," *Women's Studies Quarterly* 28, nos. 1–2 (2000): 24–45; Pamela E. Mack, "What Difference Has Feminism Made to Engineering in the Twentieth Century?" in Angela N. H. Creager, Elizabeth Lunbeck, and Londa Schiebinger, eds., *Feminism in Twentieth-Century Science, Technology, and Medicine* (Chicago, 2001) 149–64; and Ruth Oldenziel, "Gender and the Meanings of Technology: Engineering in the U.S., 1880–1945," (Ph.D. diss., Yale University, 1992), chap. 6. For a transnational perspective, see Annie Canel, Ruth Oldenziel, and Karin Zachman, eds., *Crossing Boundaries, Building Bridges: Comparing the History of Women Engineers, 1870s–1990s* (London, 2000). For British cases, see Carroll W. Pursell, "Domesticating Modernity: The Electrical Assocation for Women, 1924–1986," *British Journal of the History of Science* (1990): 47–67, and Carroll Pursell, "'Am I a Lady or an Engineer?': The Origins of the Women's Engineering Society in Britain, 1918–1940," *Technology and Culture* 34 (1993): 78–97. For German case studies, see Margot Füchs, *Wie die Väter, so die Töchter: Frauenstudium an der Technischen Hochschule München, 1899–1970* (Munich, 1994), and Barbara Duden and Hans Ebert, "Die Anfänge des Frauenstudiums an der Technischen Hochschule Berlin," in *Wissenschaft und Gesellschaft: Beiträge zur Geschichte der Technischen Universität Berlin, 1879–1979*, ed. Reinhard Rürup (Berlin, 1979). Judith S. McIlwee and J. Gregg Robinson, *Women in Engineering: Gender, Power and Workplace Culture* (Albany, 1992), is an excellent study. On women in the professions, see Joan Brumberg and Nancy Tomes, "Women in the Professions: A Research Agenda for American Historians," *Reviews in American History* 10 (1982): 275–96. Many of these topics are addressed in the conference proceedings volume from the International Symposium on Technology and Society, New Brunswick, N.J., 29–31 July 1999, *Women and Technology: Historical, Societal, and Professional Perspectives* (Piscataway, N.J., 1999).

20. Cynthia Cockburn in *Brothers: Male Dominance and Technical Change* (London, 1983) and *Machinery of Dominance: Women, Men, and Technical Know-How*, with a foreword by Ruth Schwartz Cowan (Boston, 1988 [1985]) early on described masculinity as a relationship to femininity. See also Mary Freifeld, "Technological Change and the 'Self-Acting' Mule: A Study of Skill and Sexual Division of Labour," *Social History* 11 (1986): 319–43, and Ava Baron, "Contested Terrain Revisited: Technology and Gender Definitions of Work

in the Printing Industry, 1850–1920," in Drygulski Wright, *Women, Work, and Technology* (n. 18 above). For more recent work in labor history see n. 38 below.

21. McGaw, "No Passive Victims, No Separate Spheres" (n. 15 above); Ruth Oldenziel, *Making Technology Masculine. Men, Women and Modern Machines in America, 1870–1945* (Amsterdam, 1999); Roger Horowitz, ed., *Boys and Their Toys? Masculinity, Class, and Technology* (New York and London, 2001); Carroll W. Pursell, "The Construction of Masculinity and Technology," *Polhem* 11 (1993): 206–19. This was not the first time Pursell had written about the subject; see Carroll W. Pursell, "Toys, Technology and Sex Roles in America, 1920–1940," in Trescott, *Dynamos and Virgins Revisited* (n. 3 above). See also Bruce Sinclair, "Technology on Its Toes: Late Victorian Ballets, Pageants, and Industrial Exhibitions" in Cutcliffe and Post, *In Context* (n. 15 above), and Arwen Mohun's, Ruth Oldenziel's, and Paul Edwards's articles in this reader. For general histories and sociologies on masculinity, see for example Gail Bederman, *Manliness and Civilization.: A Cultural History of Gender and Race in the United States, 1880–1917* (Chicago, 1995); Brian Easlea, *Fathering the Unthinkable* (London, 1981); Mark C. Carnes and Clyde Griffen, eds., *Meanings for Manhood: Constructions of Masculinity in Victorian America* (Chicago, 1990); Paul Willis, "Masculinity and Factory Labor," in *Working Class Culture,* eds. John Clarke et al. (London, 1979); Susan Johnson, "Bulls, Beans and Dancing Boys," *Radical History Review* 60 (1994): 4–37; E. Anthony Rotundo, *American Manhood: Transformations in Masculinity from the Revolution to the Modern Era* (New York, 1993); and David D. Gilmore, *Manhood in the Making: Cultural Concepts of Masculinity* (New Haven, Conn., 1990).

22. Robert C. Post, *High Performance: The Culture and Technology of Drag Racing, 1950–1990* (Baltimore, 1994).

23. See the work of Cynthia Cockburn, Ava Baron, Judy McGaw, Bruce Sinclair, Ruth Oldenziel, and Carroll Pursell, cited in n. 19–21.

24. For examples of social construction and actor network theory in technology studies, see Wiebe Bijker, Trevor Pinch, and Thomas Hughes, eds., *The Social Construction of Technological Systems* (Cambridge, Mass., 1987). For an argument in support of actor-network theory in studies of gender and technology, see Keith Grint and Rosalind Gill, *The Gender-Technology Relation: Contemporary Theory and Research* (London, 1995). For examples of other sociological and theoretical scholarship, see Cheris Kramarae, ed., *Technology and Women's Voices: Keeping in Touch* (London, 1988); Evelyn Fox Keller and Marianne Hirsch, eds., *Conflicts in Feminism* (New York, 1990); Gill Kirkup and Laurie Smith Keller, eds., *Inventing Women: Science, Technology and Gender* (London, 1992); Eileen Green, Jenny Owen, and Den Pain, eds., *Gendered by Design? Information Technology and Office Systems* (Bristol, U.K., 1993).

25. Bryan C. Taylor, "Register of the Repressed: Women's Voice and the Body in the Nuclear Weapons Organization," *Quarterly Journal of Speech* 97 (1993): 267–85; Jane Caputi, "The Metaphors of Radiation: or, Why a Beautiful Woman Is Like a Nuclear Power Plant," *Women's Studies International Forum* 14 (1991): 423–42; Paul N. Edwards in this volume; Carol Cohn, "Sex and Death in the Rational World of Defense Intellectuals," *Signs* 13 (1987): 687–718; Joseph O'Connell, "The Fine-Tuning of a Golden Ear: High-End Audio and the Evolutionary Model of Technology," *Technology and Culture* 33 (1992): 1–37. For an intro-

duction to work in the field of information and communication studies, see Liesbet van Zoonen, "Feminist Theory and Information Technology," *Media, Culture & Society* 14 (1992): 9–30; and Maureen Ebben and Cheris Kramarae, "Women and Information Technologies: Creating a Cyberspace of Our Own," in *Women, Information Technology, and Scholarship*, eds. H. Jeanie Taylor, Cheris Kramarae, and Maureen Ebben (Urbana, Ill., 1993).

26. Joseph J. Corn, "Making Flying 'Thinkable'," chap. 4 in *The Winged Gospel: America's Romance with Aviation, 1900–1950* (Oxford, 1983); Virginia Scharff, *Taking the Wheel: Women and the Coming of the Motor Age* (Albuquerque, N.M., 1992 [1991]).

27. Wiebe Bijker and Trevor Pinch, "The Social Construction of Artifacts," in Bijker, Pinch, and Hughes, *The Social Construction of Technology*, pp. 17–50; Scharff, *Taking the Wheel*; Rudi Volti, "Why Internal Combustion?" *American Heritage of Invention and Technology* (Fall 1990): 42–47; Michael B. Schiffer et al., *Taking Charge: The Electric Automobile in America* (Washington, D.C., 1994).

28. Cynthia Cockburn and Susan Ormrod, *Gender and Technology in the Making* (London, 1993). See also Lucille Alice Suchman, *Plans and Situated Actions: The Problem of Human-Machine Communications* (Cambridge and New York, 1987), and Danielle Chaubaud-Rychter's fine example, "Women Users in the Design Process of a Food Robot," in *Bringing Technology Home: Gender and Technology in a Changing Europe*, eds. Cynthia Cockburn and Rúza Fürst-Diliç (Buckingham, 1994). On the 'taming' of new technologies into daily practices, see also R. Silverstone and E. Hirsch, eds., *Consuming Technologies. Media and Information in Domestic Spaces* (London, 1992).

29. The following examples should point the interested reader to the broad issues here: Rosalind Petchesky, "Foetal Images: the Power of Visual Culture in the Politics of Reproduction," in *Reproductive Technologies: Gender, Motherhood and Medicine*, ed. Michelle Stanworth (Cambridge, 1987) (other essays in this volume are also useful); Lynn Spigel, *Make Room for T.V.: Television and the Family Ideal in Post-War America* (Chicago, 1992); Susan Douglas, *Where the Girls Are: Growing Up Female with the Mass Media* (New York, 1994); Lisa Cartwright, *Screening the Body: Tracing Medicine's Visual Culture* (Minneapolis, Minn., 1995); and Carole A. Stabile, *Feminism and the Technological Fix* (Manchester and New York, 1994), and "Shooting the Mother: Fetal Photography an the Politics of Disappearance," *Camera Obscura* 28 (1992): 178–205. The special issues on science and technology of this feminist film journal, vols. 28 and 29, deserve notice here. They were edited by Lisa Cartwright and Paula Treichler.

30. Unfortunately, authors contributing to the special issue of *Technology and Culture* on biomedical and behavioral technologies had apparently not encountered this literature; *Technology and Culture* 34, no. 4 (1993).

31. The earliest and classic "boon" statement dates from 1970 when Shulamith Firestone argued, in *The Dialectic of Sex: A Case for a Feminist Revolution* (New York, 1970), that the new technologies would be liberating for women. Socialist-feminists were the first to question Firestone's optimistic view of reproductive technologies, arguing that such technologies instead represented male intrusion of and control over the female body. Hilary Rose and Jalna Hanmer opened the black box of technology in "Women's Liberation: Reproduction and the Technological Fix," in *The Political Economy of Science: Ideology of/in*

the Natural Sciences, eds. Hilary Rose and Steven Rose (London, 1976). Rita Arditti, Renate Duelli Klein, and Shelley Minden, eds., *Test-Tube Women: What Future for Motherhood?* (London, 1984), includes the different trends in the discussion. Highly critical assessments include Gena Corea et al., *Man-Made Women: How New Reproductive Technologies Affect Women* (Bloomington, Ind., 1987 [1985]), and Gena Corea, *The Mother Machine: Reproductive Technologies from Artificial Insemination to Artifical Wombs* (New York, 1985). The influential and critical movement FINNRAGE (Feminist International Resistance to Reproductive and Genetic Engineering) announced itself in Patricia Spallone and D. L. Steinberg, *Made to Order: The Myth of Reproductive and Genetic Progress* (Oxford, 1987). For a response against the view of women as passive victims of male control and technology, see Stanworth, *Reproductive Technologies* (n. 29 above). See also Patricia Spallone, *Beyond Conception: The New Politics of Reproduction* (London, 1989), and Patricia H. Hynes, ed., *Reconstructing Babylon: Essays on Women and Technology* (Bloomington, Ind., 1991).

32. Linda Gordon, *Woman's Body, Woman's Right: Birth Control in America,* rev. ed. (New York, 1990). Barbara Katz Rothman, *Recreating Motherhood: Ideology and Technology in a Patriarchal Society* (New York, 1989); Petchesky, "Foetal Images" (n. 29 above); Jennifer L. Stone, "Contextualizing Biogenetic and Reproductive Technologies," *Critical Studies in Mass Communication* 8 (1991): 309–32; Laura R. Woliver, "The Influence of Technology and the Politics of Motherhood: An Overview of the United States," *Women's Studies International Forum* 14 (1991): 479–90; Maureen McNeil, Ian Varcoe, and Steven Yearley, eds., *The New Reproductive Technologies* (London, 1990); José van Dijck, *Manufacturing Babies and Public Consent* (New York, 1995). On the Dutch debate, see Marta Kirejczyk, *Met technologie gezegend? Gender en de omstreden invoering van in vitro fertilisatie in de Nederlandse gezondheidszorg* (Utrecht, 1996); Nelly Oudshoorn, *Beyond the Natural Body: An Archaeology of Sex Hormones* (London and New York, 1994); Marianne van den Wijngaard, *Reinventing the Sexes: The Biomedical Construction of Femininity and Masculinity, 1959–1985* (Bloomington, Ind., 1997).

33. For work published before 1982, see McGaw, "Women and the History of American Technology" (n. 4 above). See also Rima D. Apple, ed., *Women, Health, and Medicine in America: A Historical Handbook* (New Brunswick, N.J., 1992); Margeret J. Sandelowski, "Failures of Volition: Female Agency and Infertility in Historical Perspective," *Signs* 15 (1990): 475–99; Susan E. Klepp "Lost, Hidden, Obstructed, and Repressed: Contraceptive and Abortive Technology in the Early Delaware Valley," in Judith McGaw, ed., *Early American Technology: Making and Doing Things from the Colonial Era to 1850* (Chapel Hill, N.C.,1994); Janet Farell Brodie, *Abortion and Contraception in Nineteenth-Century America* (Ithaca, N.Y., 1994); Laura Klosterman Kidd, "Menstrual Technology in the United States, 1854–1921," (Ph.D. diss., Iowa State University, 1994); Rachel Maines in this reader and *The Technology of Orgasm: "Hysteria," the Vibrator, and Women's Sexual Satisfaction* (Baltimore, 1998); Joan Brumberg, *The Body Project: An Intimate History of American Girls* (New York, 1998); Elizabeth Siegel Watkins, *On the Pill: A Social History of Oral Contraceptives, 1950–1970* (Baltimore, 1998); and Andrea Tone, *Devices of Desire: A History of Contraceptives in America* (New York, 2001)

34. See Kathy Peiss, *Hope in a Jar: The Making of America's Beauty Culture* (New York,

1998); Claudia Kidwell and Valerie Steele, *Men and Women: Dressing the Part* (Washington, D.C., 1989); Anne Hollander, *Sex and Suits* (New York, 1994); Valerie Steele, *Fashion and Eroticism: Ideals of Feminine Beauty from the Victorian Era to the Jazz Age* (New York, 1985). See also McGaw, "Women and the History of American Technology," and articles by Judith McGaw and Rebecca Herzig in this reader. Peiss and Herzig also offer discussions of race and gender intersections in modifying appearance.

35. For a sample of this emerging literature, see Kathy Davis, *Reshaping the Female Body: The Dilemma of Cosmetic Surgery* (New York, 1995); Bernice L. Hausman, *Changing Sex: Transexualism, Technology and the Idea of Gender* (Durham, N.C., 1995); Ann Balsamo, *Technologies of the Gendered Body: Reading Cyborg Women* (Durham, N.C., 1996); and Rebecca Herzig in this volume. See also Allucquere Stone, *The War of Desire and Technology at the Close of the Mechanical Age* (Cambridge, Mass., 1995).

36. These discussions of blurring boundaries between what is "human" and what is "machine" owe much to the work of Donna Haraway (n. 15 above). For example, see Haraway, "The Promises of Monsters," in *Cultural Studies*, eds. Lawrence Grossberg, Cary Nelson, and Paula A. Treichler (New York, 1992). For further discussion, see also Paul N. Edwards, "Cyberpunks in Cyberspace: The Politics of Subjectivity in the Computer Age," in *The Cultures of Computing*, ed. Susan Leigh Star (London, 1995), and *The Closed World: Computers and the Politics of Discourse in Cold War America* (Cambridge, Mass., 1996); and Ruth Oldenziel, "Of Old and New Cyborgs: Feminist Narratives of Technology," *Letterature d'America* (1996): 95–111. See also Robbie Davis-Floyd and Joseph Dumit, *Cyborg Babies: From Techno-Sex to Techno-Tots* (London and New York, 1998).

37. One intriguing path is through the small literature on how the historical process of industrialization has reshaped conceptualizations of the human body. See Anson Rabinbach, *The Human Motor: Energy, Fatigue, and the Origins of Modernity* (New York, 1990); Mark Seltzer, *Bodies and Machines* (New York, 1992); and Barbara Duden, *Woman Beneath the Skin: A Doctor's Patients in 18th-Century Germany,* trans. Thomas Dunlap (Cambridge, Mass., 1991), "'Quick with Child': An Experience That Has Lost Its Status," *Technology in Society* 14 (1992): 335–44, and *Disembodying Women: Perspectives on Pregnancy and the Unborn,* trans. Lee Hoinacki (Cambridge, Mass., 1993).

38. For overviews of these issues, see Ava Baron's introduction to her edited *Work Engendered: Towards a New Labor History* (Ithaca, N.Y., 1991); Elizabeth Faue, "Gender and the Reconstruction of Labor History," *Labor History* 34 (1993): 169–79; and Alice Kessler-Harris, "Treating the Male as 'Other': Re-defining the Parameters of Labor History," *Labor History* 34 (1993): 190–204. For examples of work on gender and skill, see Margaret Lucille Hedstrom, "Automating the Office: Technology and Skill in Women's Clerical Work, 1940–1970" (Ph.D. diss., University of Wisconsin, 1988); Shirley Tillotson, "'We May All Soon Be First-Class Men': Gender and Skill in Canada's Early Twentieth Century Urban Telegraph Industry," *Labour* 27 (1991): 97–125; Wendy Gamber in this reader and *The Female Economy: Millinery and Dressmaking Trade, 1860–1930* (Urbana, Ill., 1997); Laura Lee Downs, *Manufacturing Inequality: Gender Division in the French and British Metalworking Industries, 1914–1939* (Ithaca, N.Y., 1995); and Sharon Hartman Strom, *Beyond the Typewriter: Gender, Class, and the Origins of Modern American Office Work, 1900–1930* (Urbana, Ill.,

1992). The telephone industry has attracted disproportionate attention: see Michèle Martin, *'Hello, Central?' Gender, Technology, and Culture of Formation of Telephone Systems* (Montreal, 1991); Kenneth Lipartito, "When Women Were Switches: Technology, Work, and Gender in the Telephone Industry, 1890–1920," *American Historical Review* 99 (1994): 1074–1111; Green, *Race on the Line* (see n 1). For older literature, see McGaw, "Women and the History of American Technology" (n. 4 above).

39. Exceptions include Judith McGaw, *Most Wonderful Machine: Mechanization and Social Change in Berkshire Paper Making, 1801–1885* (Princeton, N.J., 1987); Patricia A. Cooper in this volume and *Once a Cigar Maker: Men, Women, and Work Culture in American Cigar Factories, 1900–1919* (Urbana, Ill., 1987); and Margarete Sandelowski, *Devices and Desires. Gender, Technology, and American Nursing* (Chapel Hill, N.C., 2000). See also Daryl Hafter, ed., *European Women and Preindustrial Craft* (Bloomington, Ind., 1995); Arwen Mohun, "Why Mrs. Harrison Never Learned to Iron: Gender, Skill, and Mechanization in the Steam Laundry Industry," *Gender and History* 8 (1996): 231–51; and Roger Horowitz's article in this reader.

40. This is the argument Ruth Schwartz Cowan set forth in "The 'Industrial Revolution' in the Home" (n. 2 above).

41. Ruth Schwartz Cowan, *More Work for Mother: The Ironies of Household Technology from the Open Hearth to the Microwave* (New York, 1983) explicitly qualified conventional boosterism of household innovations as labor saving devices for women. For further discussion and examples, see Susan Strasser, *Never Done: A History of American Housework* (New York, 1982). Canadian scholarship is especially strong in this area: Tanis Day, "Substituting Capital for Labour in the Home: The Diffusion of Household Technology" (Ph.D. diss., Queen's University, 1987); Suzanne Marchand, "L'impact des innovations technologiques sur la vie quotidienne des quebeçoises du debut du XXe siècle, 1910–1940," *Material History Bulletin* 28 (1988): 1–14; and Joy Parr, *Domestic Goods. The Material, the Moral, and the Economic in the Postwar Years* (Toronto, Buffalo, and London, 1999). For the French case, see Robert L. Frost, "Machine Liberation: Inventing Housewives and Home Appliances in Interwar France," *French Historical Studies* 18 (1993): 109–30. On sewing and textiles, see Laurel Thatcher Ulrich, *The Age of Homespun: Objects and Stories in the Creation of an American Myth* (New York, 2001); Betty Ring, *"Let Virtue Be a Guide to Thee": Needlework in the Education of Rhode Island Women, 1730–1830* (Providence, R.I., 1983); Roszika Parker, *The Subversive Stitch: Embroidery and the Making of the Feminine* (London, 1984); Rachel Maines, "The Tools of the Workbasket: Needlework Technology in the Industrial Era," in *Bits and Pieces: Textile Traditions*, ed. Jeannette Lasansky (Lewisburg, Pa., 1991). On technological changes in the household and women's role in them, see also Maureen Ogle, "Domestic Reform and American Household Plumbing, 1840–1870," *Winterthur Portfolio* 28 (1993): 33–58, and *All the Modern Conveniences: American Household Plumbing, 1840–1890* (Baltimore, 1996); and discussion by Judith McGaw in this volume. For understandings of technological change and household work, see Kline in this volume. For discussions of domestic service work not explicitly focused on technology, see Faye Dudden, *Serving Women: Household Service in Nineteenth-Century America* (Middletown, Conn., 1983); Elizabeth Clark-Lewis, *Living In, Living Out: African-American Domestics and the Great Migration*

(1996); Jacqueline Jones, *Labor of Love, Labor of Sorrow* (New York 1985); and Tera Hunter, *To 'Joy My Freedom: Southern Black Women's Lives and Labor after Civil War* (Cambridge, Mass., 1997).

42. Cockburn and Fürst-Diliç, *Bringing Technology Home* (n. 28 above). On rural areas, see Joan Jensen, *Loosening the Bonds: Mid-Atlantic Farm Women, 1750–1850* (New Haven, Conn., 1986); Katherine Jellison, *Entitled to Power: Farm Women and Technology, 1913–1963* (Chapel Hill, N.C., 1993); Ronald Kline, *Consumers in the Country: Technology and Social Change in Rural America* (Baltimore, 2000); Angela E. Davis, "'Valiant Servants': Women and Technology on the Canadian Prairies, 1910–1940," *Manitoba History* 25 (1993): 33–42; and Glenda Riley, "In or Out of the Historical Kitchen? Interpretatons of Minnesota Rural Women," *Minnesota History* 52 (1990): 61–71, and *The Female Frontier: A Comparative View of Women on the Prairie and the Plains* (Lawrence, Kans., 1988).

43. Karen Calvert, *Children in the House: The Material Culture of Early Childhood, 1600–1900* (Boston, 1992); Mary Lynn Stevens Heininger et al., *A Century of Childhood, 1820–1920* (Rochester, N.Y., 1984); Pursell, "Toys, Technology, and Sex Roles in America" (n. 21 above); and Nina Lerman, "Technology," in *American Girlhood: An Encyclopedia*, ed. Miriam Forman-Brunell (ABC-Clio, 2001). See also the articles by Nina Lerman and Ruth Oldenziel in this reader.

44. See Jeanne Boydston, *Home and Work: Housework, Wages, and the Ideology of Labor in the Early Republic* (New York, 1990); Amelia Grace Preece, "Housework and American Standards of Living, 1920–1980," (Ph.D. diss., University of California, Berkeley, 1990); Arwen Palmer Mohun, *Steam Laundries: Gender, Technology, and Work in the United States and Great Britain, 1880–1940* (Baltimore, 1999); and Nina Lerman, "From 'Useful Knowledge' to 'Habits of Industry': Gender, Race, and Class in Nineteenth-Century Technical Education" (PhD diss., University of Pennsylvania, 1993). Recent work on the Home Economics Movement is also useful here; see Sarah Stage and Virginia B. Vincenti, eds., *Rethinking Home Economics: Women and the History of a Profession* (Ithaca, N.Y., 1997), and articles by Carolyn Goldstein, Joy Parr, and Ronald Kline in this volume.

45. McGaw, "Women and the History of American Technology" (n. 4 above), p. 827.

46. See Stuart Ewen, *Captains of Consciousness: Advertising and the Social Roots of Consumer Culture* (New York, 1977); T. J. Jackson Lears, "From Salvation to Self-Realization: Advertising and the Therapeutic Revolution," *The Culture of Consumption: Critical Essays in American History, 1880–1980*, eds. Richard Wightman Fox and T. J. Jackson Lears (New York, 1983); Roland Marchand, *Advertising the American Dream: Making Way for Modernity, 1920–1940* (Berkeley, 1985); Susan Strasser, *Satisfaction Guaranteed: The Making of the American Mass Market* (New York, 1989); Richard Tedlow, *New and Improved: The Story of Mass Marketing in America* (New York, 1990); William Leach, *Land of Desire: Merchants, Power, and the Rise of a New American Culture* (New York, 1993); T. J. Jackson Lears, *Fables of Abundance: A Cultural History of Advertising in America* (New York, 1994); and P. Walker Laird, *Advertising Progress: American Business and the Rise of Consumer Marketing* (Baltimore, 1998). On consumers in Europe, see Rosalind Williams, *Dream Worlds: Mass Consumption in Late Nineteenth-Century France* (Berkeley, 1982); Michael Barry Miller, *The Bon Marché: Bourgeois Culture and the Department Store, 1869–1920* (Princeton, N.J., 1981). For European

and American comparison, see Susan Strasser, Charles McGovern, and Matthias Judt, eds., *Getting and Spending. European and American Consumer Societies in the Twentieth Century* (Washington, D.C., 1998); see also B. Fine and E. Leopold, *The World of Consumption* (London and New York, 1993). For scholarship on advertising and marketing taking into account technology, gender, and consumers as active agents, see Susan Smulyan, *Selling Radio: The Commercialization of American Broadcasting, 1920–1934* (Washington, D.C., 1994); Carolyn Goldstein, "Mediating Consumption: Home Economics and American Consumers, 1900–1940" (Ph.D. diss., University of Delaware, 1994); Alison Clarke, *Tupperware: The Promise of Plastic in 1950s America* (Washington, D.C., 1999); Regina Lee Blaszcyk, *Imagining Consumers: Design and Innovation from Wedgewood to Corning* (Baltimore, 2000); and Nelly Oudshoorn and Trevor Pinch, eds., *How Users Matter: The Co-construction of Users and Technology* (Cambrigde, Mass., forthcoming).

47. For an overview, see Victoria de Grazia and Ellen Furlough, eds., *Sex of Things: Gender and Consumption in Historical Perspective* (Berkeley, 1996); Jennifer Scanlon, *The Gender and Consumer Culture Reader* (New York, 2000); and Ruth Oldenziel, "Man the Maker, Woman the Consumer: The Consumption Junction Revisited," in Creager, Lunbeck, and Schiebinger, *Feminism* (n. 19 above), pp. 128–48. For earlier work, see William Leach, "Transformations in a Culture of Consumption: Women and Department Stores, 1870–1920," *Journal of American History* 17 (1984), and Susan Porter Benson, *Counter Cultures: Saleswomen, Managers, and Customers in American Department Stores, 1890–1940* (Urbana, Ill., 1986). See also Dana Frank, *Purchasing Power: Consumer Organizing, Gender, and the Seattle Labor Movement, 1919–1929* (Cambridge, 1994); Jacqueline Dirks, "Righteous Goods: Women's Production, Reform Publicity, and the National Consumers' League, 1891–1919" (Ph.D. diss., Yale University, 1996); Elizabeth White, "Sentimental Enterprise: Sentiment and Profit in American Market Culture, 1830–1880" (Ph.D. diss., Yale University, 1996). See also Rebecca Herzig, Ruth Oldenziel, and Joy Parr in this volume.

48. On mediators, see Goldstein, "Mediating Consumption" (n. 46 above), and her article in this reader. See also Stage and Vincenti, *Rethinking Home Economics* (n. 44 above). On repair, see Kevin Borg, "From Village Blacksmith to Mr. Good Wrench: Creating Auto Mechanics in Technology's Middle Ground" (Ph.D. diss., University of Delaware, 2000). On users, see Susan J. Douglas, *Inventing American Broadcasting, 1899–1922* (Baltimore, 1987); Claude S. Fischer, *America Calling: A Social History of the Telephone to 1940* (Berkeley, 1992); Ronald Kline and Trevor Pinch, "Users as Agents of Technological Change: The Social Construction of the Automobile in the Rural United States," *Technology and Culture* 37 (1996): 763–95; Madeleine Akrich, "User Representations: Practices, Methods and Sociology," in *Managing Technology in Society*, eds. Arie Rip, Thomas Misa, and Johan Schot (London, 1995), pp. 167–84; Jellison, *Entitled to Power* (n. 42 above); and Rachel Maines, essay this volume and *The Technology of Orgasm* (n. 33 above). See also Arwen Mohun and Roger Horowitz, eds., *His and Hers: Gender, Consumption, and Technology* (Charlotte, N.C., 1998); Arwen Mohun, "Designed for Thrills and Safety: Amusement Parks and the Commodification of Risk, 1880–1929," *Journal of Design History*, November 2001; and Joy Parr's article in this volume.

49. For an elaboration on this point see Ruth Oldenziel, "Object/ions: Technology, Cul-

ture and Gender," in *Learning from Things,* eds. David W. Kingery and Steve Lubar (Washington, D.C., 1996). See also Katherine Martinez and Kenneth L. Ames, eds., *The Material Culture of Gender/The Gender of Material Culture* (Winterthur, Del., 1997), and Joan Rothschild et al., eds., *Design and Feminism, Re-Visioning Spaces, Places and Everyday Things* (New Brunswick, N.J., 1999). The culture of production framework arguably carries racial biases as well, excluding the agricultural and service work to which racialized groups in the United States have often been relegated. For discussion of racial and technological ideologies, see Nina Lerman, "New South, New North: Region, Ideology, and Access in Industrial Education," in Sinclair, *Technology and the African-American Experience* (n. 1 above).

Instructor's Notes on Organization

This collection of essays was assembled with an emphasis on the connection between gender and technological development as a mutually shaping process. In Part I the essays question conventional definitions of technology to suggest that such definitions are gendered and technology needs to be conceived broadly. Part II focuses more — though not exclusively — on the gender side of the equation, to demonstrate how technologies are involved in gender practices. These parts taken together show there is gender of technology as well as a technology of gender. Articles in Parts III and IV offer a close-up of specific technological practices and contexts in the industrial world of North America. Again, the essays are grouped to emphasize the relationship between industrial technologies and industrial social organizations.

Nevertheless, this reader may be tailored to suit different courses or interests in gender studies, cultural studies, American history, technology studies and material culture courses, to name just a few. To aid instructors and readers, we offer some alternative arrangements of the articles included in the book.

Teachers seeking to integrate a gendered history of technology into their classes may choose according to general themes. For example, a course on consumer culture focusing on the active participants in shaping the emerging consumer society, from mediating groups to end users who purchase goods and services, could draw on the articles by Parr, Goldstein, Kline, Gamber, Oldenziel, and Herzig. Assignments in labor history could easily integrate articles paying attention to specific technologies that are not limited to the industrial workplace but also include professional workplaces, the home, and the body. The articles by Maines, Gamber, Lerman, Cooper, Mohun, Horowitz, Kline, Parr, Light, and Edwards speak to those issues. Women's studies courses dealing with the body would stand to bene-fit from historical work showing that cyborgs have a history too. For those courses we suggest the work by McGaw, Gamber, Maines, Herzig, and Oldenziel. Those interested in the new field of masculinity studies will find in this collection some examples of how technology becomes a site of shoring up male identity even in the face of changing gender relations and new technological developments that defy old arrangements. The articles by Lerman, Edwards, Oldenziel, and Mohun deal with this issue, as do the essays by Gamber, Cooper, and Horowitz.

Historians of technology interested in the historical formation of gender prac-tices and technological developments during industrial capitalism may find plac-ing the articles in chronological and thematic order particularly useful. In such a course, the formation of the boundaries between gender identity and technologi-cal knowledge in the late nineteenth and early twentieth century may be mapped in the following order: McGaw, Lerman, Gamber, Maines, Mohun, Herzig, Olden-ziel. This may be followed by a second part, the formation of the industrial junction of consumption and production in the early twentieth century, when early forms of large technical systems of gas and electricity were in search of new markets and new consumers (Parr, Goldstein, Kline), and new machines were transforming the labor process (Cooper, Horowitz, Light, Edwards).

Teachers of technology studies and material culture classes may also group readings by specific technologies, such as computers (Light, Edwards), laundry (Parr, Mohun, Kline), electricity and gas technologies (Goldstein, Kline), sew-ing (Lerman, Gamber), factories (Cooper, Horowitz), automobiles (Oldenziel), and medical technologies (Maines, Herzig). By this arrangement McGaw's essay bridges several categories but would work well as either introduction or conclu-sion.

Because we firmly believe gender practices and technological developments to be historically entwined and situated in time and place, we have chosen nineteenth- and twentieth-century industrial capitalism as our point of entry. This reader is thus about industrial junctions. Of course, body and reproductive tech-nologies are also part and parcel of industrial narratives, as Herzig and Maines show. The built environment, as Roger Horowitz suggests, is also crucial; an ex-cellent review essay by Joan Rothschild and Victoria Rosner appears in *Design and Feminism: Re-Visioning Spaces, Places, and Everyday Things* (New Brunswick, N.J., 1999). Media and information technology, as suggested by Jennifer Light and Paul

Edwards in this reader, is also treated in a vast literature (for further discussion, see the historiographic essay at the end of this volume).

Graduate students and scholars who wish to survey the field, focusing on some key historiographical issues, should read our overview in this reader, and the essays by McGaw, Kline, and Light, as well as the review essay cited above.

Contributors

PATRICIA COOPER is an associate professor in the Department of History and the Women's Studies Program at the University of Kentucky. She is currently writing about the Sun Shipbuilding and Drydock Company during World War II and is also writing a book about Fair Employment in Philadelphia, 1930–1970.

PAUL N. EDWARDS is an associate professor of information at the University of Michigan, where he also directs the Science, Technology, and Society Program. He is the author of *The Closed World: Computers and the Politics of Discourse in Cold War America* (MIT Press, 1996) and co-editor of *Changing the Atmosphere: Expert Knowledge and Environmental Governance* (MIT Press, 2001). He is presently finishing a book about computer models in climate science and politics.

WENDY GAMBER is an associate professor of history at Indiana University, Bloomington, and a former associate editor of the *Journal of American History*. She is the author of *The Female Economy: The Millinery and Dressmaking Trades, 1860–1930* (University of Illinois Press, 1997). She is currently at work on *Houses Not Homes: Boardinghouses in the Nineteenth-Century America* (Johns Hopkins University Press, forthcoming).

CAROLYN M. GOLDSTEIN is curator at Lowell National Historical Park in Lowell, Massachusetts. She holds a Ph.D. in history from the University of Delaware, where she was a fellow in the Hagley Program. She is the author of *Do It Yourself: Home Improvement in 20th-Century America* (Princeton Architectural Press, 1998). She is completing a book about home economics and the mediation of consumption, which will be published by the University of North Carolina Press in 2003.

REBECCA HERZIG teaches courses on science, technology, and medicine in the Program in Women's Studies at Bates College in Lewiston, Maine. Her research centers on embodied experiences of "science" in late nineteenth-century America. She is co-editor, with Evelynn Hammonds, of a new collection of primary and secondary source material on science, race, and gender in U.S. history.

ROGER HOROWITZ is associate director of the Center for the Study of Business, Technology, and Society at the Hagley Museum and Library in Wilmington, Delaware. He is the author of *"Negro and White, Unite and Fight!": A Social History of Industrial Unionism in Meatpacking, 1930–1990* (University of Illinois Press, 1997), edited *Boys and Their Toys?: Masculinity, Technology, and Class* (Routledge, 2001), and is completing a study called *Meat in America: Taste, Technology, Transformation* (Johns Hopkins University Press, forthcoming).

RONALD R. KLINE is Professor of History of Technology in the Department of Science and Technology Studies at Cornell University. He is author of *Steinmetz: Engineer and Socialist* (Johns Hopkins University Press, 1992) and articles on the history of engineering education, electronics, the relationship between science and technology, and rural technology in the United States. His most recent book is *Consumers in the Country: Technology and Social Change in Rural America* (Johns Hopkins University Press, 2000).

NINA E. LERMAN is an associate professor of history at Whitman College in Walla Walla, Washington. She received a Ph.D. in History and Sociology of Science at the University of Pennsylvania. She co-edited the special issue of *Technology and Culture* on gender (1997) with Arwen Mohun and Ruth Oldenziel and is a founding editor of the discussion list H-SCI-MED-TECH. Her current project, *Children of Progress: Social Boundaries and Technical Education in an Industrial City*, examines gender, racial, and class dimensions of technological knowledge from 1800 to the early twentieth century.

JENNIFER LIGHT is an assistant professor in the Department of Communication Studies and a faculty fellow at the Institute for Policy Research at Northwestern University. She received her Ph.D. in History of Science from Harvard University. She is currently working on *Cities in the Information Society* (Johns Hopkins University Press, forthcoming), which explores how military techniques and technologies, and national security concerns, have shaped the intellectual and organizational history of American cities since World War II.

RACHEL P. MAINES is an independent scholar and a technical processing assistant at Cornell University's Hotel School Library. Her book *The Technology of Orgasm: "Hysteria," the Vibrator, and Women's Sexual Satisfaction* (Johns Hopkins University Press, 1998) won the 1999 Herbert Feis Award of the American Historical Association and the 2000/2001 Biennial Book Award of the American Foundation for Gender and Genital Medicine and Science.

ARWEN PALMER MOHUN is an associate professor of history at the University of Delaware. She is currently writing a book on risk and technology in industrializing America and is also the author of *Steam Laundries: Gender, Technology, and Work in Great Britain and the United States, 1880–1940,* (Johns Hopkins University Press, 1999) co-editor (with Roger Horowitz) of *His and Hers: Gender, Consumption, and Technology,* (University of Virginia Press, 1998) and co-editor (with Nina Lerman and Ruth Oldenziel) of the January 1997 special issue of *Technology and Culture* on gender and technology.

JUDITH A. McGAW is the author of *Most Wonderful Machine: Mechanization and Social Change in Berkshire Paper Making, 1801–1885* (Princeton University Press, 1987) and numerous articles on gender and technology. She is also the editor of *Early American Technology: Making and Doing Things from the Colonial Era to 1850* (University of North Carolina Press, 1994).

RUTH OLDENZIEL received her Ph.D. from Yale in American History and is Associate Professor at the University of Amsterdam. She is the author of *Making Technology Masculine: Men, Women and Modern Machines in America, 1870–1945* (Amsterdam University Press, 1999). She co-edited (with Annie Canel and Karin Zachmann) *Crossing Boundaries, Building Bridges: Comparing Women Engineers, 1870s-1990s* (Harwood Academic, 2000), (with Carolyn Bouw) *Schoon genoeg: huis-*

vrouwen en huishoudtechnologie in Nederland, 1898–1998 (Nijmegen, Netherlands: SUN, 1998), and (with Arwen Mohun and Nina Lerman) the special issue of *Technology and Culture* on gender (1997).

JOY PARR is Farley Professor at Simon Fraser University and the author of *Domestic Goods: The Material, the Moral, and the Economic in the Postwar Years* (University of Toronto Press, 1999). She is currently writing about large engineering projects, risk, and the history of the senses.

Index